LONDON MATHEMATICAL SOCIETY STUDENT TEXTS

Managing editor: Professor C.M. Series, Mathematics Institute
University of Warwick, Coventry CV4 7AL, United Kingdom

To the memory of our friends Paul Erdős,
Géza Fodor, and Eric Charles Milner

London Mathematical Society Student Texts 48

Set Theory

András Hajnal
Rutgers University

Peter Hamburger
Indiana-Purdue University

Translated by Attila Máté
Brooklyn College of CUNY

CAMBRIDGE
UNIVERSITY PRESS

PUBLISHED BY THE PRESS SYNDICATE OF THE UNIVERSITY OF CAMBRIDGE
The Pitt Building, Trumpington Street, Cambridge, United Kingdom

CAMBRIDGE UNIVERSITY PRESS
The Edinburgh Building, Cambridge, CB2 2RU, UK   http://www.cup.cam.ac.uk
40 West 20th Street, New York, NY 10011–4211, USA   http://www.cup.org
10 Stamford Road, Oakleigh, Melbourne 3166, Australia
Ruiz de Alarcón 13, 28014 Madrid, Spain

Originally published in Hungarian as *Halmazeimélet*.

First published in English 1999

Printed in the United Kingdom at the University Press, Cambridge

A catalogue record for this book is available from the British Library

Library of Congress Cataloging in Publication data
Hajnal, A.
Set Theory / András Hajnal and Peter Hamburger; translated by Attila Máté
p. cm. Includes bibliographical references and index.
ISBN 0 521 59344 1 (hc). – ISBN 0 521 59667 X (pbk.)
1. Set Theory. I. Hamburger, P. II. Title.
QA248.H235 1999
511.3'22–dc21   99–26507   CIP

ISBN   0 521 59344 1 Hardback
ISBN   0 521 59667 X paperback

CONTENTS

PREFACE

Aus dem Paradies, das Cantor uns geschaffen,
soll uns niemand vertreiben können.

(No one can chase us out of the paradise that Cantor has
created for us.)

David Hilbert

This textbook was prepared on the basis of courses and lectures by András Hajnal for mathematics majors at Roland Eötvös University in Budapest, Hungary. The first edition appeared in 1983; since then the book went through a number of new printings and editions. During each of these, new problems were added and the historical remarks were updated. A number of revisions have also been made in the present, the first English, edition. A significant one among these is that hints were added for the problems in Part II, and a completely new section (Section 20) discusses the so-called square-bracket symbol.

The book consists of two parts and an Appendix to Part I. The first part contains a detailed non-axiomatic introduction to set theory. This introduction is carried out on a quite precise, but intuitive level, initially presenting many of Cantor's original ideas, including those on defining cardinals and order types as abstract objects. Only later, in Sections 8–11, do we discuss von Neumann's definition of ordinals and prove results important even for mathematicians working in various areas other than set theory. This part is well suited for a one-semester undergraduate course, and it is generally used in Hungarian universities. As is customary in mathematics textbooks at Hungarian universities, each assertion announced in the text is accompanied by a complete and detailed proof.

The Appendix attached to Part I gives details as to how the development of set theory described in Part I can be transformed into a rigorous development of axiomatic set theory. Particular attention is paid to the clarification of conceptual difficulties encountered in the axiomatic development. The precise notion of independence proofs in set theory is discussed. This is all the more necessary, since, in order to appreciate the significance and the connections among the results presented in Part II, the discussion of a number of independence results was indispensable. This Appendix can serve as auxiliary material for an undergraduate course in mathematical logic.

The second part of the book can be used as material for a one- or two-semester early graduate course. It gives a detailed survey of the combinatorial foundations of modern set theory and of those classical results of set theory that are needed for most students of the field.

The nature of the book makes it impossible to include proofs of independence results requiring a familiarity with mathematical logic, but we include references to the most important ones among these results, since without an acquaintance of these it is impossible to find one's way in the subject.

A proportion of the problems attached to the sections in Part I are of the nature of simple exercises; the solution of these are completely left to the reader. For the more difficult problems marked with an asterisk * we include short hints at the end of Part I. These problems often build on earlier problems, and it is advisable to study them in the order given.

Certain problems in Part II are much more difficult. These are used to expand on the material presented. Most of these discuss results published in the literature, and, in addition to hints for their solutions, we also include references to the literature. Although we consider presentation of the material in the book self-contained, in certain problems marked by the symbol $^+$ we use concepts, such as, for example, that of the Riemann integral, not defined in the book.

We would like to thank the ASL Committee on Translations, and particularly to Steffen Lempp, for recommending an English translation of the book; Attila Máté, who, in addition to translating the book from Hungarian, also typeset it in $\mathcal{A}_{\mathcal{M}}\mathcal{S}$-TeX and acted as one of its scientific referees; and Dr. William Weiss, Justin Moore, and Vojkan Vuksanovic, of the University of Toronto, for carefully reading the manuscript.

Piscataway, New Jersey, December 1998

András Hajnal

Peter Hamburger

PART I
INTRODUCTION TO SET THEORY

INTRODUCTION

In creating set theory, Cantor primarily studied sets occurring in mathematics, such as sets of the integers and the real numbers and their subsets. The reader with a certain grasp of mathematics has an intuitive picture of these. If we want to study sets in general, we have to clarify, when we consider a set as given, what kind of sets exist. In the initial, "romantic" period of set theory, this problem did not arise. Every "conceivable" set was thought to exist, every collection for which it was possible to say in some way what its elements were was considered a set. It soon turned out, however, that this viewpoint is untenable, since, as we will see, the set of all elements cannot possibly exist. This turn of events eventually led to the development of axiomatic set theory. The usual axiomatic approach postulates only the existence of a single set, and other sets can be obtained from this set with the aid of the so-called conditional set existence axioms. In order to develop set theory along these lines, a number of theorems are needed which would be of no interest to the reader getting acquainted with the subject for the first time if no examples for their applications could be provided. This was one of the considerations that led us to present a half-intuitive, and half-axiomatic development of the subject.

In what follows, we consider a few sets commonly used in mathematics, such as the set of integers, of the rationals, and of the reals. Beyond these, we only consider sets whose existence can be derived from the conventions enumerated below. These conventions in effect amount to the usual Zermelo-Fraenkel axiom system of set theory. This axiom system will be formulated in full precision only in the Appendix following Part I. There we will sketch how the axiomatic development of set theory can be carried out on the basis of these axioms, how the nonnegative integers, the rational and real numbers can be defined in this development, and how proofs of such elementary results about them can be obtained as are usually proved by referring to some "well-known" property of, say, the integers.

1. NOTATION, CONVENTIONS

In what follows, we are going to use the following shorthand: $\forall$ stands for the words "for all," $\exists$, for "there exists," $\neg$, for "not," $\wedge$, for "and," $\vee$, for "or," $\implies$, for "implies," $\iff$, for "if and only if."

The symbol $\equiv$ will also be used to connect logic expressions: $F \equiv G$ will mean that F and G "say the same thing." Thus, for example, $F \equiv G \equiv H$ is used to express that F, G, and H "say the same thing," while the expression $F \iff G \iff H$ is never used (since parentheses are required to know how to read the latter).

We will also use *restricted quantifiers*:

$$(\forall x : \phi(x))\, \psi(x) \qquad \text{will mean} \qquad \forall x\, (\phi(x) \implies \psi(x))$$

and

$$(\exists x : \phi(x))\, \psi(x) \qquad \text{will mean} \qquad \exists x\, (\phi(x) \wedge \psi(x)).$$

Depending on the form of $\phi(x)$ here, restricted quantifiers can often be written in an abbreviated form. For example, $(\forall x : x < y)\, \psi(x)$ can be abbreviated to $\forall x < y\, \psi(x)$.

1. The concept of a set and that of "being an element of" will not be defined, these notions will be considered fundamental. In general, sets will be denoted by Roman capitals A, B, ..., and elements will be denoted by lower case letters x, y, ... If A is a set, then for every conceivable thing or object x it is true either that x belongs to A or that x does not belong to A. In the former case, we say that x is an element of A, and in the latter, that x is not an element of A. These possibilities will be denoted by $x \in A$ and $x \notin A$, respectively.

We will therefore imagine that a set can consist of arbitrary objects, or elements in other words, that is, of dogs, numbers, functions, and, primarily, of sets. It is expedient to agree that the statement $x \in y$ is meaningful for any two objects, and $x \in y$ is false if y is not a set. London is a city and it is not a set, therefore nothing is an element of it, and so $[0, 1] \notin$ London.

2. The set of the nonnegative integers, of the rational numbers, and of the real numbers, will be denoted by ω, $\mathbb{Q}$, and $\mathbb{R}$, respectively.

3. If A and B are two sets that have the same objects as elements, then A and B are equal, that is, if for all x, $x \in A \iff x \in B$ then $A = B$.

4. We assume that there is a set that has no elements. According to Convention **3**, there is only one such set. This set will be called the empty set and will be denoted by $\emptyset$.

5. If x_0, ..., x_{n-1} are elements, then $\{x_0, \ldots, x_{n-1}\}$ denotes the set with exactly these elements. For example, $\{x\}$ is the set whose only element is x, and $\{x, y\}$ is the set whole elements are x and y. Here x_0, ..., x_n do not all need to be distinct; e.g., $\{x, x\} = \{x\}$

We agree often to denote a set by listing its elements in braces. If it is impossible to list these elements but there is a fairly clear way to indicate what these elements are, we will also use this way of denoting sets. For example, the set of nonnegative integers can be written as $\{0, 1, \ldots, n, \ldots\}$.

Definition 1.1. *Let A and B be sets. We call the set A a subset of B if for all $x \in A$ we also have $x \in B$; in symbols, we will then write $A \subset B$.*

If $A \subset B$ and $\exists x \in B$ such that $x \notin A$, then we will call A a proper subset of B; in symbols, we will write $A \subsetneq B$.

It is easy to see in view of Convention **3** the following.

Theorem 1.1. *If $A \subset B$ and $B \subset A$ then $A = B$.*

$$* \qquad *$$
$$*$$

For the reader with a certain familiarity with mathematics, it is clear that the following convention is indispensable.

6. If A is a set and $\Phi(x)$ is a property that is either true or false for each object x, then those elements of A for which $\Phi(x)$ is true form a set; this set will be denoted by $\{x \in A : \Phi(x)\}$. The set of even numbers can, for example, be described as $\{x \in \omega : x \text{ is divisible by } 2\}$. If f is a real-valued function and $\Phi(x)$ is the property that f is defined at the place x and it is continuous there, then $\{x \in \mathbb{R} : \Phi(x)\}$ is the set of all points of continuity of f.

In this convention, we used the concept of a property but we have not defined it. In the axiomatic development, this concept can be precisely circumscribed with tools from mathematical logic. In our not completely axiomatic development, this concept may mean an arbitrary meaningful property, in the way that it is usually understood in other, not rigorously axiomatized branches in mathematics.

Using the conventions set down so far, we are able to prove that the set of all object or the set of all sets does not exist. This clearly contradicts earlier views that every set that can be "imagined" exists. This contradiction was first pointed out by the British philosopher Bertrand Russell, and for this reason it is called Russell's antinomy (or Russell's paradox).

Theorem 1.2. *There is no set A that has every set as its element.*

Proof. Assume that, on the contrary, there is a set A such that every set is an element of it. Define the property $\Phi(x)$ as follows: x is a set and

$x \notin x$. This is undoubtedly a "mathematical" property. Then, according to Convention **6**, there is a set $B = \{x \in A : \Phi(x)\}$. B consists of all sets that are not elements of themselves. If we had $B \in B$, then B would satisfy property Φ, and so we would have $B \notin B$, which is impossible. If we had $B \notin B$, then $\Phi(B)$ would be true, and, according to the assumption about A, we would have $B \in A$. Hence $B \in B$ would hold by virtue of the definition of B. This is again impossible. Thus, the assumption made above leads to a contradiction, proving the validity of the theorem.

*   *
*

Theorem 1.2 shows that in general we cannot talk about the set of all sets satisfying a property $\Phi(x)$. In spite of this, we will use the notation $\{x : \Phi(x)\}$ if it is possible to verify that, for the given Φ, there is a set A such that each object x with property $\Phi(x)$ is an element of A; indeed, in this case

$$\{x : \Phi(x)\} = \{x \in A : \Phi(x)\}.$$

If in what follows such a "dominating" set A is not explicitly given, it is because the existence of such a set is easy to establish, and it is left to the reader to do this.

The next two conventions will ensure the existence of further sets given a set A for which it would not be possible to find a "dominating" set on the basis of the conventions set down so far.

7. If A is a set, then all its subsets constitute a set; we denote it by $P(A)$ and call it the power set of A. For example, if $A = \{0, 1, 2\}$, then

$$P(A) = \{\emptyset, \{0\}, \{1\}, \{2\}, \{0, 1\}, \{0, 2\}, \{1, 2\}, \{0, 1, 2\}\}.$$

Clearly, $P(A) = \{x : x \subset A\}$. The elements of $P(A)$ are themselves sets. Those sets whose elements are also sets are often called set systems. We will not make this distinction systematically, since most sets that we will consider will have sets as their elements.

8. If A is a set then the *union* of those elements of A that are themselves sets will be denoted by $\bigcup A$, and will simply be called the union of A. That is

$$\bigcup A = \{x : \exists y (x \in y \wedge y \in A)\}.$$

For example, if $A = \{\text{Mississipi}, \{0\}, \{1\}\}$, then

$$\bigcup A = \{0, 1\}.$$

9. The concept of function can be reduced to that of set. To this end, we need first to define the notion of ordered pair.

Definition 1.2. *If x and y are two objects, then the set $\{\{x\}, \{x,y\}\}$ will be called the* ordered pair *formed by x and y, and it will be denoted by $\langle x, y \rangle$.*

That $\langle x, y \rangle$ indeed corresponds to the intuitive notion of ordered pair is shown by the following theorem.

Theorem 1.3. *For arbitrary elements x, y, x', y', the relation $\langle x,y \rangle = \langle x',y' \rangle$ holds if and only if $x = x'$ and $y = y'$.*

Proof. The condition is clearly sufficient. To show that it is also necessary, assume that

$$\{\{x\}, \{x,y\}\} = \{\{x'\}, \{x',y'\}\}.$$

The set $\{x\}$ must be equal to an element of the set on the right-hand side. From this, it follows that $\{x\} = \{x'\}$, $x = x'$, and $\{\{x\}, \{x,y\}\} = \{\{x\}, \{x,y'\}\}$. Hence we have $\{x,y\} = \{x,y'\}$, and so $y = y'$ also follows.

*   *
*

10. Next we will define the notion of function.

Definition 1.3. *A set consisting of ordered pairs is called a* function *if for every x there is at most one y such that $\langle x, y \rangle \in f$. The set of those elements x for which $\exists y (\langle x, y \rangle \in f)$ is called the* domain *of the function f and is denoted by $\mathrm{D}(f)$.*

If $x \in \mathrm{D}(f)$, then the unique y for which $\langle x, y \rangle \in f$ is called the value *of the function f assumed at the place x, and it is denoted by $f(x)$.*

That is, a function f assigns a uniquely determined value $f(x)$ to each element x of its domain $\mathrm{D}(f)$.

The set $\{f(x) : x \in \mathrm{D}(f)\}$ of values of the function f is called its range, *and is denoted by $\mathrm{R}(f)$.*

It follows from our conventions that the functions f and g are equal if and only if $\mathrm{D}(f) = \mathrm{D}(g)$ and for all $x \in \mathrm{D}(f)$ we have $f(x) = g(x)$.

The function f is called *one-to-one* if for $x \neq y$ with $x, y \in \mathrm{D}(f)$ we have $f(x) \neq f(y)$.

A function is also called a *mapping*. We say that a function f *maps* the set A *into* the set B if $\mathrm{D}(f) = A$, $\mathrm{R}(f) \subset B$. This state of affairs will sometimes be indicated symbolically as $f : A \to B$. This expression is to be distinguished from $f : x \mapsto y$; the latter signifies that $f(x) = y$.

The function f is said to *map* the set A *onto* the set B if $\mathrm{D}(f) = A$ and $\mathrm{R}(f) = B$.

If f is a one-to-one function, then its *inverse function* is the function g for which $\mathrm{D}(g) = \mathrm{R}(f)$ and for all $y \in \mathrm{R}(f)$ we have $f(g(y)) = y$.

If f and g are functions for which $\mathrm{R}(f) \subset \mathrm{D}(g)$, then $g \circ f$ is their *composition*, i.e., the function for which $\mathrm{D}(g \circ f) = \mathrm{D}(f)$ and $(g \circ f)(x) = g(f(x))$ for all $x \in \mathrm{D}(f)$.

We will also use the following notation: If f is a function and $A \subset D(f)$ is a set, the set

$$\{y : y \in R(f) \wedge \exists x \in A(f(x) = y)\} = \{f(x) : x \in A\}$$

will be denoted by $f``A$. This set is usually denoted by $f(A)$. This can, however, lead to a misunderstanding, since A can also be an element of $D(f)$. For this reason we need this notation, which is not used in other branches of mathematics. If there is no danger of misunderstanding, then, to simplify the task of the reader, we will also use the usual notation.

Given a function f and a set A, we will write

$$f^{-1}(A) = \{x \in D(f) : \exists y \in A(f(x) = y)\}.$$

We will use this notation even if the function f^{-1} is not defined.

11. There are one-to-one assignments that cannot be considered functions in the light of the above conventions. For example, for each object x, we can assign $\{x\}$. This assignment is not a function, since its domain is not a set. If this assignment were a function, then every set would be an element of its domain, which is impossible in view of Theorem 1.2. In the development of set theory, such assignments also occur. For this reason we will give a name to such assignments. If to each object x we assign another object $F(x)$ with the aid of a meaningful mathematical expression F, then this assignment F will be called an *operation*.

The notion of operation described has not been defined, just as the notion of property was not defined above (in Convention 6); the same comments that apply to the use of the notion of property also apply to the use of the notion of operation. We remark that the notion of operation can be reduced to that of property. An operation $F(x)$ can be defined with the aid of a two-variable property $\Phi(x, y)$ such that for each x there is exactly one y for which this property holds, and this unique value y can be called $F(x)$.

We stipulate that if F is an operation and A is a set, then the values $F(x)$ assigned to elements x of the set A also constitute a set; this set will be denoted by $\{F(x) : x \in A\}$.

For example, if $A = \{0, 1, 2\}$, then $\{\{x\} : x \in A\} = \{\{0\}, \{1\}, \{2\}\}$.

It often happens that an operation is meaningfully defined only for elements having a certain property. For example, in Convention 10 we defined the operations D and R, which assigned to a function f its domain $D(f)$ and its range $R(f)$, respectively. We may stipulate that the value of such an operation will be $\emptyset$ for each set for which its value has not been defined.

12. We will use the notation $\{A_\gamma : \gamma \in \Gamma\}$ (instead of the more precise notation $\langle A_\gamma : \gamma \in \Gamma \rangle$) in the sense that this is a function A for which $D(A) = \Gamma$ and $A(\gamma) = A_\gamma$ for $\gamma \in \Gamma$. We will use the customary notation

$\bigcup_{\gamma \in \Gamma} A_\gamma$ and $\bigcap_{\gamma \in \Gamma} A_\gamma$ for the union and *intersection*, respectively, of sets; these are defined as

$$\bigcup_{\gamma \in \Gamma} A_\gamma = \{x : \exists \gamma \in \Gamma \, (x \in A_\gamma)\},$$

$$\bigcap_{\gamma \in \Gamma} A_\gamma = \{x : \forall \gamma \in \Gamma \, (x \in A_\gamma)\}.$$

If A and B are two sets then $A \setminus B$ will denote the set $\{x \in A : x \notin B\}$, the *difference* of the sets A and B.

We will not enumerate the simple properties of the set operations; these properties can be found in a number of books, and we will assume that they are well known. We will also use the notation $\bigcap A = \bigcap_{a \in A} a$ for an arbitrary collection A of sets. As usual, we adopt the convention that in case $A = \emptyset$ the expression $\bigcap A$ is defined only in case we work with the subsets of an underlying set X. In this case we put $\bigcap A = X$. We will also write $A \cup B = \bigcup \{A, B\}$ and $A \cap B = \bigcap \{A, B\}$. The sets A and B are called *disjoint* if $A \cap B = \emptyset$.

At the end of this section, we would like to point out that in the spirit of Convention 6, from Convention 10 on we have defined sets without seeking a dominating set in definitions of the form $\{x : \Phi(x)\}$. The conscientious reader can fill in these gaps by solving the problems listed below.

Problems

Verify the following assertions.

1. If A is a set and $x, y \in A$, then $\{x\}, \{x, y\} \in \mathrm{P}(A)$, $\langle x, y \rangle \in \mathrm{P}(\mathrm{P}(A))$.

2. If A is a set, then the set B of all ordered pairs formed from elements of A exists.

3. If A is a set and $\{x, y\} \in A$, then $x, y \in \bigcup A$.

4. If A is a set and $\langle x, y \rangle \in A$, then $x, y \in \bigcup\bigcup A$.

5. If f is a function, then $\mathrm{D}(f), \mathrm{R}(f) \subset \bigcup\bigcup f$.

6. If f is a function that maps the set A into the set B, then $f \subset \mathrm{P}\left(\mathrm{P}(\bigcup\{A, B\})\right)$.

7. If f is a one-to-one function then $f^{-1} \subset \mathrm{P}\left(\mathrm{P}\left(\bigcup\{\mathrm{D}(f), \mathrm{R}(f)\}\right)\right)$.

8. If f is the function described in Convention 12, then

$$\bigcup_{\gamma \in \Gamma} A_\gamma = \bigcup \mathrm{R}(A).$$

2. DEFINITION OF EQUIVALENCE. THE CONCEPT OF CARDINALITY. THE AXIOM OF CHOICE

In Section 1, we enumerated all the conventions that we are going to use, with one exception. The omitted convention is the Axiom of Choice, which will be formulated in this section. Already, on the basis of our conventions so far, however, we are able to describe Cantor's key idea that led to the development of set theory. We are going to define a property called equivalence between sets. This property will express the statement that two sets have the "same number" of elements. This is the first step towards defining the concept of one set having "more" elements than another.

Definition 2.1. *The state of affairs that the function f maps the set A onto the set B in a one-to-one way will be denoted as $A \sim_f B$, and will be expressed as: A is equivalent to B according to the function f. The sets A and B will be called* equivalent *if there is a function f for which $A \sim_f B$. In symbols: $A \sim B$. If A is not equivalent to B, then we will write $A \nsim B$.*

Sets equivalent to each other will be regarded as having the "same size." This is justified in view of the following.

Theorem 2.1. *The property $\sim$ is an equivalence property, that is, it satisfies the following three conditions:*
If A, B, C are arbitrary sets then

1. $A \sim A$, *that is, $\sim$ is reflexive;*
2. *if $A \sim B$ then $B \sim A$, that is, $\sim$ is symmetric;*
3. *if $A \sim B$ and $B \sim C$ then $A \sim C$, that is, $\sim$ is transitive.*

Proof. 1. Let Id_A be the identity function on the set A, that is, the function for which $\mathrm{D}(\mathrm{Id}_A) = A$ and $\mathrm{Id}_A(x) = x$ for all x. Then $A \sim_{\mathrm{Id}_A} A$.

2. Assume that $A \sim B$, and let f be a function for which $A \sim_f B$. Then f^{-1} is a function with which $B \sim_{f^{-1}} A$ holds, and so $B \sim A$ also holds.

3. Assume that $A \sim B$ and $B \sim C$. Let f, g be functions for which we have $A \sim_f B$ and $B \sim_g C$. Then $A \sim_{g \circ f} C$ holds, and so $A \sim C$ is satisfied.

$$* \qquad *$$
$$*$$

We now know when two sets are said to have the "same size." In what follows, our goal is to see that there is an operation which assigns the same object to sets of the "same size," and assigns different things to sets not of the "same size." We will also introduce a name for such operations.

Definition 2.2. *We say that the operation F is compatible with the property $\sim$ if for any two sets A, B we have*

$$F(A) = F(B) \iff A \sim B.$$

Before pursuing the goal we will make a small detour. The example below will show that it is possible to have two equivalent sets such that one is a proper subset of the other. That is, an (infinite) set may be equivalent to one of its proper subsets.

Example. *Let $A = \omega \setminus \{0\} = \{1, 2, \ldots\}$ and let f be the function for which $D(f) = \omega$ and $f(n) = n + 1$ for $n \in \omega$. Then $\omega \sim_f A$.*

For the proof of the next theorem, acquaintance with the method of mathematical induction is necessary. This method of proof is used in several branches of mathematics. In Section 9 we will show that proofs by mathematical induction are justified. Up to that point we will use this method of proof without any reservations. We would like to point out here that some proofs below will also use the well-known fact that it is possible to define functions on ω by recursion. The justification of this fact will also be established in Section 9. We now return to our set goal. First we want to show that for finite sets, to be defined next, there is an operation that is compatible with the property $\sim$.

Definition 2.3. *For an arbitrary $n \in \omega$ denote by $\underline{n}$ the set $\{0, 1, \ldots, n - 1\}$ (for $n = 0$ put $\underline{0} = \emptyset$). The set A is called a finite set if there is an $n \in \omega$ such that $A \sim \underline{n}$. The set A is called infinite if it is not finite.*

Theorem 2.2. *If n and m are distinct elements of ω then $\underline{m} \nsim \underline{n}$.*

The assertion is easily proved by (mathematical) induction on the maximum of m and n. The details will be left to the reader.

$$* \qquad *$$
$$*$$

It follows from Theorem 2.2 that a finite set cannot be equivalent to any of its proper subsets. In view of Theorem 2.2, we can define the number of elements of A for a finite set A as the integer $n \in \omega$ such that $A \sim \underline{n}$. The operation so defined for finite sets is compatible with the property $\sim$ on finite sets. We would like to extend this operation to infinite sets.

If there were a set A consisting of all sets, then this could be accomplished easily. It is well known from algebra that the equivalence relation $\sim$ splits A up into pairwise disjoint equivalence classes, and so we would be able to assign to each set the equivalence class containing it. However, this approach is not feasible in view of Theorem 1.2.

For the formulation of the forthcoming statements we will need the Axiom of Choice. This is the axiom which elicited the most controversy among mathematicians in the course of the development of set theory. In most branches of mathematics, the Axiom of Choice and its consequences are accepted and used. It is not always pointed out explicitly when it is made use of. As we mentioned, set theory can be partially developed without the Axiom of Choice, but such a development encounters a number of unnecessary difficulties. So we do not take this approach, and we will accept and use the Axiom of Choice. In the introductory sections (up to Section 10) we will nevertheless point out explicitly on each occasion when this axiom is used.

Definition 2.4. *The function will be called a* choice function *for the set system* $\{A_\gamma : \gamma \in \Gamma\}$ *if* $D(f) = \Gamma$ *and for each element* γ *of* Γ *we have* $f(\gamma) \in A_\gamma$.

Clearly, in order that such a choice function should exist, it is necessary that the sets A_γ be nonempty. The Axiom of Choice states that this assumption is also sufficient.

Axiom of Choice. *For every system* $\{A_\gamma : \gamma \in \Gamma\}$ *of nonempty sets there is a choice function.*

We would like to point out that, for a finite index set Γ, the Axiom of Choice can easily be proved from our earlier conventions. For an infinite index set Γ, this axiom is not provable from the other usual axioms of set theory.

Theorem 2.3. *There is an operation that is compatible with the property* $\sim$.

A proof of this theorem, due to John von Neumann, uses the Axiom of Choice and is based on the theory of wellordered sets. This proof will be given later (see Theorem 10.1 below). Without using the Axiom of Choice the existence of such an operation cannot be proved on the basis of our conventions so far; if we had not assumed the Axiom of Choice, then the notion of cardinality to be defined below would have to be taken as a new primitive notion, and its compatibility with the equivalence property would have to be taken as an axiom.

Definition 2.5. *In what follows, a certain operation whose existence is claimed in Theorem 2.3 will be called the* cardinality *of A and will be denoted as* $|A|$.

According to Theorem 2.2, we may assume that if A is a finite set, then $|A| = n$ for the $n \in \omega$ for which $\underline{n} \sim A$. (Indeed, if F_1 is an operation that is compatible with $\sim$, then we can easily define an operation F_2 that does not assume any values in ω. For example, the operation

$$F_2(A) = \langle F_1(A), \omega \rangle$$

is like this. If we then define F as

$$F(A) = F_2(A) \qquad \text{if } A \text{ is not finite}$$

and

$$F(A) = \underline{n} \qquad \text{if } A \text{ is finite and } A \sim \underline{n},$$

then the operation so defined is compatible with the property $\sim$ in view of Theorem 2.2, and we have $|A| \in \omega$ if A is finite.) The values of the operation $|A|$ are called cardinals. The nonnegative integers are called finite cardinals, and the cardinalities of infinite sets are called infinite cardinals.

Infinite cardinals can be considered as generalizations of natural numbers. In the next four sections, we will show that there is an abundance of infinite cardinals, and, further, that the addition, multiplication, exponentiation, and the ordering according to size of the nonnegative integers can be extended to cardinals in such a way that a number of properties of these operations and of the ordering according to size are preserved.

3. COUNTABLE CARDINAL, CONTINUUM CARDINAL

According to Theorem 2.2 and the Example preceding it, the set ω of non-negative integers is an infinite set that is not equivalent to $\underline{n}$ for any $n \in \omega$.

Definition 3.1. *The sets that are equivalent to ω are called* countably infinite *sets.*

The cardinality of ω is denoted by $\aleph_0$. This notation was introduced by G. Cantor. ($\aleph$, called *aleph*, is the first letter of the Hebrew alphabet.)

The set A is called a *countable* set if it is either finite or countably infinite.

In this section, we will prove the well-known elementary theorems about countable sets. First we give those proofs that do not need the Axiom of Choice.

Theorem 3.1. $\aleph_0$ *is an infinite cardinal.*

Proof. ω is an infinite set.

*　　*
*

Theorem 3.2. *Every subset of a countable set is countable.*

Proof. In view of the transitivity of the property $\sim$, it is sufficient to prove that every subset of ω is countable. Let $A \subset \omega$. If A is finite, then there is nothing to prove; so we may assume that A is infinite. Define a function f on A with the stipulation that

$$f(n) = |A \cap \underline{n}|$$

for each $n \in A$. That is, $f(n)$ is the number of elements of A less than n. Since every subset of a finite set is finite, we have

$$f(n) \in \omega \quad \text{for} \quad n \in A.$$

If $m < n$ and $m, n \in A$ then

$$A \cap \underline{m} \subsetneq A \cap \underline{n},$$

and so $f(m) < f(n)$; that is, f is one-to-one. We will show that f maps A onto ω. To this end, it is sufficient to show that for every $k \in \omega$ we have

$k \in \mathrm{R}(f)$. If n is the smallest element of A, then $f(n) = 0$, and so $0 \in \mathrm{R}(f)$. Assume that $k \in \mathrm{R}(f)$. Then $f(m) = k$ for some $m \in A$. As A is infinite, it has an element that is greater than m. Let m' be the smallest among these. Then $f(m') = k + 1$, and so $k + 1 \in R(f)$.

$$*\qquad*$$
$$*$$

Corollary 3.1. *The set A is countable if and only if either $A = \emptyset$ or there is a function f that maps ω onto A. The latter statement expresses the fact that A can be written as a sequence $A = \{a_n : n \in \omega\}$.*

Proof. The "only if" part of the statement is obvious. Assume that f maps ω onto A. Define the function g on the set A such that

$$g(n) = \min\{k : f(k) = n\} \qquad \text{for} \quad n \in A.$$

Then g maps the set A onto a subset of ω in a one-to-one way. Hence A is countable, according to the preceding theorem.

$$*\qquad*$$
$$*$$

In what follows, we will prove that several sets that appear "larger" than ω are in fact countable.

Theorem 3.3. *Let $B = \{\langle n, m \rangle : n, m \in \omega\}$, that is, the set of ordered pairs formed by nonnegative integers. Then B is a countably infinite set.*

We will give two proofs.

First Proof. Define the function f as follows: If $\langle n, m \rangle \in B$, then put

$$f(\langle n, m \rangle) = 2^n \cdot 3^n.$$

According to the theorem on Unique Prime Factorization, f is one-to-one, and it maps B into ω. Hence B is countable, according to Theorem 3.2.

We remark that the theorem on Unique Prime Factorization can easily be proved by induction, and so the reader should consider the first proof just as acceptable as the next one.

Second Proof. The following arrangement shows the elements of B:

$$\langle 0,0 \rangle, \langle 0,1 \rangle, \langle 0,2 \rangle, \ldots, \langle 0,n \rangle, \ldots$$

$$\langle 1,0 \rangle, \langle 1,1 \rangle, \langle 1,2 \rangle, \ldots, \langle 1,n \rangle, \ldots$$

$$\vdots$$

$$\langle n,0 \rangle, \langle n,1 \rangle, \langle n,2 \rangle, \ldots, \langle n,n \rangle, \ldots$$

$$\vdots$$

The set B can be arranged into a sequence as follows: $a_0 = \langle 0,0 \rangle$ is the first element of the sequence. $a_1 = \langle 0,1 \rangle$ and $a_2 = \langle 1,0 \rangle$ are the next two elements of the sequence. If the first $\binom{n+1}{2}$ elements of the sequence a_n are given, then in the next $n+1$ steps we enumerate the elements of the $(n+1)$st secondary diagonal in the diagram, proceeding in the direction of the arrow. In this diagonal, we find those pairs $\langle k, l \rangle$ for which $k + l = n + 1$, and their order in the enumeration is determined by stipulating that those with smaller first elements come earlier.

It may be worth pointing out that there is a simple explicit formula for the mapping defined above. Denoting this function by f, we have

$$f(\langle m,n \rangle) = \frac{1}{2}(m+n)(m+n+1) + m.$$

It is easy to show directly that this function maps the set of all ordered pairs of nonnegative integers onto the set of all nonnegative integers in a one-to-one way. The function f is sometimes called the Cantor pairing function.

$$* \quad *$$
$$*$$

In the proof of the next theorem, we need to use the Axiom of Choice.

Theorem 3.4. *The union of countably many countable sets is also countable. That is, if $\{A_n : n \in \omega\}$ is a sequence of sets such that A_n is countable for each $n \in \omega$, then $\bigcup_{n \in \omega} A_n$ is also a countable set.*

Proof. We may assume that $A_n \neq \emptyset$ for $n \in \omega$, since by adding elements to the sets A_n their union can only increase. For an arbitrary $n \in \omega$ let

$$B_n = \{f : f \text{ is a function} \wedge \mathrm{D}(f) = \omega \wedge \mathrm{R}(f) = A_n\}.$$

According to our assumption, the set system $\{B_n : n \in \omega\}$ consists of nonempty sets. Therefore, by the Axiom of Choice, there is a choice function

f for it. Denote by f_n the value of the function f assumed at n. Then we have

$$A_n = \{f_n(k) : k \in \omega\}$$

for each $n \in \omega$. Let g be a function with domain the set of ordered pairs formed by elements of ω, and write

$$g(\langle n, k \rangle) = f_n(k) \qquad \text{for} \quad n, k \in \omega.$$

Then

$$R(g) = \bigcup_{n \in \omega} A_n.$$

Thus $\bigcup_{n \in \omega} A_n$ is countable according to Theorems 3.2 and 3.3.

*   *
*

We mentioned above that for an arbitrary finite system of nonempty sets, we can find a choice function without assuming the Axiom of Choice. In virtue of this remark, the proof of Theorem 3.4 shows the following: the fact that the union of finitely many countable sets is countable can be proved without the Axiom of Choice.

Theorem 3.5. *The set of rational numbers is countable.*

Proof. Let $\mathbb{Q}^+$ and $\mathbb{Q}^-$ be the set of the positive and of the negative rational numbers, respectively. Then

$$\mathbb{Q} = \mathbb{Q}^+ \cup \{0\} \cup Q^-.$$

Each $r \in \mathbb{Q}^+$ can be uniquely written in the form $\frac{p}{q}$ of an irreducible fraction, where $p, q \in \omega$. Hence Q^+ is countable according to Theorems 3.2 and 3.3. Similarly, Q^- is also countable, and so Q is countable in view of Theorem 3.4.

*   *
*

Theorem 3.6. *If A is an infinite set, then it has a subset of cardinality $\aleph_0$.*

Proof. Let f be a function that to each set $X \in P(A)$ with $X \neq \emptyset$ assigns one of its elements $f(X)$. Such a function f exists according to the Axiom of Choice.

We are going to define a function g on ω by specifying the values $g(n) \in R(g)$ for each $n \in \omega$ via recursion. Let $g(0) = f(A)$. This is well defined, since A is not empty; indeed, it is not even finite. Assume that $n > 0$, and that $g(i)$ has already been defined for each $i < n$. The set $\{g(0), \ldots, g(n-1)\}$ is a finite set, and so

$$A \setminus \{g(0), \ldots, g(n-1)\}$$

is not empty, since otherwise A would be finite. Let

$$g(n) = f(A \setminus \{g(0), \ldots, g(n-1)\}).$$

Thus $g(n)$ has been defined for all $n \in \omega$. If $i < n$ then $g(n) \neq g(i)$, since

$$g(n) \notin \{g(0), \ldots, g(n-1)\}.$$

Therefore, g is a one-to-one function. Put $A' = \mathrm{R}(g)$. According to the definition of g, we have $A' \subset A$ and, further, $\omega \sim_g A'$; consequently

$$|A'| = \aleph_0.$$

*   *
*

So far, we have discussed only results pertaining to finite and countably infinite sets. The next theorem is the first one that establishes the existence of an *uncountable*, i.e., non-countable, set. In the course of proving this theorem, Cantor discovered one of the most fundamental ideas of set theory, the "method of diagonalization."

Theorem 3.7. $\mathbb{R}$ *is not countable.*

Proof. It is sufficient to prove that for an arbitrary nonempty countable set $A \in \mathbb{R}$ there is a real number $y \in \mathbb{R} \setminus A$. As A is countable, it can be written in the form

$$A = \{a_n : n \in \omega\}.$$

The real number a_n has a unique representation as a decimal fraction that does not consist purely of the digit nine from a certain point on. Denote by $a_{n,k}$ the kth digit of a_n after the decimal point in this representation. Define the sequence y_n by stipulating that

$$y_n = \begin{cases} 0 & \text{if } a_{n,n} \neq 0, \\ 1 & \text{if } a_{n,n} = 0. \end{cases}$$

Then $0.y_0 \ldots y_n \ldots$ is the decimal fraction representation of a real number y that does not consist of the digit nine from a certain point on. Furthermore, for an arbitrary $n \in \omega$, the nth digits of the numbers y and a_n after the decimal point are different; thus we have $y \neq a_n$ in view of the uniqueness of the decimal fraction representation. Thus $y \notin A$.

*   *
*

The cardinality of $\mathbb{R}$ is denoted by $\mathfrak{c}$ and is called the cardinal *continuum*.

Theorem 3.8. *If $(a, b) \subset \mathbb{R}$ is an arbitrary nonempty open interval, then*

$$|(a, b)| = \mathfrak{c}.$$

Proof. If f is an arbitrary strictly increasing continuous function on (a, b), then, by the well-known theorem in elementary analysis, f maps (a, b) in a one-to-one way onto the open interval (α, β) with

$$\alpha = \lim_{x \to a+0} f(x), \qquad \beta = \lim_{x \to b-0} f(x).$$

Hence

$$(a, b) \sim_f \left(-\frac{\pi}{2}, \frac{\pi}{2}\right),$$

where f is an appropriate linear function; in fact,

$$f(x) = \frac{\pi}{b - a} x - \frac{\pi}{2} \cdot \frac{b + a}{b - a}.$$

Furthermore, $\left(-\frac{\pi}{2}, \frac{\pi}{2}\right) \sim_{\tan} \mathbb{R}$, where tan denotes the tangent function.

$$* \qquad *$$
$$*$$

According to the theorems proven so far, the cardinals $n \in \omega$, $\aleph_0$, and $\mathfrak{c}$ are all distinct. It would be interesting to calculate the cardinalities of various simple sets at this point. We will do this later, in Section 6, when we are in possession of some general results. As we have seen, there are at least two distinct infinite cardinalities. We are now going to carry out our program of extending the arithmetic operations and the ordering according to size to cardinals.

4. COMPARISON OF CARDINALS

We start the extension of ordering according to size from numbers to cardinals by stipulating when a cardinal is less than or equal to another one.

Definition 4.1. Let a and b be arbitrary cardinals. We say that the cardinal a is less than or equal to the cardinal b if there are sets A and B such that $|A| = a$, $|B| = b$, and there is a one-to-one mapping of A onto a subset of B. This state of affairs will be denoted as $a \leq b$.

In order to show the soundness of this definition we have to show that the property $\leq$ does not depend on the choice of the sets A and B.

Theorem 4.1. Let a and b be arbitrary cardinals and let A, A', B, B' be sets for which $|A| = |A'| = a$, $|B| = |B'| = b$. If there is a one-to-one function f for which $D(f) = A$ and $R(f) \subset B$, then there is a one-to-one function g such that $D(g) = A'$ and $R(g) \subset B'$.

Proof. As $|A| = |A'|$ and $|B| = |B'|$, there are mappings h, k such that $A' \sim_h A$ and $B \sim_k B'$.

Let $g = k \circ (f \circ h)$. Then $D(g) = A'$, $R(g) \subset B'$. The function g is one-to-one, since the functions f, h, k are one-to-one.

$$* \qquad *$$
$$*$$

It is easy to verify the following.

Corollary 4.1. For arbitrary cardinals a, b we have $a \leq b$ if and only if there are sets $A \subset B$ for which $|A| = a$ and $|B| = b$.

$$* \qquad *$$
$$*$$

On the basis of this remark, we already know that $m \leq n$ if $m, n \in \omega$ and m is not greater than n in the ordering of integers; further, $n \leq \aleph_0$ if $n \in \omega$; finally, $\aleph_0 \leq c$. Indeed, these assertions follow from the relations $\underline{m} \subset \underline{n}$, $\underline{n} \subset \omega$, $\omega \subset \mathbb{R}$, respectively.

Theorem 4.2. *The property $\leq$ defined for cardinals is reflexive and transitive.*

Proof. The proof of reflexivity is obvious, since we have $A \subset A$ for an arbitrary set A. For the proof of transitivity, assume that the cardinals a, b, c are such that

$$a \leq b \quad \text{and} \quad b \leq c.$$

Choose sets A, B, C and one-to-one functions f, g such that

$$|A| = a, \quad |B| = b, \quad |C| = c,$$

$$\mathrm{D}(f) = A, \quad \mathrm{R}(f) \subset B, \quad \mathrm{D}(g) = B, \quad \text{and} \quad \mathrm{R}(g) \subset C.$$

Then the function $h = g \circ f$ maps the set A into C in a one-to-one way.

*   *
*

We recall a lemma from algebra. This lemma is needed since it gives a condition to convert a property "of type $\leq$" into a property "of type $<$."

Lemma 4.1. *Let $\leq$ be a reflexive, transitive, and anti-symmetric property. The last attribute means that $x \leq y$ and $y \leq x$ imply $x = y$. Define the property $x < y$ as saying that $x \leq y$ and $x \neq y$. Then the property $<$ is irreflexive and transitive.*

Proof. We have $\neg(x < x)$, since $x = x$.

For the proof of transitivity, assume that we have $x < y$ and $y < z$ for the elements x, y, z. Then

$$x \leq y \quad \text{and} \quad y \leq z.$$

From this, we can conclude that $x \leq z$ by the transitivity of the property $\leq$. If we had $z = x$, then we would also have

$$y \leq x$$

in view of $y \leq z$. Taking this together with $x \leq y$, we would obtain $x = y$ in view of anti-symmetry; this would contradict the assumption $x < y$. Thus $x \neq z$, and so $x < z$.

*   *
*

In view of this lemma, we will need the following theorem.

Bernstein's Equivalence Theorem 4.3. *The property $\leq$ defined for cardinals is anti-symmetric.*

For this, according to the definition of the property $\leq$, we need to prove the following assertion about sets:

If A, B are sets and f, g are one-to-one functions such that

$$D(f) = A, \quad R(f) \subset B, \quad D(g) = B, \quad \text{and} \quad R(g) \subset A,$$

then

$$A \sim B,$$

that is, there is a function h such that $A \sim_h B$.

Before we formulate a somewhat more general theorem that will give this assertion as a consequence, we will introduce some notation concerning functions.

Definition 4.2. *If f is a function and A is a set, then the restriction of f to A, denoted as $f|A$, is the function whose domain is $D(f) \cap A$, and which has the property that for every element x of $D(f) \cap A$ we have $(f|A)(x) = f(x)$.*

A similar notation will be used even if F is an operation; we point out that $F|A$ is a function even in this case, according to Convention 11 in Section 1.

Instead of Theorem 4.3 we will prove the following theorem.

Theorem 4.4. *Assume that A, B, f, g satisfy the conditions in the assertion given after Theorem 4.3. Then there are sets A', B', A'', B'' such that*

$$A = A' \cup A'', \quad A' \cap A'' = \emptyset,$$

$$B = B' \cup B'', \quad B' \cap B'' = \emptyset,$$

and such that

$$A' \sim_{f|A'} B', \quad B'' \sim_{g|B''} A''.$$

As a consequence, we have $A \sim B$. Indeed, we can define a one-to-one mapping h of A onto B as follows:

$$h(x) = \begin{cases} f(x) & \text{for } x \in A', \\ g^{-1}(x) & \text{for } x \in A''. \end{cases}$$

Proof. For the reader with a certain familiarity with mathematics, the following facts are well known: If f is an arbitrary function and $\{A_\gamma : \gamma \in \Gamma\}$ is a set system for which $A_\gamma \subset D(f)$ for each $\gamma \in \Gamma$, then

(1)
$$f``\bigcup_{\gamma \in \Gamma} A_\gamma = \bigcup_{\gamma \in \Gamma} f``A_\gamma.$$

Furthermore, if f is one-to-one, then

(2)
$$f``\bigcap_{\gamma \in \Gamma} A_\gamma = \bigcap_{\gamma \in \Gamma} f``A_\gamma.$$

also holds. The proofs of assertions (1) and (2) are simple and they are left to the reader.

If the sets A', B', A'', B'' satisfy the requirements of the theorem, then the following relations hold:

$$B' = f``A',$$
$$B'' = B \setminus f``A',$$
$$A'' = g``(B \setminus f``A'),$$
$$A' = A \setminus g``(B \setminus f``A').$$

The last relation suggests that we consider the mapping $h(X) = A \setminus g``(B \setminus f``X)$, which sends an arbitrary subset X of A to another subset $h(X) \subset A$, and look for the "fixed point" of the mapping h, i.e., for a subset $X \subset A$ such that $h(X) = X$.

We will do this via "iteration." By recursion, we define the sets $A_n \subset A$ as follows:

$$A_0 = A, \quad A_{n+1} = h(A_n) \quad \text{for } n \in \omega.$$

Let $A' = \bigcap_{n \in \omega} A_n$ and $B' = f``A'$. Then

$$B' = \bigcap_{n \in \omega} f``A_n$$

in virtue of (2). Put $B'' = B \setminus B'$. Then we have

$$B'' = \bigcup_{n \in \omega} B \setminus f``A_n$$

according to the De Morgan identity. Let $A'' = g``B''$. Then we have

$$A'' = \bigcup_{n \in \omega} g``(B \setminus f``A_n)$$

by (1). Using the De Morgan identity again, we obtain

$$A \setminus A'' = A \setminus g``B'' = h(A') = \bigcap_{n \in \omega} A \setminus g``(B \setminus f``A_n) = \bigcap_{n \in \omega} A_{n+1}.$$

Taking into account that $A_0 = A$, we obtain that

$$\bigcap_{n \in \omega} A_{n+1} = \bigcap_{n \in \omega} A_n = A'.$$

That is, we have

$$A \setminus A'' = A' = h(A').$$

As $B' = f``A'$ and $A'' = g``B''$, the sets so constructed satisfy the requirements of the theorem.

*   *
*

Definition 4.3. *We say that cardinal a is less than cardinal b, or, in symbols, $a < b$, if $a \leq b$ and $a \neq b$.*

The corollaries below show that these definitions are appropriate, and the property $<$ is a generalization of the ordering of numbers.

Corollary 4.2. *For arbitrary cardinals a, b the relation $a < b$ is true if and only if for any two sets A, B with $|A| = a$, $|B| = b$ there is a $B' \subset B$ with $A \sim B'$ and $A \nsim B$.*

Proof. The first half of the condition is equivalent to the statement $a \leq b$, and the second half, to the statement $a \neq b$.

*   *
*

Corollary 4.3. *The property $<$ is irreflexive and transitive on cardinals.*

Proof. The property $\leq$ is reflexive and transitive according to Theorem 4.2, and it is anti-symmetric according to the Equivalence Theorem 4.3. Hence the assertion follows from Lemma 4.1 above.

*   *
*

Corollary 4.4. *The property $<$ on cardinals is an extension of the ordering by size of numbers.*

Proof. Taking Theorem 2.2 into account, it is easy to see that in both orderings the statement "n is less than m" is equivalent to $\underline{n} \subsetneq \underline{m}$ for arbitrary $n, m \in \omega$.

*   *
*

For the cardinals we met so far, we have

$$0 < 1 < 2 < \cdots < n < \cdots < \aleph_0 < \mathfrak{c}.$$

The question arises whether any two cardinals are comparable. With the aid of the Axiom of Choice, we will show (in Corollary 10.3) that the answer is affirmative, that is, for cardinals the property $<$ is trichotomous. We mention here without proof that this trichotomy implies the Axiom of Choice. For the present, we will prove the special case of trichotomy saying that $\aleph_0$ is comparable to every cardinal:

Theorem 4.5. $\aleph_0$ *is the smallest infinite cardinal.*

Proof. Let a be an infinite cardinal and let A be a set with $|A| = a$. According to Theorem 3.6, ω is equivalent of a subset of A, and so $\aleph_0 \leq a$.

$$* \qquad *$$
$$*$$

We would like to point out that in the proof of Theorem 3.6 we used the Axiom of Choice; in fact, neither Theorem 3.6 nor Theorem 4.5 can be proved from the conventions of Section 1 alone; we will express this by saying that these results cannot be proved without assuming the Axiom of Choice (although in certain cases some related weaker assumptions might also suffice).

The following theorem of G. Cantor ensures that there are infinitely many infinite cardinals, and, in fact, for every cardinal there is a cardinal that is larger.

Theorem 4.6 *(Cantor's Theorem). For every set A we have*

$$|A| < |\mathrm{P}(A)|.$$

Proof. We have $|A| \leq |\mathrm{P}(A)|$, since the function $\{x\}|A$ maps A into $\mathrm{P}(A)$ in a one-to-one way. For this reason, it is sufficient to show that $A \nsim \mathrm{P}(A)$. We use reduction ad absurdum. Assume, on the contrary, that there is a function f for which $A \sim_f \mathrm{P}(A)$. .

Let $B = \{a \in A : a \notin f(a)\}$. Then $B \subset A$, and so $B \in \mathrm{P}(A)$. Thus, there is an element $b \in A$ for which $f(b) = B$, Then we have

$$a \in f(b) \iff a \notin f(a)$$

for an arbitrary $a \in A$, according to the definition of B. As $b \in A$, we may replace a with b here; we obtain

$$b \in f(b) \iff b \notin f(b).$$

This is a contradiction. Hence our assumption must be false; therefore $A \nsim \mathrm{P}(A)$.

$$* \qquad *$$
$$*$$

A consequence of this theorem is that

$$|\omega| < |\mathrm{P}(\omega)| < \cdots < |\mathrm{P}^n(\omega)| < \ldots,$$

where the sets $\mathrm{P}^n(\omega)$ are defined by recursion on n:

$$\mathrm{P}^0(\omega) = \omega \quad \text{and} \quad \mathrm{P}^{n+1}(\omega) = \mathrm{P}(\mathrm{P}^n(\omega)) \quad \text{for} \quad n \in \omega.$$

We will state one more theorem in this section. We would like to mention that this cannot be proved without the Axiom of Choice either (see the remark after Theorem 4.5 for the precise meaning of this phrase).

Theorem 4.7. *If the function f maps the set A onto the set B, then* $|B| \le |A|$.

Proof. For each $y \in B$ let $A_y = \{x \in A : f(x) = y\}$. According to the assumption, the set system $\{A_y : y \in B\}$ consists of nonempty sets. By the Axiom of Choice, there is a choice function g with $g(y) \in A_y$ for each $y \in B$. It is obvious that g maps the set B onto a subset of A in a one-to-one way.

5. OPERATIONS WITH SETS AND CARDINALS

We now turn to the extension to cardinals of the operations defined for numbers. To this end, we need some results involving set operations.

Definition 5.1. *Let* $\{A_\gamma : \gamma \in \Gamma\}$ *be a set system. The* direct product *or* Cartesian product, *denoted as* $\mathsf{X}_{\gamma \in \Gamma} A$, *is defined as the set*

$$\{f : f \text{ is a function } \quad and \quad D(f) = \Gamma \quad and \quad \forall \gamma \in \Gamma (f(\gamma) \in A_\gamma)\},$$

that is the set of all choice functions for the given set system.

Our definition is legitimate, since all elements of the set $\mathsf{X}_{\gamma \in \Gamma} A_\gamma$ belong to the set

$$P\left(P\left(P(\Gamma \cup \bigcup_{\gamma \in \Gamma} A_\gamma)\right)\right).$$

As a special case, the above definition also gives the direct product of two sets. If A_0 and A_1 are sets, then we write $A_0 \times A_1 = \mathsf{X}_{i \in 2} A_i$. This is isomorphic, but not identical to the set of ordered pairs taken from A_0 and A_1. If there is no danger of misunderstanding, we will, in what follows, use the notation $A_0 \times A_1$ also for the set formed by ordered pairs.

In verifying certain properties of cardinal operations, the following "associative" law will be very useful.

Theorem 5.1. *Let* $\{A_\gamma : \gamma \in \Gamma\}$ *be a set system. Assume that*

$$\Gamma = \bigcup_{\delta \in \Delta} \Gamma_\delta,$$

where the sets Γ_δ *are pairwise without common elements, i.e., pairwise disjoint. Then*

$$\mathsf{X}_{\gamma \in \Gamma} A_\gamma \sim \mathsf{X}_{\delta \in \Delta} \left(\mathsf{X}_{\gamma \in \Gamma_\delta} A_\gamma \right).$$

Proof. Denote the set on the left-hand side by T, and the one on the right-hand side by S. Define a mapping Φ on T by putting

$$D(\Phi(f)) = \Delta \qquad \text{for} \qquad f \in T,$$

and

$$\Phi(f)(\delta) = f|\Gamma_\delta \qquad \text{if} \qquad \delta \in \Delta.$$

It is easy to check that $T \sim_\Phi S$.

*   *
*

The mapping Φ is usually called the "canonical" correspondence between the sets T and S.

If there is no danger of misunderstanding, then the two isomorphic sets featured in Theorem 5.1 will be considered "identical."

Definition 5.2. *Define the set A raised to the power B (or the set A to the B) as follows:*

$$\{f : \mathrm{D}(f) = B \wedge \mathrm{R}(f) \subset A\}.$$

This set is denoted as $^B A$. The reason for this peculiar notation is that below we will have to distinguish between the *exponentiation* of sets just introduced from the exponentiation of cardinals to be introduced later. From the definition it follows immediately that $^B A = \bigtimes_{b \in B} A_b$ with $A_b = A$ for all $b \in B$.

Definition 5.3. *A set operation is said to be* compatible *with the property of equivalence if, when replacing its arguments with equivalent sets, the result of the operation will be equivalent to its original result.*

Theorem 5.2. *The direct product, the exponentiation, and the disjoint union (i.e., the union of pairwise disjoint sets) are compatible with the equivalence property.*

Proof. Let

$$\{A_\gamma : \gamma \in \Gamma\} \quad \text{and} \quad \{A'_\gamma : \gamma \in \Gamma\}$$

be set systems such that

$$A_\gamma \sim A'_\gamma \quad \text{for all} \quad \gamma \in \Gamma.$$

Using the Axiom of Choice, for each $\gamma \in \Gamma$ we can pick a function ϕ_γ for which

$$A_\gamma \sim_{\phi_\gamma} A'_\gamma.$$

We will now verify the assertion about *direct products*. Let

$$A = \bigtimes_{\gamma \in \Gamma} A_\gamma, \qquad A' = \bigtimes_{\gamma \in \Gamma} A'_\gamma.$$

Define a mapping Φ of A onto A' as follows: for $f \in A$ let $\Phi(f)$ be the element f' of A' such that

$$f'(\gamma) = \phi_\gamma(f(\gamma))$$

holds for every $\gamma \in \Gamma$. Let $f_1, f_2 \in A$ with $f_1 \neq f_2$, and let $f_1' = \Phi(f_1)$, $f_2' = \Phi(f_2)$. Then there is a $\gamma \in \Gamma$ such that $f_1(\gamma) \neq f_2(\gamma)$. As ϕ_γ is one-to-one, we then have $f_1'(\gamma) \neq f_2'(\gamma)$, and so $f_1' \neq f_2'$. Thus Φ is one-to-one. Furthermore, $R(\Phi) = A'$ holds; indeed, if $f' \in A'$, and f is the element of A such that

$$f(\gamma) = \phi^{-1}(f'(\gamma))$$

for all $\gamma \in \Gamma$, then $\Phi(f) = f'$. Thus

$$A \sim_\Phi A'.$$

To verify the assertion for *disjoint union*, assume that the sets A_γ are pairwise disjoint, and that the sets A_γ' are so, as well. Denote by A and A' the sets $\bigcup_{\gamma \in \Gamma} A_\gamma$ and $\bigcup_{\gamma \in \Gamma} A_\gamma'$, respectively. Then we can define a mapping Φ satisfying the relation $A \sim_\Phi A'$ as follows: If $x \in A$, then let $\Phi(x) = \phi_\gamma(x)$ for the unique γ for which $x \in A_\gamma$. The verification of the requisite properties of Φ will be left to the reader. (We remark that Φ could also have been defined by the relation $\Phi = \bigcup_{\gamma \in \Gamma} \Phi_\gamma$.)

Finally assume that A, A', B, B' are sets and ϕ, ψ are functions such that

$$A \sim_\phi A', \qquad B \sim_\psi B'.$$

Define a mapping Φ of the set $^B A$ as follows: For $f \in {}^B A$ let

$$\Phi(f) = \psi^{-1} \circ f \circ \phi.$$

It is easy to verify that

$$^B A \sim_\Phi {}^{B'} A'.$$

*   *
*

We remark that for operations with finitely many arguments the proof of Theorem 5.2 does not need the Axiom of Choice.

Definition 5.4. *Let* $\{a_\gamma : \gamma \in \Gamma\}$ *be a system of cardinals.*

1. *Choose sets* $\{A_\gamma : \gamma \in \Gamma\}$ *such that* $|A_\gamma| = a_\gamma$ *for each* $\gamma \in \Gamma$. *The cardinality of the set* $\bigtimes_{\gamma \in \Gamma} A_\gamma$ *is denoted as* $\prod_{\gamma \in \Gamma} a_\gamma$ *and is called the* product *of the cardinals* $\{a_\gamma : \gamma \in \Gamma\}$.

2. *Choose pairwise disjoint sets* $\{A_\gamma' : \gamma \in \Gamma\}$ *such that* $|A_\gamma'| = a_\gamma$ *for each* $\gamma \in \Gamma$. *The cardinality of the set* $\bigcup_{\gamma \in \Gamma} A_\gamma'$ *is denoted as* $\sum_{\gamma \in \Gamma} a_\gamma$ *and is called the* sum *of the cardinals* $\{a_\gamma : \gamma \in \Gamma\}$.

3. *Let* a *and* b *be cardinals. Choose sets* A, B *such that* $|A| = a$, $|B| = b$. *The cardinality of the set* $^B A$ *is denoted as* a^b, *and is called the* bth power *of the cardinal* a.

We will be able to verify the soundness of these definitions only later, with the aid of a theorem of Section 10. At that point, we will show that a cardinality operation described in Definition 2.5 can be chosen in such a way that cardinal a is a set of cardinality a itself. This will ensure that for an arbitrary set $\{a_\gamma : \gamma \in \Gamma\}$ of cardinals, we can find a system $\{A_\gamma : \gamma \in \Gamma\}$ of sets such that $|A_\gamma| = a_\gamma$ for each $\gamma \in \Gamma$. (This statement would follow directly from the Axiom of Choice only if for every cardinal a there existed a set consisting of all sets of cardinality a. One can, however, prove with the aid of Theorem 1.2 that such a set does not exist.) If for an arbitrary system $\{a_\gamma : \gamma \in \Gamma\}$ of cardinals we can find sets $\{A_\gamma : \gamma \in \Gamma\}$ such that $|A_\gamma| = a_\gamma$ for each $\gamma \in \Gamma$, then the sets $A'_\gamma = A_\gamma \times \{\gamma\}$ are pairwise disjoint and $A_\gamma \sim A'_\gamma$ for every $\gamma \in \Gamma$.

Finally, Theorem 5.2 ensures that the value of none of these operations depends on the choice of the sets in question.

If a_1 and a_2 are cardinals, then we will also use the notation

$$a_0 + a_1 = \sum_{i \in \underline{2}} a_i,$$

$$a_0 \cdot a_1 = a_0 a_1 = \prod_{i \in \underline{2}} a_i.$$

Naturally, we do not need the Axiom of Choice for the definition of $a + b$ and $a \cdot b$, but we do need it for the definition of operations with infinitely many arguments. We will, however, not call any special attention to this fact any more.

Examples:

a) $\aleph_0 + \aleph_0 = \aleph_0$.
b) $\aleph_0 \aleph_0 = \aleph_0$.
a) and b) follow from Theorems 3.3 and 3.4.

c) $2^{\aleph_0} > \aleph_0$. To verify this, we need to point out only that $2^{\aleph_0} = |{}^\omega 2|$ and, further, ${}^\omega \underline{2} \sim P(\omega)$; indeed, there is a canonical one-to-one correspondence between the so-called characteristic functions, i.e., functions assuming only the values 0 and 1, on ω, and the set of all subsets of ω.

We can verify the following theorem without any difficulty.

Theorem 5.3. *The addition, the multiplication, and the exponentiation are extensions of the corresponding operations for nonnegative integers.*

Proof. Let A, B be finite sets such that $|A| = n$, $|B| = m$. Then

$$|A \times B| = m \cdot n, \qquad |{}^B A| = n^m.$$

If $A \cap B = \emptyset$ is also satisfied then

$$|A \cup B| = m + n.$$

$$* \qquad *$$
$$*$$

Theorem 5.4.
1. *The addition and the multiplication are commutative operations.*
2. *The addition and the multiplication are associative operations.*
3. *The operations addition, multiplications, and exponentiation are increasing in the weak sense.*

Proof. Assertion 1 for addition follows from the commutativity of the union; for multiplication it follows from the fact that there is a canonical one-to-one mapping between $A \times B$ and $B \times A$. Assertion 2 for addition follows from the associativity of union, for multiplication it follows from Theorem 4.1 used to make up for the non-associativity of the direct product. To verify Assertion 3, let a_γ, a'_γ ($\gamma \in \Gamma$) and a, a', b, b' be cardinals such that

$$a_\gamma \leq a'_\gamma, \quad a \leq a', \quad b \leq b'.$$

We have to prove that

$$\sum_{\gamma \in \Gamma} a_\gamma \leq \sum_{\gamma \in \Gamma} a'_\gamma,$$

$$\prod_{\gamma \in \Gamma} a_\gamma \leq \prod_{\gamma \in \Gamma} a'_\gamma,$$

and

$$a^b \leq a'^{b'}.$$

We carry out the proof only for addition, the other two operations can be handled similarly. We choose the sets A_γ pairwise disjoint, the sets A'_γ likewise, such that

$$|A_\gamma| = a_\gamma, \qquad |A'_\gamma| = a'_\gamma$$

for each $\gamma \in \Gamma$.

Making use of the inequalities $a_\gamma \leq a'_\gamma$, we may pick sets $A''_\gamma \subset A'_\gamma$ such that $A_\gamma \sim A''_\gamma$. According to Theorem 5.2, we have

$$\sum_{\gamma \in \Gamma} a_\gamma = \left| \bigcup_{\gamma \in \Gamma} A''_\gamma \right| \leq \left| \bigcup_{\gamma \in \Gamma} A'_\gamma \right| = \sum_{\gamma \in \Gamma} a'_\gamma.$$

$$* \qquad *$$
$$*$$

We would like to mention the following consequence of the Axiom of Choice:

If $A = \bigcup_{\gamma \in \Gamma} A_\gamma$, then $|A| \leq \sum_{\gamma \in \Gamma} |A_\gamma|$.

Indeed, choose pairwise disjoint sets A'_γ and functions ϕ_γ such that $A'_\gamma \sim_{\phi_\gamma} A_\gamma$, and denote the set $\bigcup_{\gamma \in \Gamma} A'_\gamma$ by A'. The mapping $\phi = \bigcup_{\gamma \in \Gamma} \phi_\gamma$ maps the set A' onto A, and

$$|A'| = \bigcup_{\gamma \in \Gamma} |A'_\gamma| = \bigcup_{\gamma \in \Gamma} |A_\gamma|.$$

Thus $|A| \leq |A'|$ according to Theorem 4.7.

The operations described in Definition 5.4 are not strictly increasing. This is shown by the following *examples*:

a) $\aleph_0 + 0 = \aleph_0 + 1 = \aleph_0 + \aleph_0 = \aleph_0$,

b) $\aleph_0 \cdot 1 = \aleph_0 \cdot \aleph_0 = \aleph_0$.

An example involving exponentiation will be given in Section 6.

In order to prove the distributivity of multiplication over addition we need the distributive law for the corresponding set operations. This is stated in the following theorem.

Theorem 5.5 *(General Distributive Law). Given arbitrary set systems* $\{A_{\gamma,\delta} : \delta \in \Delta_\gamma\}$ *for each* $\gamma \in \Gamma$, *we have*

$$\underset{\gamma \in \Gamma}{\times} \left(\bigcup_{\delta \in \Delta_\gamma} A_{\gamma,\delta} \right) = \bigcup_{\phi \in \times_{\gamma \in \Gamma} \Delta_\gamma} \left(\underset{\gamma \in \Gamma}{\times} A_{\gamma,\phi(\gamma)} \right).$$

Furthermore, if the set systems consist of pairwise disjoint sets for each γ, *then the summands occurring on the right-hand side are pairwise disjoint.*

Proof. Denote the set on the left-hand side by P, that on the right-hand side by R. We have

$$f \in P \equiv \forall \gamma \in \Gamma \left(f(\gamma) \in \bigcup_{\delta \in \Delta_\gamma} A_{\gamma,\delta} \right)$$

$$\equiv \forall \gamma \in \Gamma \, \exists \delta \, (f(\gamma) \in A_{\gamma,\delta}).$$

(The difference between the use of $\equiv$ and $\Longleftrightarrow$ was explained at the beginning of Section 1.) Using the Axiom of Choice, we can see this to be equivalent to

$$(\exists \phi \in \underset{\gamma \in \Gamma}{\times} \Delta_\gamma) \forall \gamma \in \Gamma \, (f(\gamma) \in A_{\gamma,\phi(\gamma)}).$$

This in turn in equivalent to the statement $f \in R$.

Now assume that

$$A_{\gamma,\delta} \cap A_{\gamma,\delta'} = \emptyset \quad \text{for} \quad \delta, \delta' \in \Delta_\gamma, \quad \delta \neq \delta', \quad \text{and} \quad \gamma \in \Gamma.$$

Let

$$\phi, \phi' \in \underset{\gamma \in \Gamma}{\times} \Delta_\gamma \quad \text{with} \quad \phi \neq \phi'.$$

Then there is a $\gamma_0 \in \Gamma$ such that

$$\phi(\gamma_0) \neq \phi'(\gamma_0).$$

If we now have $f \in \bigtimes_{\gamma \in \Gamma} A_{\gamma,\phi(\gamma)}$ and $f' \in \bigtimes_{\gamma \in \Gamma} A_{\gamma,\phi'(\gamma)}$, then

$$f(\gamma_0) \in A_{\gamma_0,\phi(\gamma_0)}, \qquad f'(\gamma_0) \in A_{\gamma_0,\phi'(\gamma_0)},$$

and so $f \neq f'$. Thus the summands of the set R are pairwise disjoint.

$$* \qquad *$$
$$*$$

A similar general distributive law can be proved if the pair $\bigtimes$, $\bigcup$ is replaced with the pair $\bigcap$, $\bigcup$, or with the one $\bigcup$, $\bigcap$. These, however, will not be needed for now.

Corollary 5.1. *Given arbitrary systems $\{a_{\gamma,\delta} : \delta \in \Delta_\gamma\}$ of cardinals for each $\gamma \in \Gamma$, we have*

$$\prod_{\gamma \in \Gamma} \left(\sum_{\delta \in \Delta_\gamma} a_{\gamma,\delta} \right) = \sum_{\phi \in \bigtimes_{\gamma \in \Gamma} \Delta_\gamma} \left(\prod_{\gamma \in \Gamma} a_{\gamma,\phi(\gamma)} \right);$$

that is, there is a general distributive law of multiplication over addition for cardinals.

Proof. Choose pairwise disjoint subsets $A_{\gamma,\delta}$ such that

$$|A_{\gamma,\delta}| = a_{\gamma,\delta} \qquad \text{for} \qquad \delta \in \Delta_\gamma \text{ and } \gamma \in \Gamma.$$

Using Theorem 5.5, the assertion follows.

$$* \qquad *$$
$$*$$

Finally, we will show that the well-known identities of exponentiation for nonnegative integers also remain valid.

Theorem 5.6. *Let a, b, c, a_γ, b_γ ($\gamma \in \Gamma$) be arbitrary cardinals. Then the following identities hold:*
1. $a^{bc} = (a^b)^c$.
2. *If $b = \sum_{\gamma \in \Gamma} b_\gamma$, then $a^b = \prod_{\gamma \in \Gamma} a^{b_\gamma}$.*
3. *If $a = \prod_{\gamma \in \Gamma} a_\gamma$, then $a^b = \prod_{\gamma \in \Gamma} a_\gamma^b$.*

Proof. Pick the sets A, B, C such that $|A| = a$, $|B| = b$, $|C| = c$.
1. $a^{bc} = |{}^{B \times C}A|$. According to Theorem 5.1, we have

$$^{B \times C}A \sim \bigtimes_{\langle y,z \rangle \in B \times C} A \sim \bigtimes_{z \in C} \left(\bigtimes_{y \in B} A \right) = {}^C({}^B A).$$

The cardinality of the last set is $(a^b)^c$.

2. Let B_γ be pairwise disjoint sets such that

$$B = \bigcup_{\gamma \in \Gamma} B_\gamma, \qquad \text{and} \qquad |B_\gamma| = b_\gamma \quad \text{for} \quad \gamma \in \Gamma.$$

According to Theorem 5.1, we have

$$ {}^B\!A \sim \underset{x \in B}{\times} A \sim \underset{\gamma \in \Gamma}{\times} \left(\underset{x \in B_\gamma}{\times} A \right) = \underset{\gamma \in \Gamma}{\times} {}^{B_\gamma}\!A. $$

The cardinality of the last set is $\prod_{\gamma \in \Gamma} a^{b_\gamma}$.

3. Choose the sets A_γ $(\gamma \in \Gamma)$ such that

$$A \sim \underset{\gamma \in \Gamma}{\times} A_\gamma \qquad \text{and} \qquad |A_\gamma| = a_\gamma \quad \text{for} \quad \gamma \in \Gamma.$$

Then

$$a^b = |{}^B\!A|.$$

Further, according to Theorem 5.1, we have

$$ {}^B\!A = \underset{x \in B}{\times} A \sim \underset{x \in B}{\times} \left(\underset{\gamma \in \Gamma}{\times} A_\gamma \right) \sim \underset{\gamma \in \Gamma}{\times} \left(\underset{x \in B_\gamma}{\times} A_\gamma \right) = \underset{\gamma \in \Gamma}{\times} \left({}^B\!A_\gamma \right). $$

The cardinality of the last set is

$$\prod_{\gamma \in \Gamma} a_\gamma^b.$$

$$* \qquad *$$
$$*$$

Corollary 5.2. *Multiplication is "repeated" addition. Exponentiation is "repeated" multiplication. By this we mean that if a is a cardinal and $|B| = b$, then*

$$\sum_{x \in B} a = ab \qquad \text{and} \qquad \prod_{x \in B} a = a^b.$$

Proof. Using the distributive law, we obtain

$$\sum_{x \in B} a = a \cdot \sum_{x \in B} 1 = a \cdot b.$$

The second assertion follows from the equality

$$\prod_{x \in B} a = a^{\sum_{x \in B} 1} = a^b.$$

6. EXAMPLES

In fulfilling our promise made at the end of Section 3, we are now going to use the general theorems obtained so far to establish several simple results. It would be instructive if the reader established these results also directly (without the use of general theorems).

Statement 6.1. *If $k \in \omega$ and $k \neq 0$ then*

$$\aleph_0^k = \aleph_0.$$

Proof. We use induction. For $k = 1$ the assertion is obvious. Assume $k > 1$. According to the induction hypothesis, we have

$$\aleph_0^k = \aleph_0^{k-1}\aleph_0 = \aleph_0^2.$$

We proved earlier that $\aleph_0^2 = \aleph_0$.

$$* \qquad *$$
$$*$$

Statement 6.2. $\mathfrak{c} = 2^{\aleph_0}$.

Proof. We need to show that $|\mathbb{R}| = 2^{\aleph_0}$. We have $|\mathbb{R}| \leq |[0,1)|$ according to Theorem 3.8. Given an arbitrary $x \in [0,1)$, let $x = [0.x_1x_2\ldots x_i\ldots]_2$ be its binary expansion such that it does not consist purely of the digit 1 from a certain point on.

This expansion induces a one-to-one mapping of $[0,1)$ into $^\omega 2$. Hence $|\mathbb{R}| \leq 2^{\aleph_0}$. On the other hand, given an arbitrary function $f \in {^\omega 2}$, assign to it the real number x_f whose decimal fraction expansion is $x_f = 0.f(0)f(1)\ldots$. In this way, $f \in {^\omega 2}$ is mapped into $\mathbb{R}$ in a one-to-one way. Thus $2^{\aleph_0} \leq |\mathbb{R}|$ also holds.

$$* \qquad *$$
$$*$$

Statement 6.3. *If a is an infinite cardinal then $a + \aleph_0 = a$.*

Proof. Let A be a set with $|A| = a$. According to Theorem 3.6, there is a set $A' \subset A$ for which $|A'| = \aleph_0$. Let $B = A \backslash A'$ and $b = |B|$. Then $a = b + \aleph_0$. Thus

$$a + \aleph_0 = (b + \aleph_0) + \aleph_0 = b + \aleph_0 \cdot 2 = b + \aleph_0 = a.$$

$$* \quad *$$
$$*$$

Statement 6.4. *For the n-dimensional Euclidean space E^n we have $|E^n| = \mathfrak{c}$.*

Proof. $|E^n| = |^n\mathbb{R}| = \mathfrak{c}^n = (2^{\aleph_0})^n = 2^{\aleph_0 \cdot n} = 2^{\aleph_0} = \mathfrak{c}$.

$$* \quad *$$
$$*$$

Statement 6.5. *The set of all sequences of real numbers has cardinality $\mathfrak{c}$.*

Proof.

$$|^{\omega}\mathbb{R}| = \mathfrak{c}^{\aleph_0} = (2^{\aleph_0})^{\aleph_0} = 2^{\aleph_0^2} = \mathfrak{c}.$$

$$* \quad *$$
$$*$$

Statement 6.6. *Denote by $\mathcal{F}$ the set of all real functions. Then*

$$|\mathcal{F}| = 2^{2^{\aleph_0}} > \mathfrak{c}.$$

Proof.

$$|\mathcal{F}| = |^{\mathbb{R}}\mathbb{R}| = \mathfrak{c}^{\mathfrak{c}} = (2^{\aleph_0})^{\mathfrak{c}} = 2^{\aleph_0 \cdot \mathfrak{c}} = 2^{2^{\aleph_0}} > \mathfrak{c},$$

since according to Statement 6.4, we have $\mathfrak{c} \leq \aleph_0 \mathfrak{c} \leq \mathfrak{c}^2 = \mathfrak{c}$.

$$* \quad *$$
$$*$$

Statement 6.7.[+] *Let $C(\mathbb{R})$ be set of continuous real functions. Then $|C(\mathbb{R})| = \mathfrak{c}$.*

The symbol [+] after the number of the above statement indicates that the statement relies on mathematical concepts undefined in this book. As explained in the preface, this symbol is used to indicate problems relying on mathematical concepts undefined here. Such concepts do not usually occur in the main text of the book; the present section is an exception, since it discusses only illustrative examples rather than develops the theory that is the subject of the present book.

Proof. According to a well-known theorem of elementary analysis, if f is a real function, then f is determined by its values assumed at rational points. Hence

$$|C(\mathbb{R})| \leq |^{\mathbb{Q}}\mathbb{R}| = \mathfrak{c}^{\aleph_0} = \mathfrak{c}.$$

On the other hand, the constant functions are continuous; hence

$$\mathfrak{c} \le |C(\mathbb{R})|$$

is also satisfied.

*   *

*

Statement 6.8. *Define the sequence* $\{\mathfrak{c}_n : n \in \omega\}$ *of cardinals by recursion as follows:*

$$\mathfrak{c}_0 = \aleph_0, \qquad \mathfrak{c}_{n+1} = 2^{\mathfrak{c}_n} \quad \text{for} \quad n \in \omega.$$

By Cantor's Theorem, we have

$$\mathfrak{c}_0 < \mathfrak{c}_1 < \cdots < \mathfrak{c}_n < \ldots.$$

We claim that

$$\mathfrak{c}_n^2 = \mathfrak{c}_n \quad \text{for} \quad n \in \omega.$$

Proof. We use induction. For $n = 0$ we have

$$\mathfrak{c}_0^2 = \aleph_0^2 = \aleph_0 = \mathfrak{c}_0.$$

Let $n \in \omega$, and assume the assertion is true for n, that is,

$$\mathfrak{c}_n^2 = \mathfrak{c}_n.$$

Then

$$\mathfrak{c}_n \le \mathfrak{c}_n \cdot 2 \le \mathfrak{c}_n \cdot \mathfrak{c}_n = \mathfrak{c}_n.$$

Hence

$$\mathfrak{c}_{n+1}^2 = \left(2^{\mathfrak{c}_n}\right)^2 = 2^{\mathfrak{c}_n \cdot 2} = 2^{\mathfrak{c}_n} = \mathfrak{c}_{n+1}.$$

*   *

*

Each of the infinite cardinals that we have met so far is one of the above $\mathfrak{c}_n$'s. Therefore the question arises on the basis of Statement 6.8 whether the relation $a^2 = a$ is true for every infinite cardinal a? The answer is affirmative. This important statement is customarily called the Fundamental Theorem of Cardinal Arithmetic (see Theorem 10.3). This statement is also one of those results that can be proved only with the aid of the Axiom of Choice. We would like to mention without proof that this statement is equivalent to the Axiom of Choice.

Statement 6.9. *Let the cardinal d be the sum of the c_n's above. Then we have $d^2 = d$.*

Proof.

$$d^2 = \left(\sum_{n\in\omega} c_n\right)^2 = \sum_{\langle i,j\rangle\in\omega\times\omega} c_i c_j = \sum_{k\in\omega}\left(\sum_{\max\langle i,j\rangle=k} c_i c_j\right)$$
$$= \sum_{k\in\omega} c_k(2c_0 + \cdots + 2c_{k-1} + c_k) = \sum_{k\in\omega} c_k^2 = \sum_{k\in\omega} c_k = d.$$

In the course of this calculation, we used the distributive law and Statement 6.8.

*   *
*

The cardinal d is clearly greater than every c_n, since $c_n < c_{n+1} \le d$ holds.

Statement 6.10. *For the cardinal d defined in the preceding passage, we have the relation $d^{\aleph_0} = 2^d$.*

Proof.

$$2^d = 2^{\sum_{n\in\omega} c_n} = \prod_{n\in\omega} 2^{c_n} = \prod_{n\in\omega} c_{n+1} \le \prod_{n\in\omega} d = d^{\aleph_0}.$$

Thus $2^d \le d^{\aleph_0}$. Further
$$(2^d)^d = 2^{d^2} = 2^d,$$

and so, noting that cardinal operations are increasing in the wider sense, we obtain
$$d^{\aleph_0} \le (2^d)^d = 2^d.$$

*   *
*

With Statement 6.10 we have an example saying that exponentiation is not monotonic in the strict sense.

Finally, we show the following generalization of Cantor's Theorem 4.6.

Theorem 6.1. *For any system $\{a_\gamma : \gamma \in \Gamma\}$ of cardinals there is a cardinal that is larger than each of these cardinals.*

Proof. Let $a' = \sum_{\gamma\in\Gamma} a_\gamma$ and $a = 2^{a'}$. If $\gamma \in \Gamma$ then $a_\gamma \le a' < a$.

*   *
*

Theorem 6.1 shows that cardinals are very numerous, since there is no set that contains all of them.

Here we would like to summarize the statements whose proofs we have yet to give. First of all, we need to establish assertions justifying proofs by induction and definitions by recursion. Secondly, we need to prove the existence of the cardinality operation; this will be accomplished with the aid of the Axiom of Choice.

Since we have used these statements already, in establishing them we have to be careful not to use those results that were obtained by their application.

Finally, we still owe proofs of the Trichotomy Theorem and of the Fundamental Theorem of Cardinal Arithmetic; these proofs will require the Axiom of Choice.

A key tool in these proofs is the theory of ordered sets.

Problems

1. Prove that the set of all finite subsets of ω is countable.

2. Prove that if the cardinal a satisfies $a \leq \mathfrak{c}$ then $\mathfrak{c} + a = \mathfrak{c}$.

3. Prove that if $a^2 = a$ for each infinite cardinal a then $b + c = bc$ for any two infinite cardinals b, c.

4. Prove that the cardinality of the set of all permutations of ω has cardinality $2^{\aleph_0}$.

5. Prove that the set of irrational numbers has cardinality $\mathfrak{c}$.

6.[+] Prove that the cardinality of every nonempty perfect set of reals has cardinality $\mathfrak{c}$.

7.[+] Prove that the set of Lebesgue measurable functions on the interval $[0, 1]$ has cardinality $\mathfrak{c}^{\mathfrak{c}}$.

8.[+] Prove that the set of Riemann integrable functions on the interval $[0, 1]$ has cardinality $\mathfrak{c}^{\mathfrak{c}}$.

7. ORDERED SETS. ORDER TYPES. ORDINALS

We now turn to the discussion of the theory of ordered sets. For this we need the following definition.

Definition 7.1. *Let $k \in \omega$. A function whose domain is $\underline{k}$ will be called an* ordered k-tuple. *We will use the customary notation $\langle a_0, \ldots, a_{k-1} \rangle$ to denote ordered k-tuples.*

Although for $k = 2$ this notation is not identical to the ordered pair $\langle a_0, a_1 \rangle$, this will not cause any misunderstanding in what follows. If A is a set, then the set of ordered k-tuples formed by elements of A are denoted by $\underline{k}A$.

Definition 7.2. *If A is a set, then we say that the set R is a k-place relation on the set A if $R \subset \underline{k}A$.*

Let R be a k-place relation on A and let $\langle a_0, \ldots, a_{k-1} \rangle \in \underline{k}A$. We say that R is true *on the ordered k-tuple $\langle a_0, \ldots, a_{k-1} \rangle$, or that the elements $a_0, \ldots, a_{k-1}$ are related by R if $\langle a_0, \ldots, a_{k-1} \rangle \in R$. If $\langle a_0, \ldots, a_{k-1} \rangle \notin R$, then we say that R is* false *on the ordered k-tuple $\langle a_0, \ldots, a_{k-1} \rangle \in R$, or that the elements $a_0, \ldots, a_{k-1}$ are* not related *by R.*

The two-place relations are called relations, *in short.*

Examples: a) Let A be the set of all points of the plane and put

$$R = \{\langle P_0, P_1, P_2 \rangle \in \underline{3}A : \text{the points } P_0, P_1, P_2 \text{ lie on the same line}\}.$$

Then R is a three-place relation on the plane.

b) An example for a two-place relation on the set of the nonnegative integers is the ordering by size; that is, if $A = \omega$, and $<$ is the set $\{\langle m, n \rangle \in \underline{2}\omega : n$ is less than $m\}$, then $<$ is a two-place relation on A.

If R is a relation on A, then the fact that $\langle x, y \rangle \in R$ is often denoted as xRy. For example, $\langle m, n \rangle \in <$ is usually written as $m < n$.

Definition 7.3. *The pair $\langle A, \prec \rangle$ is called an* ordered set *if $\prec$ is a relation that orders A, that is, the relation $\prec$ is irreflexive, transitive, and trichotomous on A. That is,*

$$\forall x \, \neg \, x \prec x,$$

$$\forall x, y, z \, [(x \prec y \wedge y \prec z) \implies x \prec z],$$

and

$$\forall x, y \, [x \prec y \lor x = y \lor y \prec x],$$

where the variables x, y, and z run over elements of A.

We remark that if $\prec$ is only irreflexive and transitive on A, then we call $\langle A, \prec \rangle$ a partially ordered set.

For example, the pairs $\langle \omega, < \rangle$ and $\langle \mathbb{R}, < \rangle$ are ordered sets, where $<$ is the well-known ordering by size of the nonnegative integers and of the real numbers, respectively. If $A \subset \mathbb{R}$, then we can regard the pair $\langle A, < \rangle$ also as an ordered set, with the ordering inherited from $\mathbb{R}$. This notation is imprecise in that $<$ is a relation not on A, but on the larger set $\mathbb{R}$. For situations when we need such fine distinctions, we are going to introduce a symbol:

Definition 7.4. *Given a k-place relation R on the set A and $B \subset A$, the relation $\mathrm{S} = \mathrm{R} \cap {}^{\underline{k}}B$ defined on B is denoted as $\mathrm{R} \restriction B$, and it is called the restriction of R to B.*

Definition 7.5. *Given a two-place property $\Phi(x, y)$ and a set A, the relation*

$$\mathrm{S} = \{\langle x, y \rangle \in {}^{2}A : \Phi(x, y)\}$$

on A is denoted as $\Phi \restriction A$ and is called the restriction of Φ to A.

In the same way as in the case of the property $\sim$, we will describe what we mean by saying that the ordering of two ordered sets are "alike," in other words, that two ordered sets are *similar*.

Definition 7.6. *Let $\langle A, \prec \rangle$ and $\langle A', \prec' \rangle$ be ordered sets. We say that the function f is a monotonic (isomorphic) mapping of the ordered set $\langle A, \prec \rangle$ onto the ordered set $\langle A', \prec' \rangle$ if*

1. $A \sim_f A'$;
2. *For all $x, y \in A$ we have $x \prec y \iff f(x) \prec' f(y)$.*

This state of affairs will be denoted as

$$\langle A, \prec \rangle \simeq_f \langle A', \prec' \rangle.$$

We say that the ordered set $\langle A, \prec \rangle$ is similar to the ordered set $\langle A', \prec' \rangle$ if there is an f for which

$$\langle A, \prec \rangle \simeq_f \langle A', \prec' \rangle.$$

(In symbols: $\langle A, \prec \rangle \simeq \langle A', \prec' \rangle$.)

For example, ω is similar to $\omega \setminus \{\emptyset\}$, and $\left(-\frac{\pi}{2}, \frac{\pi}{2}\right)$ is similar to $\mathbb{R}$ in the ordering according to size. We can write this in our notation as follows:

$$\langle \omega, \prec \rangle \simeq_f \langle \omega \setminus \{\emptyset\}, < \restriction \omega \setminus \{\emptyset\} \rangle,$$

where $\mathrm{D}(f) = \omega$ and $f(n) = n + 1$ for $n \in \omega$, and

$$\left\langle \left(-\frac{\pi}{2}, \frac{\pi}{2}\right), < \restriction \left(-\frac{\pi}{2}, \frac{\pi}{2}\right) \right\rangle \simeq_{\tan \restriction (-\frac{\pi}{2}, \frac{\pi}{2})} \langle \mathbb{R}, < \rangle.$$

Theorem 7.1. *The property of similarity is an equivalence property.*

Proof. Reflexivity is satisfied, since

$$\langle A, \prec \rangle \simeq_{\mathrm{Id}_A} \langle A, \prec \rangle.$$

To show symmetry, assume that for a function f we have

$$\langle A, \prec \rangle \simeq_f \langle A', \prec' \rangle;$$

then

$$\langle A', \prec' \rangle \simeq_{f^{-1}} \langle A, \prec \rangle.$$

Thus

$$\langle A, \prec \rangle \simeq \langle A', \prec' \rangle$$

implies

$$\langle A', \prec' \rangle \simeq \langle A, \prec \rangle.$$

To establish transitivity, assume that

$$\langle A, \prec \rangle \simeq \langle A', \prec' \rangle \quad \text{and} \quad \langle A', \prec' \rangle \simeq \langle A'', \prec'' \rangle$$

hold. Choose functions f and g such that

$$\langle A, \prec \rangle \simeq_f \langle A', \prec' \rangle \quad \text{and} \quad \langle A', \prec' \rangle \simeq_g \langle A'', \prec'' \rangle.$$

We know that for the composition function $h = f \circ g$ we have $A \sim_h A''$. Furthermore, h is monotonic, since for $x, y \in A$ we have

$$x \prec y \equiv f(x) \prec' f(y) \equiv g(f(x)) \prec'' g(f(y)).$$

Hence

$$\langle A, \prec \rangle \simeq_h \langle A'', \prec'' \rangle,$$

that is

$$\langle A, \prec \rangle \simeq \langle A'', \prec'' \rangle.$$

$$* \qquad *$$
$$*$$

If $\langle A, \prec \rangle$ is an ordered set, and it is clear from the context what the ordering $\prec$ is, then instead of $\langle A, \prec \rangle$ we will call the set A itself an ordered set. In the same spirit, to say that $\langle A, \prec \rangle$ is a finite ordered set means, in particular, that A is a finite set. The next theorem says that on a finite set there is only one "kind" of ordering.

Theorem 7.2. *If* $\langle A, \prec \rangle$ *is a finite ordered set such that* $|A| = n$, *then* $\langle A, \prec \rangle \simeq \langle \underline{n}, < \rangle$, *where* $<$ *is the ordering of the set* $\underline{n}$ *according to size.*

Proof. The assertion of the theorem easily follows from Theorem 2.2 by induction on n. The details are left to the reader.

*   *
*

Definition 7.7. *We say that the operation F is* compatible *with the* (similarity) *property* $\simeq$ *if for any two ordered sets* $\langle A, \prec \rangle$, $\langle A', \prec' \rangle$ *we have*

$$F(\langle A, \prec \rangle) = F(\langle A', \prec' \rangle) \iff \langle A, \prec \rangle \simeq \langle A', \prec' \rangle.$$

Theorem 7.3. *There is an operation that is compatible with the property* $\simeq$.

Theorem 7.3 will be proved in Section 10, along with the analogous Theorem 2.3 concerning the property $\sim$ (see Theorem 10.2).

Definition 7.8. *In what follows,* type *will denote a fixed operation compatible with the property* $\simeq$; *such an operation exists according to Theorem 7.3. The value of this operation on the ordered set* $\langle A, \prec \rangle$ *will be denoted as* type$\langle A, \prec \rangle$ *or* type $A(\prec)$, *and it will be called the* order type *of the ordered set* $\langle A, \prec \rangle$.

Just as we did in the case of the cardinality operation, according to Theorem 7.2, we may assume that if A is finite and $|A| = n$, then type $A(\prec) = n$. The order types of finite sets will be called *finite order types*, the other ones will be called *infinite order types*. The finite order types are the nonnegative integers. In what follows, order types will be denoted by Greek letters. We will, in particular, use the following notation:

$$\text{type}\, \omega(<) = \omega_0, \qquad \text{type}\, \mathbb{Q}(<) = \eta_0, \qquad \text{type}\, \mathbb{R}(<) = \lambda_0.$$

Here $<$ always denotes the ordering by size of the sets in question. We would like to add that the symbols ω and ω_0 are firmly associated with the meaning defined above, so we will never use these symbols to mean anything else; however, the symbols η_0 and λ_0 are only loosely associated with the meanings defined above, and if there is no danger of misunderstanding, these symbols may be reused with other meanings.

We remark that the similarity property is a "refinement" of the equivalence property on ordered sets. If $\langle A, \prec \rangle \simeq \langle A', \prec' \rangle$, then $A \sim A'$, but the converse is not true. For example, $\omega \sim \mathbb{Q}$ since both sets are countable, but $\omega_0 \neq \eta_0$, since in the ordering by size of the set of rational numbers $\mathbb{Q}$ has no first element.

Order types can also be regarded as a generalization of the nonnegative integers. We can define extensions to order types of the usual operations, and we can study the properties of these operations. In this book, we will study this question only to the extent it is absolutely necessary for what follows.

Before we do this, we introduce another concept, the concept of wellordered sets. These sets will be ordered sets that possess an important property of the nonnegative integers.

Definition 7.9. *The ordered set* $\langle A, \prec \rangle$ *is called* wellordered *if every nonempty subset* A' *of* A *has a least element* a *in the ordering* $\prec$*, that is, there is an* $a \in A'$ *such that for each* $x \in A'$ *we have either* $a \prec x$ *or* $a = x$.

It is easy to see that every subset of a wellordered set is also wellordered by the inherited ordering. Furthermore, if $\langle A, \prec \rangle \simeq \langle A', \prec' \rangle$, then the set A is wellordered if and only if A' is so. This last remark justifies the next definition.

Definition 7.10. *The order types of wellordered sets are called* ordinals.

For example, the sets $\langle \omega, < \rangle$ and $\langle \underline{n}, < \rangle$ for $n \in \omega$ are wellordered, and so n and ω_0 are ordinals; on the other hand, the order types η_0, λ_0 are not ordinals.

We need to specify some of the usual notation. If $\prec$ is an ordering, then $x \preceq y$ is an abbreviation of $x \prec y$ or $x = y$. If $\langle A, \prec \rangle$ is an ordered set, $A' \subset A$, and the set A' has a least element in the ordering $\prec$, then this least element is denoted by $\min_{\prec} A'$.

Definition 7.11. *Let* $\{\langle A_\gamma, \prec_\gamma \rangle : \gamma \in \Gamma\}$ *be a system of ordered sets, and let* $\prec_\Gamma$ *be an ordering of the set* Γ.

Assume that the sets A_γ *are pairwise disjoint. The ordered set* $\langle A, \prec \rangle$, *where*

$$A = \bigcup_{\gamma \in \Gamma} A_\gamma$$

and for each $x, y \in A$ *we put*

$$x \prec y \iff \exists \gamma, \delta \, (x \in A_\gamma, \, y \in A_\delta \quad \text{and}$$
$$\text{either} \quad \gamma \prec_\Gamma \delta \quad \text{or} \quad (\gamma = \delta \quad \text{and} \quad x \prec_\gamma y)),$$

is called the ordered union with respect to the ordering $\prec_\Gamma$ *of the set system*

$$\{\langle A_\gamma, \prec_\gamma \rangle : \gamma \in \Gamma\}.$$

In words: In the ordering $\prec$ *"we put the sets* A_γ *one after the other" according to the ordering* $\prec_\Gamma$*, and within each* A_γ *we "keep the old ordering."*

To show that this definition is sound, we are about to prove that $\prec$ is an ordering of A. For an arbitrary element $x \in A$, there is exactly one $\gamma \in \Gamma$ for which $x \in A_\gamma$. Denote this γ by $\gamma(x)$. Then

$$x \prec y \iff \gamma(x) \prec_\Gamma \gamma(y) \quad \text{or} \quad (\gamma(x) = \gamma(y) \quad \text{and} \quad x \prec_{\gamma(x)} y).$$

Thus one can easily conclude that the relation $\prec$ is irreflexive, from the relations $\prec_\gamma$ and $\prec_\Gamma$ being irreflexive.

Next we verify trichotomy. If $x, y \in A$ then

$$\text{either} \quad \gamma(x) \prec_\Gamma \gamma(y) \quad \text{or} \quad \gamma(y) \prec_\Gamma \gamma(x) \quad \text{or} \quad \gamma(x) = \gamma(y)$$

holds. In the first or second case, we have

$$x \prec y \quad \text{or} \quad y \prec x.$$

In the third case,

$$x \preceq_{\gamma(x)} y \quad \text{or} \quad y \preceq_{\gamma(x)} x$$

holds, and then either $x \preceq y$ or $y \preceq x$ is valid.

To verify transitivity, assume that $x \prec y$ and $y \prec z$ hold for the elements x, y, z of A. Then

$$\gamma(x) \preceq_\Gamma \gamma(y) \quad \text{and} \quad \gamma(y) \preceq_\Gamma \gamma(z)$$

also hold. If in one of these two relations we actually have $\prec_\Gamma$ in place of $\preceq_\Gamma$, then

$$\gamma(x) \prec_\Gamma \gamma(z)$$

holds, and so $x \prec z$ holds. Hence we may assume that

$$\gamma = \gamma(x) = \gamma(y) = \gamma(z).$$

Then, according to the definition of $\prec$, we have

$$x \prec_\gamma y \quad \text{and} \quad y \prec_\gamma z.$$

Therefore the assertion follows from the transitivity of $\prec_\gamma$.

Theorem 7.4. 1. *The operation of ordered union with respect to a fixed ordered set $\langle \Gamma, \prec_\Gamma \rangle$ is compatible with the similarity property.*

2. *If $\langle A_\gamma, \prec_\gamma \rangle$ are disjoint wellordered sets ($\gamma \in \Gamma$) and $\langle \Gamma, \prec_\Gamma \rangle$ is a wellordered set, then the ordered union of the sets A_γ with respect to $\prec_\Gamma$ is wellordered.*

Proof. 1. Assume that for all $\gamma \in \Gamma$ we have

$$\langle A_\gamma, \prec_\gamma \rangle \simeq \langle A'_\gamma, \prec'_\gamma \rangle$$

and all of the sets A_γ and A'_γ are pairwise disjoint. Denote the ordered union of these sets with respect to $\langle \Gamma, \prec_\Gamma \rangle$ by

$$\langle A, \prec \rangle \quad \text{and} \quad \langle A', \prec \rangle',$$

respectively. According to the Axiom of Choice, there is a system $\{\phi_\gamma : \gamma \in \Gamma\}$ of functions such that

$$\langle A_\gamma, \prec_\gamma \rangle \simeq_{\phi_\gamma} \langle A'_\gamma, \prec'_\gamma \rangle$$

holds for each $\gamma \in \Gamma$. Define a mapping f of A onto A' by stipulating that $f(x) = \phi_\gamma(x)$ if $x \in A_\gamma$ and $\gamma \in \Gamma$. From the proof of Theorem 5.2 we know that $A \sim_f A'$. It is easy to see that f is monotonic; we leave the details to the reader.

2. Denote by $\langle A, \prec \rangle$ the ordered union described in the assertion being proved. Let $A' \subset A$, $A' \neq \emptyset$. Write

$$\Gamma' = \{\gamma \in \Gamma : A' \cap A_\gamma \neq \emptyset\}.$$

Then $\Gamma' \neq \emptyset$, and so, $\langle \Gamma, \prec_\Gamma \rangle$ being wellordered, the element

$$\gamma_0 = \min_{\prec_\gamma} \Gamma'$$

is well defined.

As $A_{\gamma_0} \cap A' \neq \emptyset$ and A_{γ_0} is wellordered, there is an element

$$a_0 = \min_{\prec_{\gamma_0}} A_{\gamma_0} \cap A'.$$

It is clear that a_0 is the least element of A' in the ordering $\prec$.

$$* \qquad *$$
$$*$$

After this preparation we can define the sum of order types.

Definition 7.12. *Let $\langle \Gamma, \prec_\Gamma \rangle$ be an ordered set, and let $\{\Theta_\gamma : \gamma \in \Gamma\}$ be a system of order types. Choose pairwise disjoint ordered sets $\langle A_\gamma, \prec_\gamma \rangle$ such that we have type $A_\gamma(\prec_\gamma) = \Theta_\gamma$.*

The order type of the ordered union with respect to $\langle \Gamma, \prec_\Gamma \rangle$ of the sets $\langle A_\gamma, \prec_\gamma \rangle$ is denoted as $\sum_{\gamma \in \Gamma(\prec_\Gamma)} \Theta_\gamma$, and is called the sum with respect to $\langle \Gamma, \prec_\Gamma \rangle$ *of the order types $\{\Theta_\gamma : \gamma \in \Gamma\}$.*

The $\sum$ notation is needed so that we can distinguish between the addition of order types and of cardinals.

That this definition is sound can only be verified with the aid of a theorem to be established in Section 10. There we will show that the operation type $A(\prec)$ can be defined in such a way that type $A(\prec)$ is a set that consists of ordered sets similar to $\langle A, \prec \rangle$. Then we can prove with the aid of the Axiom of Choice that, given an arbitrary system $\{\Theta_\gamma : \gamma \in \Gamma\}$ of order types, it is possible to choose a system

$$\{\langle A_\gamma, \prec_\gamma \rangle : \gamma \in \Gamma\}$$

of ordered sets such that type $A_\gamma(\prec_\gamma) = \Theta_\gamma$ holds for every $\gamma \in \Gamma$. We can then see that the A_γ's can be chosen pairwise disjoint in the same way as in the case of the definition of sums of cardinals. Finally, this sum does not depend on the specific choice of the sets $\langle A_\gamma, \prec_\gamma \rangle$ according to Theorem 7.4. This same theorem implies the following.

Corollary 7.1. *A sum of ordinals with respect to a wellordered set* $\langle \Gamma, \prec_\Gamma \rangle$ *is an ordinal.*

The definition naturally gives the sum of two order types. If Θ_0 and Θ_1 are ordinals, then their sum $\Theta_0 \dotplus \Theta_1$ is defined as

$$\Theta_0 \dotplus \Theta_1 = \sum_{i \in \underline{2}(<)}^{\cdot} \Theta_i.$$

It is of course not necessary to use the Axiom of Choice to see that the definition of the sum of two order types is sound.

Examples:

$$\omega_0 \dotplus 1 = \text{type}\left\{0, \frac{1}{2}, \ldots, 1 - \frac{1}{n}, \ldots, 1\right\}(<)$$
$$= \text{type}(\omega \cup \{+\infty\})(<).$$

$$\omega_0 \dotplus \omega_0 = \text{type}\left\{0, \frac{1}{2}, \ldots, 1 - \frac{1}{n}, \ldots, 1, \frac{3}{2}, \ldots, 2 - \frac{1}{n}, \ldots\right\}(<).$$

$$\text{type}(0,1)\,(< \restriction (0,1)) \dotplus 1 = \text{type}(0,1]\,(< \restriction (0,1]).$$

In this examples, $<$ is the ordering of the reals according to size.

$\sum_{n \in \omega}^{\cdot} \omega_0$, which may also be denoted as $\omega_0 \dotplus \cdots \dotplus \omega_0 \dotplus \ldots$, represents the order type of an ordered set obtained by "placing ordered sets of order type ω_0 one after another" in a sequence of type ω_0.

Theorem 7.5. 1. *The addition of order types is an extension of the addition of nonnegative integers.*

2. *The addition of order types is associative, but it is not commutative even for ordinals.*

Proof. 1. If n, m are finite order types, then

$$n \dotplus m = \text{type}\, A(\prec)$$

for a set A of $n + m$ elements. Then, according to our conventions,

$$\text{type}\, A(<) = n + m.$$

Hence

$$n \dotplus m = n + m.$$

2. The first part of the second assertion follows from the associativity of ordered union; the verification of this latter will be left to the reader.

The lack of commutativity is shown by the following example:

$$1 \dotplus \omega_0 \neq \omega_0 \dotplus 1.$$

In fact, $1 \dotplus \omega_0 = \omega_0$, as $\mathrm{type}(\omega \setminus \{0\})(<) = \omega_0$; on the other hand,

$$\omega_0 \dotplus 1 \neq \omega_0,$$

since a set of order type $\omega_0 \dotplus 1$ has a last element, while one of order type ω_0 has no last element.

$$* \qquad *$$
$$*$$

In what follows we will define the product of two order types. To this end, we need the following definition.

Definition 7.13. *Let $\langle A', \prec' \rangle$, $\langle A'', \prec'' \rangle$ be ordered sets. Their anti-lexicographic product will be defined as the ordered set $\langle A, \prec \rangle$, such that $A = A' \times A''$, and the ordering $\prec$ of the set A satisfies the following condition: If*

$$x = \langle x', x'' \rangle \in A \qquad and \qquad y = \langle y', y'' \rangle \in A,$$

then

$$x \prec y \iff x'' \prec'' y'' \qquad or \qquad x'' = y'' \quad and \quad x' \prec' y'.$$

To verify the soundness of this definition, we need to show that $\prec$ is an ordering of the set A. The irreflexitivy is obvious.

Trichotomy can be verified as follows. Let

$$x = \langle x', x'' \rangle, \qquad y = \langle y', y'' \rangle$$

two arbitrary elements of the set A. Since $\prec''$ is an ordering of the set A'', we have

$$\text{either} \quad x'' \prec'' y'' \quad \text{or} \quad y'' \prec'' x'' \quad \text{or} \quad x'' = y''.$$

In the first two cases, we have $x \prec y$ or $y \prec x$. If $x'' = y''$, then we have $x \preceq y$ or $y \preceq x$ according as $x' \preceq' y'$ or $y' \preceq' x'$.

To verify transitivity, assume that we have

$$x = \langle x', x'' \rangle \prec y = \langle y', y'' \rangle$$

and

$$y \prec z = \langle z', z'' \rangle.$$

Then we certainly have $x'' \preceq'' y''$ and $y'' \preceq'' z''$. If in either of these relations we also have $\prec''$ instead of $\preceq''$, then $x \prec z$ holds. We may therefore assume that $x'' = y'' = z''$. In this case, however, we have

$$x' \prec' y' \qquad and \qquad y' \prec' z'$$

in view of our assumptions. Hence $x' \prec' y'$ also holds. Thus we again have

$$x \prec z.$$

Theorem 7.6. 1. *The anti-lexicographic product of ordered sets is compatible with the similarity property.*

2. *If the sets* $\langle A', \prec' \rangle$, $\langle A'', \prec'' \rangle$ *are wellordered, then their anti-lexicographic product* $\langle A, \prec \rangle$, *is also wellordered.*

Proof. 1. Assume that

$$\langle A', \prec_0' \rangle \simeq \langle B', \prec_1' \rangle \quad \text{and} \quad \langle A'', \prec_0'' \rangle \simeq \langle B'', \prec_1'' \rangle.$$

Choose monotonic mappings f and g witnessing these relations. Let $\langle A, \prec_0 \rangle$, $\langle B, \prec_1 \rangle$ be the corresponding anti-lexicographic products. Define a mapping h of A into B as follows: For an arbitrary $x = \langle x', x'' \rangle \in A$ put

$$h(x) = \langle f(x'), g(x'') \rangle.$$

It can immediately be seen from the definition that h is one-to-one, it maps onto B, and is monotonic.

2. Let $X \subset A' \times A''$, $X \neq \emptyset$. Put

$$X'' = \{ x'' \in A'' : \exists x' (\langle x', x'' \rangle \in X) \}.$$

Then $X'' \neq \emptyset$. A'' is wellordered, so we can consider the element $a'' = \min_{\prec''} X''$. Put

$$X' = \{ x' \in A' : \langle x', a'' \rangle \in X \}.$$

The set X' is again nonempty, and A' is nonempty. So there is an element

$$a' = \min_{\prec'} X'.$$

Let

$$a = \langle a', a'' \rangle.$$

It is clear that a is the least element of X in the ordering $\prec$.

$$* \qquad *$$
$$*$$

Definition 7.14. *Let* Θ', Θ'' *be arbitrary order types. Choose ordered sets for which* type $A'(\prec') = \Theta'$ *and* type $A''(\prec'') = \Theta''$. *The order type of the anti-lexicographic product of these sets is called the* product of the order types Θ' *and* Θ'', *and is denoted as* $\Theta' \times \Theta''$.

This definition is sound, since according to Theorem 7.6, the product it defines does not depend on the specific choice of the ordered sets A', A''.

The following result is valid in view of Theorem 7.6.

Corollary 7.2. *The product of ordinals is also an ordinal.*

Examples: $\omega_0 \times 2 = \text{type} \, \omega \times \underline{2}(\prec)$, where the ordering $\prec$ is given by the following enumeration:

$$\omega \times \underline{2} = \{\langle 0,0 \rangle, \langle 1,0 \rangle, \ldots, \langle n,0 \rangle, \ldots,$$
$$\langle 0,1 \rangle, \langle 1,1 \rangle, \ldots, \langle n,1 \rangle, \ldots\}.$$

From this, it is clear that $\omega \times 2 = \omega_0 \dot{+} \omega_0$. On the other hand, $2 \times \omega_0$ is the order type of the set $\underline{2} \times \omega$ given by the following enumeration:

$$\underline{2} \times \omega = \{\langle 0,0 \rangle, \langle 1,0 \rangle, \langle 0,1 \rangle, \langle 1,1 \rangle, \ldots\}.$$

Hence, clearly, $2 \times \omega_0 = \omega_0$.

Theorem 7.7. 1. *The multiplication of order types is an extension of the multiplication of nonnegative integers.*

2. *The multiplication of order types is associative, but it is not commutative even for ordinals.*

Proof. 1. If n, m are finite order types, then $n \times m$ is the order type of a set of cardinality $n \cdot m$. The order type of this set is $n \cdot m$, and so $n \times m = n \cdot m$.

2. The first part of the assertion can be established by following the proof of Theorem 5.1, which was designed as a substitute of the associative law for Cartesian products, and then using the definition of ordered union. The details are left to the reader.

The examples mentioned above show that $\omega_0 \times 2 \neq 2 \times \omega_0$, and so multiplication is not commutative even for ordinals.

$$* \qquad *$$
$$*$$

As multiplication is not commutative, there may be two different distributive laws. The following example shows that multiplication on the right is not distributive

$$\omega_0 = 2 \times \omega_0 = (1 \dot{+} 1) \times \omega_0 \neq \omega_0 \dot{+} \omega_0.$$

On the other hand, the following is true.

Theorem 7.8. *Given an arbitrary ordered set $\langle \Gamma, \prec_\Gamma \rangle$ and given order types $\Theta, \Theta_\gamma \, \gamma \in \Gamma$, we have*

$$\Theta \times \left(\sum_{\gamma \in \Gamma(\prec_\Gamma)} \Theta_\gamma \right) = \sum_{\gamma \in \Gamma(\prec_\Gamma)} \Theta \times \Theta_\gamma;$$

that is, multiplication on the left is distributive.

Proof. Choose pairwise disjoint ordered sets $\{\langle A_\gamma, \prec_\gamma \rangle : \gamma \in \Gamma\}$ and an ordered set $\langle A', \prec' \rangle$ such that

$$\text{type}\, A_\gamma(\prec_\gamma) = \Theta_\gamma \qquad \text{for} \qquad \gamma \in \Gamma$$

and

$$\text{type}\, A'(\prec') = \Theta$$

hold. Then, according to the distributive law for sets, we have

$$A' \times \left(\bigcup_{\gamma \in \Gamma} A_\gamma \right) = \bigcup_{\gamma \in \Gamma} A' \times A_\gamma.$$

Denote this set by P; further, denote by π the order type on the left-hand side of the assertion of the theorem, and by ρ the one on the right-hand side. For each $x \in \bigcup_{\gamma \in \Gamma} A_\gamma$, denote by $\gamma(x)$ the unique element $\gamma \in \Gamma$ for which $x \in A_{\gamma(x)}$. Then

$$\pi = \text{type}\, P(\prec^*), \qquad \rho = \text{type}\, P(\prec^{**})$$

for the appropriate orderings $\prec^*$ and $\prec^{**}$ of the set P. We can see from the definitions of ordered union and of anti-lexicographic product that the relations

$$\langle x, y \rangle \prec^* \langle x', y' \rangle$$

and

$$\langle x, y \rangle \prec^{**} \langle x', y' \rangle$$

are both equivalent to the following:

$$\begin{array}{lll} \text{either} & \gamma(y) \prec_\Gamma \gamma(y') & \\ \text{or} & \gamma(y) = \gamma(y') & \text{and} \quad y \prec_{\gamma(y)} y' \\ \text{or} & y = y' & \text{and} \quad x \prec' x'. \end{array}$$

Hence $\pi = \rho$.

$$* \qquad *$$
$$*$$

Example:

$$\sum_{n \in \omega_0(<)}^{\cdot} \omega_0 = \omega_0 \times \left(\sum_{n \in \omega_0(<)}^{\cdot} 1 \right) = \omega_0 \times \omega_0.$$

For the sake of completeness, we introduce one more piece of notation.

Definition 7.15. *Given an ordered set $\langle A, \prec \rangle$, we denote by $\succ$ the relation defined on A for which we have*

$$x \succ y \iff y \prec x$$

whenever $x, y \in A$.

Clearly, $\langle A, \succ \rangle$ is an ordered set, and for two arbitrary ordered sets A, A' we have

$$\langle A, \prec \rangle \simeq \langle A', \prec' \rangle \iff \langle A, \succ \rangle \simeq \langle A', \succ' \rangle.$$

Hence the following definition makes sense. .

Definition 7.16. *Given an arbitrary order type Θ, we denote by Θ^*, and call the* reverse *of Θ, the order type $\Theta^* = \text{type } A(\succ)$, where $\langle A, \prec \rangle$ is an ordered set of order type Θ.*

Example: $\omega_0^* = \text{type}\{-n : m \in \omega\}(<)$, where $<$ is the ordering of integers according to size.

Problems

1. Using the Axiom of Choice, prove that a set $\langle A, \prec \rangle$ is wellordered if and only if it has no subset of order type ω^*.

2. Prove that $\eta_0 = \eta_0^*$ and $\lambda_0 = \lambda_0^*$, where η_0 is the order type of the rationals, and λ_0 is that of the real numbers.

3. Define an "ordering" of the order types as follows: we put

$$\Theta_0 \preceq \Theta_1$$

if there are sets

$$\langle A_0, \prec_0 \rangle, \qquad \langle A_1, \prec_1 \rangle$$

of order type Θ_0, Θ_1, respectively, such that A_0 is similar to a subset of A_1 Prove that

  a) the definition is sound;
  b) $\preceq$ is reflexive and transitive;
  c) $\preceq$ is not trichotomous;
  d) $\preceq$ is not antisymmetric.

4.[*] Prove that if Θ is an arbitrary countable order type then $\Theta \preceq \eta_0$.

5.[*] The ordered set $\langle A, \prec \rangle$ is said to have a *dense ordering* if it is not empty, if it has an element preceding, and an element succeeding, any of its elements, and if it has an element between any two of its elements. Prove that if $\langle A, \prec \rangle$ is countable and has a dense ordering, then $\text{type}(\langle A, \prec \rangle) = \eta_0$.

6. Prove that the set of irrational numbers is not similar to the set of real numbers.

7.[*] Prove that for an arbitrary infinite order type Θ we have either $\omega \preceq \Theta$ or $\omega \preceq \Theta^*$.

8.[*] Prove that if $A \subset \mathbb{R}$ and the set $\langle A, < \rangle$ or the set $\langle A, > \rangle$ is wellordered in the ordering inherited from $\mathbb{R}$, then $|A| \leq \aleph_0$.

8. PROPERTIES OF WELLORDERED SETS.
GOOD SETS. THE ORDINAL OPERATION

We are now setting out on the course of giving proofs of theorems that we still "owe." It will be easy to see that we use here only a few elementary definitions and theorems, and that for the formulations or proofs of these we do not need any results that we are yet going to establish. In this section, we will not use the Axiom of Choice.

Definition 8.1. *Let* $\langle A, \prec \rangle$ *be an ordered set and let* $B \subset A$. *We say that* B *is an* initial segment *of* A *if, whenever* $x \in B$ *and* $y \prec x$, *we also have* $y \in B$.

A is clearly an initial segment of A. If B is an initial segment of A and $B \neq A$, we call B a proper initial segment of A.
Example: Let $B = \{r \in \mathbb{Q} : r < \sqrt{2}\}$. Then B is an initial segment of $\mathbb{Q}$.

Definition 8.2. *Let* $\langle A, \prec \rangle$ *be an ordered set and* $x \in A$. *The set* $\{y \in A : y \prec x\}$ *is called the* initial segment *of* A *determined by* x. *It is denoted as* $A| \prec x$.

It is clear that $A| \prec x$ is an initial segment of A. As shown by the above example, there is an ordered set that has a proper initial segment that is not determined by any of its elements.
The next three theorems summarize the simplest fundamental properties of wellordered sets.

Theorem 8.1. *Every proper initial segment of a wellordered set is determined by one of its elements.*

Proof. Assume $\langle A, \prec \rangle$ is wellordered, and let B be one of its proper initial segments. Then $A \setminus B \neq \emptyset$, and so there is an element

$$x = \min_{\prec} A \setminus B.$$

We claim that we have

$$B = A| \prec x.$$

Indeed, if $y \in B$ then $y \prec x$, since otherwise we would have $x \preceq y$, and so, B being an initial segment, we would have $x \in B$. Thus $y \in B$ implies

$y \in A| \prec x$. Assume now that $y \in A| \prec x$. Then $y \in A$ and $y \prec x$. In view of the minimality of x, we therefore have $y \notin A \setminus B$. Thus $y \in B$.

$$* \quad *$$
$$*$$

Theorem 8.2. *If f is a monotonic mapping of a wellordered set $\langle A, \prec \rangle$ into itself, then we have $x \preceq f(x)$ for $x \in A$.*

Proof. Denote by B the set $\{x \in A : f(x) \prec x\}$. Assume that, on the contrary to the assertion, we have $B \neq \emptyset$. Then, A being wellordered, we can consider the element $x_0 = \min_{\prec} B$. Put $y_0 = f(x_0)$. According to the definition of B, we have $y_0 \prec x_0$. In view of the monotonicity of f, we have $f(y_0) \prec f(x_0) = y_0$. Thus $f(y_0) \prec y_0$, and so y_0 is an element of B that is less than x_0. This is a contradiction; thus $B = \emptyset$, and so $x \preceq f(x)$ holds for every $x \in A$.

$$* \quad *$$
$$*$$

Theorem 8.3. *If $\langle A, \prec \rangle$ and $\langle A', \prec' \rangle$ are similar wellordered sets, then there is exactly one monotonic mapping of A onto A'.*

Proof. Assume f and g are mappings such that

$$\langle A, \prec \rangle \simeq_f \langle A', \prec' \rangle \quad \text{and} \quad \langle A, \prec \rangle \simeq_g \langle A', \prec' \rangle.$$

Put $h = g^{-1} \circ f$. Then $h^{-1} = f^{-1} \circ g$. Both these mappings are mappings of A into itself. So, according to the preceding theorem, we have

$$x \preceq h(x) \quad \text{and} \quad x \preceq h^{-1}(x),$$

that is, according to the second relation, we also have

$$h(x) \preceq x.$$

Thus h is the identity mapping. Hence it follows that $f = \text{Id}_A \circ g = g$.

$$* \quad *$$
$$*$$

To understand the following definitions, we need some explanation. When we discussed sets above, we were usually concerned with subsets of ω, $\mathbb{R}$, or of some given structure, or, in a "very abstract situation," with sets of subsets of such sets. Whereas we never made an explicit restriction to this extent, we usually separated elements and sets. Even if we allowed for elements of a set to be sets themselves, we have never been concerned with the question of whether these sets occurring as elements were elements of each other or not.

Definition 8.3. *Given an arbitrary set A, the relation $\in_A$ denotes the restriction of the property $\in$ to A, that is,*

$$\in_A = \in \restriction A = \{\langle x, y \rangle : x, y \in A \text{ and } x \in y\}.$$

The idea of studying this relation does not originate with Cantor. It originates with E. Zermelo and J. von Neumann. In retrospect, it appears natural that we need to study this relation, since in Section 1 we imposed stipulations precisely on the property $\in$. We were unable to prove a number of our assertions exactly because we disregarded this relation, and so were unable to exploit our assumptions about the property $\in$.

Before continuing, we give some *examples* that may help the reader to get accustomed to the idea.

If A does not consist of sets, then the relation $\in_A$ is of no use. For example, if $A = \{\text{Sylvester P. Pussycat}, 7\}$ then

$$\in_A = \emptyset.$$

If $A = \emptyset$ then

$$\in_A = \emptyset.$$

If $A = \{\emptyset\}$ then

$$\in_A = \emptyset.$$

If $A = \{\emptyset, \{\emptyset\}\}$ then

$$\in_A = \{\langle \emptyset, \{\emptyset\} \rangle\}.$$

If $A = \{\emptyset, \{\emptyset\}, \{\emptyset, \{\emptyset\}\}\}$ then

$$\in_A = \{\langle \emptyset, \{\emptyset\} \rangle, \langle \emptyset, \{\emptyset, \{\emptyset\}\} \rangle, \langle \{\emptyset\}, \{\emptyset, \{\emptyset\}\} \rangle\}.$$

In the last three examples $\in_A$ is an ordering of the set A. If, on the other hand, $A = \{\emptyset, \{\emptyset\}, \{\{\emptyset\}\}\}$, then

$$\in_A = \{\langle \emptyset, \{\emptyset\} \rangle, \langle \{\emptyset\}, \{\{\emptyset\}\} \rangle\},$$

but since $\emptyset \notin \{\{\emptyset\}\}$, the relation $\in_A$ is not an ordering of A.

We will define a kind of sets that we will temporarily call "good sets." These are intended to be wellordered sets; further, for every wellordered set we want there to be exactly one good set that is similar to it. Having done this, we can define the order type operation in such a way that it should assign a good set to every wellordered set. That is, the good sets will be the ordinals.

Definition 8.4. *We call a set A* transitive *if all its elements are sets, and for each element v of every element u of A, we also have v ∈ A. This latter requirement can also be formulated by saying that each element of A must also be one of its subsets.*

We remark that if A is a transitive set then every element of its elements is also a set, etc. We may express this by saying that A consists of sets hereditarily.

Definition 8.5. *The set A is called* good *if it is transitive and it is wellordered by the relation ∈_A; that is, if A is transitive and ⟨A, ∈_A⟩ is a wellordered set.*

For example, $\emptyset$, $\{\emptyset\}$, $\{\emptyset, \{\emptyset\}\}$, $\{\emptyset, \{\emptyset\}, \{\emptyset, \{\emptyset\}\}\}$ are all good sets.

However surprising it may be, it will turn out that there are many good sets.

The next three theorems will enumerate the basic properties of good sets.

Theorem 8.4. *If A and B are good sets, then A ∩ B is also a good set.*

Proof. If A and B are transitive sets then $A \cap B$ is also transitive. Indeed, $A \cap B \subset A$, and so the elements of $A \cap B$ are also sets. If $u \in A \cap B$, then $u \in A$, $u \in B$, and so $u \subset A$, $u \subset B$, and so $u \subset A \cap B$ as well. $\langle A \cap B, \in_{A \cap B} \rangle$ is wellordered, since $\in_{A \cap B}$ is the ordering inherited from the wellordering $\in_A$.

$$* \quad *$$
$$*$$

Theorem 8.5. *Each element of a good set is also a good set.*

Proof. Assume that A is a good set and $x \in A$. A is transitive, so $x \subset A$. Hence it follows that x consists of sets and is wellordered by $\in_x$. Assume that $u \in x$ and $v \in u$. Using the transitivity of A twice, we obtain that $u \in A$, $v \in A$. As $\in_A$ wellorders A, $\in_A$ is a transitive relation on A, and so $v \in u$ and $u \in x$ implies $v \in x$. That is, for any $u \in x$ we also have $u \subset x$. Thus x is a good set.

$$* \quad *$$
$$*$$

Before enunciating the next theorem, we would like to consider the following.

Lemma 8.1. *Let A be a transitive set and assume that ∈_A orders A. A set B ⊂ A is then an initial segment of the ordering ⟨A, ∈_A⟩ if and only if B is a transitive set.*

Proof. Assume that $B \subset A$ and B is an initial segment of A. Then, clearly, B also consists of sets only. If $u \in B$ then $u \in A$ as well, so u is a subset of A. If we now have $v \in u$, then v is an element of A that precedes u in the ordering $\in_A$; thus, B being an initial segment, we have $v \in B$. Hence, if $u \in B$ then also $u \subset B$, and so B is transitive.

Conversely, assume that B is a subset of A and it is transitive. If $u \in B$ and $v \in u$, then we have $v \in B$ in virtue of $u \subset B$; that is, B is an initial segment of A.

*   *
*

Theorem 8.6. *If A is a good set and B is a transitive subset of A, then either $B \in A$ or $B = A$; thus B is also a good set.*

Proof. According to Lemma 8.1, B is an initial segment of the wellordered set $\langle A, \in_A \rangle$. We may assume that B is a proper initial segment, since otherwise we would have $B = A$. According to Theorem 8.1, there is an $x \in A$ such that $B = A| \in_A x$, the initial segment determined by x. We claim that $B = x$. If $v \in B$, then we have $v \in x$ according to the definition of $A| \in_A x$. If $v \in x$, then we also have $v \in A$ in view of the transitivity of A, and so $v \in A| \in_A x = B$. Thus $B = x \in A$. It then follows from Theorem 8.5 that B is a good set.

*   *
*

From now on, we will use α, β, γ, δ to denote *good sets*. We are going to define an ordering of good sets.

Definition 8.6. *For arbitrary good sets α, β, we put*

$$\alpha < \beta \iff \alpha \in \beta.$$

Theorem 8.7. *The property $<$ just defined is a wellordering of good sets. By this we mean that the three usual properties of orderings are satisfied, and, furthermore, if for some property Φ we have $\exists \alpha \Phi(\alpha)$, then there is a least α for which $\Phi(\alpha)$ holds. If there is such an α, we will denote it by $\min_<\{\alpha : \Phi(\alpha)\}$.*

We had to add the explanation because, as it will turn out later, the collection of all good sets does not form a set.

Proof. The property $<$ is irreflexive on good sets: Indeed, assume $\alpha < \alpha$; then $\alpha \in \alpha$. Thus we would also have $\alpha \in \alpha \in \alpha$. This, however, is not possible since $\in_\alpha$ is irreflexive on α.

The property $<$ is transitive on good sets: Indeed, assume that $\alpha < \beta$ and $\beta < \gamma$. Then $\alpha \in \beta$ and $\beta \in \gamma$. Now γ is a transitive set; so we have $\alpha \in \gamma$, that is, $\alpha < \gamma$.

The property $<$ is trichotomous on good sets: Indeed, assume that $\alpha \neq \beta$. According to Theorem 8.4, $\alpha \cap \beta$ is a good set, and so it is transitive;

furthermore, $\alpha \cap \beta$ is a subset of both α and β. According to Theorem 8.6, the following two assertions hold: on the one hand,

$$\alpha \cap \beta = \alpha \quad \text{or} \quad \alpha \cap \beta \in \alpha,$$

and on the other,

$$\alpha \cap \beta = \beta \quad \text{or} \quad \alpha \cap \beta \in \beta.$$

If the second part of both assertions hold, then

$$\alpha \cap \beta \in \alpha \cap \beta,$$

which contradicts the already established irreflexivity.

Hence, either

$$\alpha \cap \beta = \alpha$$

or

$$\alpha \cap \beta = \beta.$$

As $\alpha \neq \beta$, exactly one of these relations must be true, and so either

$$\alpha \cap \beta = \beta \in \alpha$$

or

$$\alpha \cap \beta = \alpha \in \beta$$

holds.

The property $<$ is a wellordering on good sets: Indeed, assume that $\Phi(\alpha)$ is true for some property Φ and for some good set α. Put

$$A = \{x \in \alpha : \Phi(x)\}.$$

If $A = \emptyset$, then α is the least good set with property Φ. If $A \neq \emptyset$, then, noting that $\in_\alpha$ wellorders α, the element $\min_{\in_\alpha} A$ exists. According to Theorem 8.5, this element is a good set. Denote $\min_{\in_\alpha} A$ by α_0. α_0 has property Φ, according to its definition. If $\beta < \alpha_0$, then, α being transitive, we have $\beta \in \alpha$. Hence, β does not have property Φ, according to the definitions of α_0 and A.

*　　*
*

We now formulate some important consequences of our results.

Corollary 8.1. *If A consists of good sets and is transitive, then A is a good set.*

Proof. $\in_A$ wellorders the set A, since, according to Theorem 8.7, $<$ is a wellordering property on good sets. As A is also transitive, it is a good set.

*   *
*

Corollary 8.2. *There is no set that has each good set as an element.*

Proof. Assuming the assertion is false, there is a set A consisting of all good sets. This A is transitive since, according to Theorem 8.5, each element of a good set is also a good set. This means, according to the preceding corollary, that the set A itself is a good set; thus we have $A \in A$. This contradicts the fact that the property $<$ is irreflexive on good sets.

*   *
*

Corollary 8.3. *If $\alpha < \beta$ then $\langle \beta, \in_\beta \rangle$ cannot be mapped monotonically onto a subset of $\langle \alpha, \in_\alpha \rangle$; hence, if α and β are distinct good sets, then $\langle \alpha, \in_\alpha \rangle$ is not similar to $\langle \beta, \in_\beta \rangle$.*

Proof. If, on the contrary, there is an f and an $A \subset \alpha$ such that

$$\langle \beta, \in_\beta \rangle \simeq_f \langle A, \in_\alpha \restriction A \rangle,$$

then f is a monotonic mapping that maps β into β and

$$f(\alpha) \in_\beta \alpha.$$

This is impossible according to Theorem 8.2.

The second part of the assertion follows from the first part, since if $\alpha \neq \beta$ then either $\alpha < \beta$ or $\beta < \alpha$.

*   *
*

The next theorem says that indeed there are many good sets.

Existence Theorem 8.8. *For every wellordered set $\langle A, \prec \rangle$ there is exactly one good set α such that*

$$\langle A, \prec \rangle \simeq \langle \alpha, \in_\alpha \rangle.$$

Proof. According to Corollary 8.3, given an arbitrary wellordered set $\langle A, \prec \rangle$, there is at most one such good set. Let A' consist of those elements x of A that determine initial segments for which such good sets exist; that is,

$$A' = \{x \in A : \exists \beta \, (A \restriction \prec x \simeq \langle \beta, \in_\beta \rangle)\}.$$

For an arbitrary $x \in A'$, denote by β_x the unique good set β for which

$$\langle A| \prec x, \prec \rangle \simeq \langle \beta, \in_\beta \rangle).$$

According to Theorem 8.3, there is exactly one function f that maps the initial segment $A| \prec x$ monotonically onto β_x; denote this function by f_x $(x \in A')$.

We need to make two simple observations. First, if

$$\langle B, \prec \rangle \simeq_g \langle B', \prec' \rangle,$$

then g maps an initial segment onto an initial segment, and an initial segment determined by an element onto an initial segment determined by an element. Second, for arbitrary good sets $\beta \in \alpha$ we have

$$\langle \alpha, \in_\alpha \rangle| \prec \beta = \beta.$$

Therefore, if $x \in A'$ and $y \prec x$ then $f_x|(A| \prec y)$ maps the initial segment $A| \prec y$ monotonically onto a good set; thus it follows that A' is an initial segment of A. In fact, the range of this mapping is β_y, and

$$f_x|(A| \prec y) = f_y.$$

Furthermore, we also have

$$\beta_y < \beta_x.$$

Consider the set $\{\beta_x : x \in A'\}$. This set consists of good sets. If $\gamma \in \beta_x$, then it follows from what was said before that $\gamma = \beta_y$ for some $y \in A'$; in fact, it is easy to see this with

$$y = \min_{\prec}\{z \in A| \prec x : \gamma \leq f_x(z)\}$$

($u \leq v$ abbreviates $u < v \vee u = v$). Thus the set $\{\beta_x : x \in A'\}$ is transitive, and so it is a good set in virtue of Corollary 8.1. Denote the set $\{\beta_x : x \in A'\}$ by α. Define a mapping f of A' by the stipulation

$$f(x) = \beta_x \qquad (x \in A').$$

As we have seen, this mapping is monotonic, and so

$$\langle A', \prec \restriction A' \rangle \simeq \langle \alpha, \in_\alpha \rangle.$$

It is sufficient to show that $A' = A$. If this were not true, then according to Theorem 8.1 we would have

$$A' = A| \prec x$$

for some $x \in A$. This is impossible, since in this case we would have $x \in A'$ by the definition of A', and so $A' = A| \prec x$ would imply $x \prec x$.

$$* \qquad *$$
$$*$$

According to Theorem 8.8, there is an operation $\mathcal{F}_0$ defined on wellordered sets such that for a wellordered set $\langle A, \prec \rangle$ its value $\mathcal{F}_0(\langle A, \prec \rangle)$ is the unique good set α satisfying

$$\langle A, \prec \rangle \simeq \langle \alpha, \in_\alpha \rangle.$$

According to Theorem 8.8 and Corollary 8.3, this operation is compatible with the similarity property on wellordered sets.

Therefore we may modify Definition 7.8 of the order type operation as follows.

Definition 8.7. *From now on,* type $A(\prec)$ *will denote an operation compatible with similarity such that*

$$\text{type } A(\prec) = \mathcal{F}_0(\langle A, \prec \rangle)$$

if $\langle A, \prec \rangle$ *is wellordered.*

According to Definition 7.8, the ordinals are the values of the order type operation on wellordered sets, that is, the good sets.

We summarize what we know so far about ordinals.

The property $\alpha < \beta$ (i.e., $\alpha \in \beta$) is a *wellordering property* on ordinals (Theorem 8.7).

Each element and every transitive subset of an ordinal is an ordinal (Theorems 8.5, 8.6).

$$\alpha < \beta \iff \alpha \subsetneq \beta$$

and

$$\alpha \leq \beta \iff \alpha \subset \beta$$

(Theorem 8.6).

According to the stipulation made after Definition 7.8, if A is a finite set and $|A| = n$, then, for the order type of A in an arbitrary ordering $\prec$, we have type $A(\prec) = n$. We will maintain this stipulation, so in what follows the nonnegative integers will be identified with the finite ordinals.

$$0 = \emptyset, 1 = \{\emptyset\} = \{0\}, 2 = \{\emptyset, \{\emptyset\}\} = \{0, 1\}, \ldots.$$

This convention will make the symbol $\underline{n}$ used thus far superfluous, since for $n \in \omega$, we have

$$\underline{n} = \{0, 1, \ldots, n-1\} = n.$$

According to the new notation,

$$\omega_0 = \text{type } \omega(<) = \omega = \{0, 1, \ldots, n, \ldots\}.$$

In what follows, we will use the symbol ω.

Corollary 8.4. *For arbitrary wellordered sets $\langle A, \prec \rangle$, $\langle A', \prec' \rangle$ we have*

$$\text{type } A(\prec) < \text{type } A'(\prec')$$

if and only if the set $\langle A, \prec \rangle$ is similar to an initial segment $\langle A', \prec' \rangle$ determined by an element.

Proof. Let type $A(\prec) = \alpha$, type $A'(\prec') = \beta$. We know that

$$\langle A, \prec \rangle \simeq \langle \alpha, \in_\alpha \rangle,$$

$$\langle A', \prec' \rangle \simeq \langle \beta, \in_\beta \rangle.$$

On the other hand,

$$\alpha < \beta \iff \alpha = \beta| \in_\beta \alpha \wedge \alpha \in \beta.$$

Thus the assertion follows from the transitivity of the similarity property.

$$* \qquad *$$
$$*$$

Next we will establish a few simple theorems for ordinals.

Theorem 8.9. *The addition and multiplication of ordinals is weakly increasing in the first argument, and strictly increasing in the second argument.*

Proof. The assertions concerning weak monotonicity can be proved similarly as Assertion 3 of Theorem 5.4; we will not go into details. As for strict monotonicity, assume that for the ordinals α, β, γ we have $\beta < \gamma$. We claim that

$$\alpha \dotplus \beta < \alpha \dotplus \gamma,$$

$$\alpha \times \beta < \alpha \times \gamma.$$

Choose disjoint ordered sets $\langle A, \prec \rangle$, $\langle C, \prec' \rangle$ for which

$$\alpha = \text{type } A(\prec), \qquad \gamma = \text{type } C(\prec').$$

According to Corollary 8.4, there is an $x \in C$ such that, writing $B = C| \prec' x$, we have

$$\text{type } B(\prec') = \beta.$$

Then $A \cup B$ is the initial segment determined by x of the ordered sum $A \cup C$, and $A \times B$ is the initial segment determined by $\langle a_0, x \rangle$ of the anti-lexicographic product $A \times C$, where $a_0 = \min_{\prec} A$. Then, using Corollary 8.4 again, we can see that

$$\alpha \dot{+} \beta < \alpha \dot{+} \gamma, \qquad \alpha \dot{+} \beta < \alpha \dot{+} \gamma.$$

*   *
*

Strict monotonicity in the first argument is not true. This is illustrated by the following examples:

$$0 < 1, \qquad \text{but} \qquad \omega = 0 \dot{+} \omega = 1 \dot{+} \omega,$$

$$1 < 2, \qquad \text{but} \qquad \omega = 1 \times \omega = 2 \times \omega,$$

as we saw in Section 7 above.

Theorem 8.10. *For an arbitrary ordinal α, the ordinal $\alpha \dot{+} 1$ is the least ordinal that is greater than α, and*

$$\alpha \dot{+} 1 = \alpha \cup \{\alpha\}.$$

Proof. By the preceding theorem, we have $\alpha < \alpha \dot{+} 1$. On the other hand, $\alpha \cup \{\alpha\}$ is a transitive set that consists of ordinals, and so it is an ordinal. According to the definition of ordered sum, we have

$$\text{type}(\alpha \cup \{\alpha\})(\in) = \alpha \dot{+} 1,$$

and so

$$\alpha \dot{+} 1 = \alpha \cup \{\alpha\}.$$

Finally, if $\alpha < \beta$ then $\alpha \in \beta$; hence

$$\alpha \cup \{\alpha\} \subset \beta, \qquad \alpha \cup \{\alpha\} \leq \beta.$$

*   *
*

Theorem 8.11. *An arbitrary set A of ordinals has an upper bound, that is an ordinal that is greater than or equal to each element of A. The least upper bound is the ordinal $\bigcup A$.*

Proof. According to Convention 8 in Section 1, we have $\bigcup A = \bigcup_{\alpha \in A} A$. We claim that $\bigcup A$ is an ordinal. Indeed, $\bigcup A$ consists of ordinals, since all elements of ordinals are also ordinals. Further, $\bigcup A$ is transitive, since it is a union of transitive sets. Thus $\bigcup A$ is an ordinal as claimed. If $\alpha \in A$, then $\alpha \subset \bigcup A$, and so $\alpha \le \bigcup A$; so $\bigcup A$ is an upper bound. Assume now that β is an upper bound of A. Then, for each $\alpha \in A$ we have $\alpha \le \beta$; and so we also have $\alpha \subset \beta$. Thus we have $\bigcup A \subset \beta$, and so $\bigcup A \le \beta$ also holds.

$$* \qquad *$$
$$*$$

If A is a set consisting of ordinals, then the ordinal $\bigcup A$ is denoted by $\sup A$.

Definition 8.8. *The ordinal α is called an ordinal of the first kind or a successor ordinal if $\alpha = \beta \dotplus 1$ for some β. α is said to be of the second kind if it is not of the first kind. If $\alpha > 0$ and is of the second kind, then α is called a* limit ordinal.

For example: ω is a limit ordinal, 0 is an ordinal of the second kind, $1, 2, \ldots, n, \ldots$, and $\omega \dotplus 1$ are successor ordinals.

Problems

1. Show that
$$\sup\{\xi \dotplus 1 : \xi \in A\}$$
is the least ordinal that is greater than each element of A.

2. Prove that if $\alpha \le \beta$ then the equation $\alpha \dotplus \xi = \beta$ has a unique solution.

3. Prove that
$$\sup\{\alpha \dotplus \eta \dotplus 1 : \eta < \xi\} = \alpha \dotplus \xi.$$

4. Prove that if ξ is a limit ordinal then
$$\sup\{\alpha \times \eta : \eta < \xi\} = \alpha \times \xi.$$

5. Show that there are limit ordinals α and ξ such that
$$\sup\{\eta \dotplus \alpha : \eta < \xi\} \ne \xi \dotplus \alpha,$$
$$\sup\{\eta \times \alpha : \eta < \xi\} \ne \xi \times \alpha.$$

6. * Show that for arbitrary ordinals $\beta \ge 0$ and α there are uniquely determined ordinals ξ, ρ such that
$$\alpha = \beta \times \xi \dotplus \rho \qquad \text{and} \qquad \rho < \beta.$$
(This equation is a generalization of division with remainder for positive integers.)

7. * Without using the Axiom of Choice, prove that for every set A there is an ordinal α such that $\alpha \not\sim A'$ holds for every $A' \subset A$. *(Hartog's Theorem.)*

8. * Without using the Axiom of Choice, prove that if
$$A \cap \alpha = \emptyset \qquad \text{and} \qquad A \times \alpha \sim A \cup \alpha$$
then there is an $A' \subset A$ such that $\alpha \sim A'$ or a $B \subset \alpha$ such that $A \sim B$.

9. * Prove that if for every infinite cardinal we have $a^2 = a$, then the Axiom of Choice holds. *(Tarski's Theorem.)*

9. TRANSFINITE INDUCTION AND RECURSION. SOME CONSEQUENCES OF THE AXIOM OF CHOICE, THE WELLORDERING THEOREM

In this section we will pay off two of our debts. Namely, we will prove the theorems on transfinite induction and transfinite recursion. After this, we will list the most familiar statements equivalent to the Axiom of Choice. As applications, we will establish some results of set theory that have frequent uses in the second part of the book and in other branches of mathematics, such as algebra, topology, etc.

Transfinite Induction Theorem 9.1. *Let $\Phi(\alpha)$ be an arbitrary property defined for ordinals. Assume that for every ordinal α the following assertion holds: If $\Phi(\beta)$ is true for every $\beta < \alpha$, then $\Phi(\alpha)$ is also true. In these circumstances, $\Phi(\alpha)$ is true for every α. Formally:*

$$\forall \alpha \left(\forall \beta < \alpha \, \Phi(\beta) \implies \Phi(\alpha) \right) \implies \forall \alpha \Phi(\alpha).$$

This theorem is a generalization of mathematical induction on nonnegative integers. The theorem remains true, with appropriate changes, if instead of ordinals we say nonnegative integers or any wellordered collection.

Induction on nonnegative integers is usually formulated as follows. If $\Phi(0)$ and for every $n < \omega$ we have $\Phi(n) \implies \Phi(n+1)$, then $\Phi(n)$ is true for every $n \in \omega$.

This cannot be translated to ordinals literally, since, for example, 0 is finite, and if α is finite then $\alpha \overset{.}{+} 1$ is finite, and yet there is an ordinal (e.g. ω) that is not finite; that is, the literal translation fails for ordinals with the property $\Phi(\alpha)$ meaning "α is finite." The formulation of Theorem 9.1 has the advantage of being equally valid for ordinals and for nonnegative integers. It is also worth pointing out that the validity of $\Phi(0)$ need not be required separately, since the statement $\forall \beta < 0 \, \Phi(\beta)$ is true for any property Φ.

Proof. Assume, on the contrary, that there is a β for which $\Phi(\beta)$ is false. Then, according to Theorem 8.7 there is a smallest ordinal α for which $\Phi(\alpha)$ is false. Then Φ true for every $\beta < \alpha$, and so it is also true for α, according to our assumptions. This is a contradiction, showing that $\Phi(\alpha)$ is true for all α.

$$* \quad *$$
$$*$$

The Transfinite Recursion Theorem will say that there always exists an operation F such that $F(\alpha)$ depends in a given way on how F was defined for ordinals β less than α. This is a much deeper assertion than the Transfinite Induction Theorem. Theorem 9.2 will also justify definition by recursion on nonnegative integers.

Transfinite Recursion Theorem 9.2. *Let G be an operation that assigns a set $G(f)$ to each function f. Then there is a unique operation defined for ordinals such that we have*

$$F(\alpha) = G(F|\alpha)$$

for every ordinal α.

Before we start out with the proof, we recall that the restriction of an operation to a set is a function (see Section 4). Theorem 9.2 indeed reflects our intentions expressed above, as can be seen by the following considerations. The way F is defined up to α is described by the function $F|\alpha$; this function does not only give the values $F(\beta)$ for $\beta < \alpha$, it also specifies how $F(\beta)$ depends on β. We are looking for an F such that, for all α, $F(\alpha)$ depends "in a way expressed by the operation G" on how F has been been specified up to α; that is, we want an F such that $F(\alpha)$ depends on $F|\alpha$ "in a way expressed by the operation G." The proof that follows fits in with the present intuitive framework; in the Appendix, we will outline how a more formal proof can be given along the same lines (see Theorem A5.1 below).

Proof. First we show that for every ordinal α there is at most one function f defined on α such that

(1) $$\forall \beta < \alpha \, (f(\beta) = G(f|\beta)),$$

that is, "an f that up to α satisfies our defining equation."

Assume that there are functions f_1 and f_2 on α that satisfy (1). By transfinite induction on β, we are going to show that

$$f_1(\beta) = f_2(\beta)$$

for every ordinal $\beta < \alpha$. In view of the Transfinite Induction Theorem, for this it is enough to show that if $\beta < \alpha$ and $f_1(\gamma) = f_2(\gamma)$ for every $\gamma \in \beta$, then $f_1(\beta) = f_2(\beta)$. If

$$f_1(\gamma) = f_2(\gamma)$$

for every $\gamma < \beta$, then

$$f_1|\beta = f_2|\beta,$$

and so
$$f_1(\beta) = G(f_1|\beta) = G(f_2|\beta) = f_2(\beta).$$

Thus $f_1 = f_2$.

Next we will show by transfinite induction that for every α there is a function satisfying (1). Assume that for every $\gamma \leq \alpha$ there is a function that satisfies (1). Since we have already seen that there is at most one such function for each γ, we may denote the function satisfying (1) with γ replacing α by f_γ. (This is a somewhat vague description; more precisely, what is meant is that there is a function that to every γ assigns f_γ.)

Note that if $\delta < \gamma < \alpha$ then f_γ clearly satisfies (1) up to δ, and so, in view of the uniqueness of f_δ, we have

$$f_\delta = f_\gamma|\delta \qquad \text{for arbitrary ordinals} \quad \delta < \gamma < \alpha.$$

We now distinguish two cases according as $a)$ $\alpha = \gamma \dotplus 1$ is of the first kind, or $b)$ α is of the second kind.

$a)$ Define f as follows:
$$D(f) = \alpha,$$
$$f(\delta) = \begin{cases} f_\gamma(\delta) & \text{for } \delta < \gamma, \\ G(f_\gamma) & \text{for } \delta = \gamma. \end{cases}$$

Then, for every $\delta \leq \gamma$ we have
$$f|\delta = f_\gamma|\delta,$$

and so for every $\delta < \alpha$
$$f(\delta) = G(f_\gamma|\delta) = G(f|\delta)$$

holds.

$b)$ For arbitrary $\gamma < \alpha$ we have
$$\gamma \dotplus 1 < \alpha$$

(note that this is true even in the case $\alpha = 0$). Define f as follows:
$$D(f) = \alpha$$

and
$$\forall \gamma < \alpha \quad f(\gamma) = f_{\gamma \dotplus 1}(\gamma).$$

Then we have
$$f(\gamma) = f_{\gamma \dotplus 1}(\gamma) = f_\delta(\gamma)$$

for all $\gamma < \delta < \alpha$, and so

$$f|\gamma = f_\delta|\gamma$$

holds for any arbitrary $\gamma \leq \delta$. Thus

$$f(\gamma) = f_{\gamma+1}(\gamma) = G(f_{\gamma+1}|\gamma) = G(f|\gamma)$$

also holds.

Thus f satisfies (1) in both cases. Hence, by the Transfinite Induction Theorem, for every α there is a (unique) function satisfying (1).

Denoting this function by f_α for each α, we can define the operation F by

$$F(\alpha) = f_{\alpha+1}(\alpha)$$

for every ordinal α. The claim that F satisfies the equation $F(\alpha) = G(F|\alpha)$ can be established in a way similar to what was done in case $b)$ when proving the existence of f_α. Finally, for any two operations F_1, F_2 satisfying the recursion equation $F(\alpha) = G(f|\alpha)$ of the theorem, we have $F_1(\alpha) = F_2(\alpha)$. This can be proved by transfinite induction in the same way as the uniqueness of f_α.

*　　*
*

Theorem 9.3. *The following five assertions are equivalent:*

a) AXIOM OF CHOICE. *For each system $\{A_\gamma : \gamma \in \Gamma\}$ of nonempty sets there is a choice function.*

b) WELLORDERING THEOREM (ZERMELO'S THEOREM). *For every set A there is a relation $\prec$ that wellorders A.*

c) TEICHMÜLLER–TUKEY LEMMA. *Let A be a set and Φ a property defined on all finite subsets of A. Assume that B is a subset of A such that each finite subset of B has property Φ. Then B can be extended to a maximal subset M of A such that each finite subset of M has property Φ. (Here the maximality of M means that no subset M' of A with $M \subsetneq M'$ is such that each finite subset of M' has property Φ.)*

d) HAUSDORFF'S MAXIMAL CHAIN THEOREM. *An arbitrary partially ordered subset $\langle P, \prec \rangle$ has a maximal ordered subset M; i.e., no subset M' of P with $M \subsetneq M'$ is ordered. (Partially ordered sets were described in Definition 7.3.)*

e) ZORN'S LEMMA. *Assume that in the partially ordered set $\langle P, \prec \rangle$, every ordered subset $R \subset P$ has an upper bound, that is, there exists an $x_R \in P$ such that $y \preceq x_R$ for each $y \in R$. Then there is an $x \in P$ that is a maximal element, that is, an element x such that there is no element of P greater than x. (Zorn's Lemma was discovered by K. Kuratowski in 1922 and rediscovered by M. Zorn in 1935.)*

The proof of the equivalence of these assertions will of course not rely on the Axiom of Choice.

Proof. We will show that the implications $a) \Longrightarrow b)$, $b) \Longrightarrow c)$, $c) \Longrightarrow d)$, $d) \Longrightarrow e)$, and $e) \Longrightarrow a)$ hold.

I. $a) \Longrightarrow b)$. It is enough to show that there is a function f and an ordinal such that $\alpha \sim_f A$. Indeed, a wellordering of A can then be defined as follows:

$$x \prec y \iff f^{-1}(x) < f^{-1}(y)$$

for every $x, y \in A$.

Let g be a choice function defined on all nonempty subsets of A, i.e., let g be such that

$$g(X) \in X \qquad \text{for} \qquad X \subset A \quad \text{with} \quad X \neq \emptyset.$$

We may extend g so that it is defined also for the empty set by putting $g(\emptyset) = a$ for some $a \notin A$. Such an a exists, since according to Theorem 1.2, A cannot contain every set.

Define an operation F by transfinite recursion on α as follows: Assume that $F(\beta)$ has been defined for all ordinals $\beta < \alpha$. Put

$$F(\alpha) = g(A \setminus \{F(\beta) : \beta < \alpha\}).$$

Considering that the argument of g is a subset of A, this definition is meaningful.

This being the first application of the Transfinite Recursion Theorem, as an illustration, we will describe an operation G for which $F(\alpha) = G(F|\alpha)$ holds. Since such an operation is usually easy to find, we will not go into such detail in the future. In the present case,

$$G(f) = g(A \setminus \mathrm{R}(f))$$

is an appropriate choice, since for an arbitrary operation F, we have

$$\mathrm{R}(F|\alpha) = \{F(\beta) : \beta < \alpha\}.$$

According to the Transfinite Recursion Theorem, $F(\alpha)$ is defined for every α. In view of the choice of g, we have either $F(\alpha) \in A$ or $F(\alpha) = a \notin A$. We are now going to see that if $F(\alpha) \in A$ and $\gamma < \alpha$, then

$$F(\gamma) \in A \qquad \text{and} \qquad F(\gamma) \neq F(\alpha).$$

Indeed,

$$F(\alpha) = g(A \setminus \{F(\beta) : \beta < \alpha\}),$$

and so, if $F(\alpha) \in A$ then

$$A \setminus \{F(\beta) : \beta < \alpha\} \neq \emptyset$$

and

$$F(\alpha) \in A \setminus \{F(\beta) : \beta < \alpha\}.$$

Thus, $F(\alpha) \neq F(\gamma)$, since

$$F(\gamma) \notin A \setminus \{F(\beta) : \beta < \alpha\}.$$

On the other hand, as

$$A \setminus \{F(\beta) : \beta < \alpha\} \neq \emptyset,$$

we have

$$A \setminus \{F(\beta) : \beta < \gamma\} \neq \emptyset$$

a fortiori, and so

$$F(\gamma) \in A \setminus \{F(\beta) : \beta < \gamma\} \in A.$$

We are next going to show that there is an α such that

$$F(\alpha) = a.$$

Informally, this means that if we select elements $F(0)$, $F(1)$, ... of A one after the other, we will eventually "run out of elements" of A. Put

$$B = \{x \in A : \exists \alpha \, (F(\alpha) = x)\}.$$

According to what we showed above, for every $x \in B$ there is exactly one α for which $F(\alpha) = x$; thus we may denote this α by $F^{-1}(x)$. Then

$$C = \{F^{-1}(x) : x \in B\}$$

is a set according to Convention 11 in Section 1. On the other hand,

$$C = \{\beta : F(\beta) \in A\}.$$

We saw that if $\gamma < \beta \in C$, then $\gamma \in C$. C consists of ordinals and is transitive, and so it is an ordinal. From now on, denote this ordinal by α:

$$\alpha = \{\beta : F(\beta) \in A\}.$$

Then $F(\alpha) = a$, since otherwise we would have $F(\alpha) \in A$, and so $\alpha \in \alpha$. We have

$$F(\alpha) = g(A \setminus \{F(\beta) : \beta < \alpha\}),$$

and so

$$A \setminus \{F(\beta) : \beta < \alpha\} = \emptyset.$$

Thus, using the notation $f = F|\alpha$, we have

$$\alpha \sim_f A.$$

Thus we established the implication *a)* $\implies$ *b)*.

We can now easily see without any reference to the other implications that the Wellordering Theorem is equivalent to the Axiom of Choice. Indeed, if

$$\{A_\gamma : \gamma \in \Gamma\}$$

is a system of nonempty sets, then, in possession of the Wellordering Theorem, we can directly specify a choice function:

Let

$$A = \bigcup_{\gamma \in \Gamma} A_\gamma$$

and let $\prec$ be a wellordering of A. Then

$$f(\gamma) = \min_{\prec} A_\gamma$$

is a choice function.

II. *b)* $\implies$ *c)*. For the proof, consider the sets A, B described in the statement of *c)*. According to the Wellordering Theorem, we can write the set $A \setminus B$ as a sequence

$$A \setminus B = \{a_\alpha : \alpha < \eta\},$$

where the elements a_α are distinct. We can now specify the elements of M. Naturally, B must be a subset of M. Using Transfinite Recursion, we will decide which elements of $A \setminus B$ will also belong to M. In precise terms, we should say that we are going to define the characteristic function of M, but this would make our notation too complicated.

Assume that $\alpha < \eta$, and for all $\beta < \alpha$ we have already decided whether $a_\beta \in M$ or $a_\beta \notin M$. Write

$$M_\alpha = B \cup \{a_\beta : a_\beta \in M \quad \text{and} \quad \beta < \alpha\}.$$

Put $a_\alpha \in M$ if and only if every finite subset of $M_\alpha \cup \{a_\alpha\}$ has property Φ. This completes the definition of the set M. Let X be an arbitrary finite subset of M. We claim that the set X has property Φ. This is certainly true if $X \subset B$. If $X \not\subset B$, then, X being finite, there is a largest α such that $a_\alpha \in X$. Then $X \subset M_\alpha \cup \{a_\alpha\}$, and so X has property Φ, since otherwise we would not have put a_α in M.

We have yet to show that M is maximal. For this, we have to show that if $a_\alpha \notin M$ then $M \cup \{a_\alpha\}$ has a finite subset X that does not have property Φ. This is, however, true, since the reason we did not put a_α in M was that $M_\alpha \cup \{a_\alpha\}$ had such a subset X, and

$$M_\alpha \cup \{a_\alpha\} \subset M \cup \{a_\alpha\}.$$

This completes the proof of $b) \Longrightarrow c)$.

III. $c) \Longrightarrow d)$. Let $\langle P, \prec \rangle$ be a partially ordered set. Define the property $\Phi(X)$ for each finite subset X of P by saying that $\Phi(X)$ is false if and only if X has exactly two elements, that is, $X = \{u, v\}$ with $u \neq v$, and $u \not\prec v$, $v \not\prec u$. As the empty set has property Φ, according to $c)$ there is a maximal set R such that each of its finite subsets has property Φ. Clearly, R is a maximal ordered subset.

IV. $d) \Longrightarrow e)$. Let $\langle P, \prec \rangle$ be a partially ordered set satisfying the assumptions of Assertion $e)$. According to $d)$, P has a maximal ordered subset R. Let x_R be an upper bound of R. We claim that x_R is a maximal element of the set P. Indeed, if there were a $u \in P$ such that $x_R \prec u$, then $R \cup \{u\}$ would be an ordered subset of P larger than R. This contradicts the maximality of R.

V. $e) \Longrightarrow a)$. Let $\{A_\gamma : \gamma \in \Gamma\}$ be a system of nonempty sets. Let P be the set of partial choice functions for this system, that is,

$$P = \{f : f \text{ is a function } \text{ and } D(f) \subset \Gamma \text{ and } \forall \gamma \in D(f)\, f(\gamma) \in A_\gamma\}.$$

Define a relation on P by the stipulation

$$f \prec g \iff f \subsetneq g;$$

clearly, this is a partial ordering. If $R \subset P$ is ordered, then $\bigcup R$ is a partial choice function, and, in fact, it is an upper bound of the set R. Thus, according to $e)$, P has a maximal element, say f. We claim that f is a choice function for the set system $\{A_\gamma : \gamma \in \Gamma\}$. Indeed, if this is not the case, then there is a $\gamma \in \Gamma$ such that $\gamma \notin D(f)$. Then there is a function g defined on the set $D(f) \cup \{\gamma\}$ such that g agrees with f on $D(f)$ and $g(\gamma) \in A_\gamma$. Then g is a partial choice function with $f \prec g$, contradicting the maximality of f.

$$* \quad *$$
$$*$$

The Axiom of Choice and the assertions equivalent to it can neither be proved nor disproved from the other usual axioms of set theory. We will not go into details at this point.

As we mentioned in Section 2, from now on *we will use the Axiom of Choice without necessarily pointing this out explicitly.*

Definition 9.1. 1. Let X be a *nonempty set* and let $\mathcal{F} \subset \mathrm{P}(X)$. We say that $\mathcal{F}$ is a filter if it is not empty, $\emptyset \notin \mathcal{F}$, and the following two properties are satisfied:

(I) $A, B \in \mathcal{F} \implies A \cap B \in \mathcal{F}$;

(II) $A \in \mathcal{F}$ and $A \subset B \implies B \in \mathcal{F}$.

2. $\mathcal{F}$ is called an ultrafilter on X if it is a filter and for an arbitrary $A \in \mathrm{P}(X)$ we have either $A \in \mathcal{F}$ or $X \setminus A \in \mathcal{F}$.

3. $\mathcal{F}$ is said to have the Finite Intersection Property if for each finite $\mathcal{F}' \in \mathcal{F}$ we have $\bigcap \mathcal{F}' \neq \emptyset$.

As examples, we mention the following filters:

a) $\mathcal{F} = \{X\}$, where X is an arbitrary nonempty set.

b) If X is an arbitrary set and $\emptyset \neq D \subset X$, then $\mathcal{F}_D = \{Y \subset X : D \subset Y\}$ is also a filter. Filters of this type are called *principal filters*. It is clear that a principal filter is an ultrafilter if and only if D is a *singleton*, that is, a set having exactly one element.

Ultrafilters not of this type can be obtained only with the aid of the following theorem.

Theorem 9.4. Let X be a nonempty set and assume $\mathcal{F} \subset X$ has the Finite Intersection Property. Then $\mathcal{F}$ can be extended to an ultrafilter; i.e., there is an ultrafilter $\mathcal{U} \subset \mathrm{P}(X)$ with $\mathcal{F} \subset \mathcal{U}$.

Proof. According to the Teichmüller–Tukey Lemma, there is a system $\mathcal{U} \subset \mathrm{P}(X)$ of sets such that $F \subset \mathcal{U}$ and $\mathcal{U}$ is a maximal set with the Finite Intersection Property. We claim that $\mathcal{U}$ is an ultrafilter.

Clearly, $\mathcal{U} \neq \emptyset$ and $\emptyset \notin \mathcal{U}$. Assume that $A, B \in \mathcal{U}$. If $\mathcal{U}' \subset \mathcal{U}$ and $\mathcal{U}'$ is finite, then we have

$$\emptyset \neq \bigcap(\mathcal{U}' \cup \{A, B\}) = \bigcap \mathcal{U}' \cap (A \cap B),$$

since $\mathcal{U}$ has the Finite Intersection Property. Thus $\mathcal{U} \cup \{A \cap B\}$ also has the Finite Intersection Property. In view of $\mathcal{U}$ being maximal, we have $A \cap B \in \mathcal{U}$.

Assume that $A \in \mathcal{U}$, $A \subset B$. If $\mathcal{U}' \subset \mathcal{U}$ and $\mathcal{U}'$ is finite, then, using the Finite Intersection Property of $\mathcal{U}$, we have

$$\emptyset \neq \bigcap \mathcal{U}' \cap A \subset \bigcap \mathcal{U}' \cap B,$$

and so $\mathcal{U} \cup \{B\}$ has the Finite Intersection Property. Again using the maximality of $\mathcal{U}$, we have $B \in \mathcal{U}$. Thus, so far we have shown that $\mathcal{U}$ is a filter.

Assume now that $A \subset X$ and $X \setminus A \notin \mathcal{U}$. Let $\mathcal{U}' \subset \mathcal{U}$ be finite. Then, in view of the already verified intersection property, that is, Property (I) for $\mathcal{U}$, we have $\bigcap \mathcal{U}' \in \mathcal{U}$, and by Property (II) we have $\bigcap \mathcal{U}' \not\subset X \setminus A$. Thus, for an arbitrary finite $\mathcal{U}' \subset \mathcal{U}$ we have $\bigcap \mathcal{U}' \cap A \neq \emptyset$; therefore, the set $\mathcal{U} \cup \{A\}$ has the Finite Intersection Property. Hence $A \in \mathcal{U}$ follows by the maximality of $\mathcal{U}$.

*   *

*

Corollary 9.1. *If X is infinite, then there is an ultrafilter $\mathcal{U} \subset P(X)$ that is not a principal filter.*

Proof. Let $\mathcal{F} = \{X \setminus \{a\} : a \in X\}$. As X is infinite, $\mathcal{F}$ has the Finite Intersection Property. According to Theorem 9.4, there is an ultrafilter $\mathcal{U}$ on X for which $\mathcal{F} \subset \mathcal{U}$. If $\mathcal{U}$ were a principal filter, then there would exist an $a \in X$ for which $a \in G$ for every $G \in \mathcal{U}$. This is, however, not possible since $X \setminus \{a\} \in \mathcal{U}$.

*   *

*

We would like to point out that we used the Axiom of Choice in the proof of Theorem 9.4. It is known that this theorem cannot be proved just by using the other axioms. It is known, further, that Theorem 9.4 even together with the other axioms does not imply the Axiom of Choice.

Problems

1. Let $\langle P, \prec \rangle$ be a partially ordered set. $A \subset P$ is called an independent set if for any two of its elements neither $a \prec b$ nor $b \prec a$ holds. Prove that every partially ordered set has a maximal independent subset.

2. Prove that the assertion of the Teichmüller–Tukey Lemma is no longer true if we say countable set instead of finite set in its formulation.

3. Show that the following assertion is equivalent to the Axiom of Choice: Every set has an ordering, and in every partially ordered set there is a maximal independent subset.

4. For an arbitrary set X put

$$H(X) = \{f : f \text{ is a function } \wedge |f| < \aleph_0 \wedge \mathrm{D}(f) \subset X \wedge \mathrm{R}(f) \subset 2\}.$$

Show that

$$|H(X)| = |X| \qquad \text{whenever} \qquad |X| \geq \aleph_0.$$

5.* For an ordinal $\kappa \geq \omega$, write

$$\tilde{H}(\kappa) = H(H(\kappa))$$

with the function H defined in the preceding problem. For an arbitrary $\phi \in {}^\kappa 2$ and $i < 2$ put

$$A_{\phi,i} = \{g \in \bar{H}(\kappa) : \forall f \in D(g) \, (f \subset \phi \implies g(f) = i)\}.$$

Show that for an arbitrary $k \in H({}^\kappa 2)$ the set

$$\bigcap\{A_{\phi,k(\phi)} : \phi \in D(k)\}$$

is not empty.

6.* Let κ be an infinite cardinal. Show that there are 2^{2^κ} many ultrafilters on κ.

7. Define the "exponentiation α^β of ordinals" by transfinite recursion as follows:

$$\alpha^0 = 1, \qquad \alpha^{\beta+1} = \alpha^\beta \times \alpha, \qquad \alpha^\xi = \sup\{\alpha^\gamma : \gamma < \xi\}$$

for ξ a limit ordinal. Prove that
a) $\alpha^\beta \times \alpha^\gamma = \alpha^{\beta+\gamma}$;
b) $(\alpha^\beta)^\gamma = \alpha^{\beta \times \gamma}$.

8.* *(Rado's Selection Lemma)* Let $\{A_\alpha : \alpha < \kappa\}$ with $\kappa \geq \omega$ be a sequence of nonempty finite sets. Denoting by $[\kappa]^{<\omega}$ the set of finite subsets of κ (for the definition of this symbol, see Section 12), for each $V \in [\kappa]^{<\omega}$ let $f_V \in \bigtimes_{\alpha \in V} A_\alpha$ be a choice function for $\{A_\alpha : \alpha \in V\}$. Then there is a choice function $f \in \bigtimes_{\alpha \in \kappa} A_\alpha$ such that for each $V \in [\kappa]^{<\omega}$ there is a V' with $V \subset V' \in [\kappa]^{<\omega}$ for which $f|V = f_{V'}|V$.

10. DEFINITION OF THE CARDINALITY
OPERATION. PROPERTIES OF CARDINALITIES.
THE COFINALITY OPERATION

In this section, we will present the remaining proofs that were omitted earlier. In the second part of the section, we will prove the Fundamental Theorem of Cardinal Arithmetic, and then we will prove some further results about cardinals.

Theorem 10.1. *The operation* $\min_<\{\alpha : \alpha \sim A\}$ *is compatible with the property* $\sim$.

Proof. According to the Wellordering Theorem, the expression $\min_<\{\alpha : \alpha \sim A\}$ is well defined for every set A. Denote this operation briefly by $F(A)$. According to its definition, for arbitrary sets A, B we have $A \sim F(A)$, $B \sim F(B)$. Therefore we have $A \sim B$ whenever $F(A) = F(B)$. On the other hand, if $A \sim B$ then for every α we have $\alpha \sim A \iff \alpha \sim B$, and so

$$F(A) = \min_<\{\alpha : \alpha \sim A\} = \min_<\{\alpha : \alpha \sim B\} = F(B)$$

holds.

*　　*
*

Thus we have completed the proof of Theorem 2.3. Next we are going to modify Definition 2.5:

Definition 10.1. *The operation*

$$\min_<\{\alpha : \alpha \sim A\}$$

is denoted by $|A|$. *In words, this means that the cardinality of a set is the least ordinal that is equivalent to it.*

With this notation, we now have three different symbols for ω, as $\omega = \omega_0 = \aleph_0$. Furthermore, for an arbitrary $n \in \omega$ we have $|n| = \underline{n} = n$. This definition is the reason that we needed different symbols for the cardinal and the ordinal operations; namely,

$$\omega + \omega = \omega \qquad (\aleph_0 + \aleph_0 = \aleph_0),$$

but

$$\omega \dotplus \omega \neq \omega,$$

and

$$\omega \cdot \omega = \omega \quad (\aleph_0 \cdot \aleph_0 = \aleph_0), \qquad \text{but} \qquad \omega \times \omega \neq \omega.$$

As cardinals are also ordinals, in what follows we will also use lower case Greek letters (usually κ, λ) to denote cardinals. This may not be easy to get used to, but it is well worth doing, since identifying cardinals with ordinals often significantly simplifies the presentation of proofs.

Corollary 10.1. *For an arbitrary set A we have $\|A\| = |A|$, that is, $|A|$ is a set of cardinality $|A|$.*

Proof. We have $|A| \sim A$, hence both sets have the same cardinality.

*   *
*

We remark that this property of the cardinality operation was used in Definition 5.4 when defining cardinal operations with infinitely many arguments.

Corollary 10.2. *An ordinal ξ is a cardinal if and only if for every $\eta < \xi$ we have $\eta \not\sim \xi$.*

Proof. ξ is a cardinal if and only if $|\xi| = \xi$. Furthermore, $|\xi|$ is the least ordinal that is equivalent to ξ.

*   *
*

Corollary 10.3. *The partial ordering by size of the cardinals is the same as the ordering of ordinals. Hence, the partial ordering of cardinals is trichotomous, i.e., it is an ordering, and this ordering is also a wellordering.*

Proof. For the duration of this proof, denote by $<^*$ the partial ordering by size of cardinals described in Definition 4.1, and by $<$, the ordering of ordinals. Let κ, λ be cardinals. If $\kappa < \lambda$ then $\kappa \subset \lambda$, and, λ being a cardinal, we have $\kappa \not\sim \lambda$ according to Corollary 10.2. Hence we have $\kappa <^* \lambda$ also. Conversely, assume that $\kappa <^* \lambda$. Then there is an $A \subset \lambda$ such that $\kappa \sim A$. According to Corollary 8.3, type $A(\in) \leq \lambda$. On the other hand, $\kappa \leq$ type $A(\in)$ by Corollary 10.2. Thus $\kappa \leq \lambda$. Since $\kappa = \lambda$ is impossible by our assumption, we have $\kappa < \lambda$.

*   *
*

In the proof of Corollary 10.3 given above, we only used the definition of $<^*$. Considering that $\leq$ is antisymmetric for ordinals, we obtained a new proof of Bernstein's Equivalence Theorem, albeit one that uses the Axiom of Choice.

We are going to show next that there is an operation that is compatible with the property $\simeq$. This is the last result that we owe the proof of.

Theorem 10.2. *Define the operation $F(\langle A, \prec \rangle)$ for an arbitrary ordered set $\langle A, \prec \rangle$ as follows:*

$$F(\langle A, \prec \rangle) = \{\langle |A|, \prec' \rangle : \prec' \in P(^2|A|) \wedge \langle |A|, \prec' \rangle \simeq \langle A, \prec \rangle\}.$$

That is, $F(\langle A, \prec \rangle)$ is the set of all ordered sets of form $\langle |A|, \prec' \rangle$ that are similar to $\langle A, \prec \rangle$. The operation so defined is compatible with the relation $\simeq$.

Proof. If $\langle A, \prec \rangle$ is an arbitrary ordered set, then

$$F(\langle A, \prec \rangle) \neq \emptyset,$$

since there is an f for which $A \sim_f |A|$, and then $\langle |A|, \prec' \rangle \in F(\langle A, \prec \rangle)$ for the ordering $\prec'$ defined by the stipulation

$$x \prec' y \iff f^{-1}(x) \prec f^{-1}(y).$$

Furthermore, if

$$\langle A, \prec \rangle \not\simeq \langle A_1, \prec_1 \rangle,$$

and $\langle |A|, \prec' \rangle \in F(\langle A, \prec \rangle)$, then

$$\langle |A|, \prec' \rangle \notin F(\langle A_1, \prec_1 \rangle).$$

and so

$$F(\langle A, \prec \rangle) \neq F(\langle A_1, \prec_1 \rangle).$$

On the other hand, if

$$\langle A, \prec \rangle \simeq \langle A_1, \prec_1 \rangle,$$

then

$$|A| = |A_1|,$$

and the relation $\langle |A|, \prec' \rangle \in F(\langle A, \prec \rangle)$ implies

$$\langle |A|, \prec' \rangle \in F(\langle A_1, \prec_1 \rangle).$$

Therefore,

$$F(\langle A, \prec \rangle) = F(\langle A_1, \prec_1 \rangle)$$

holds if and only if

$$\langle A, \prec \rangle \simeq \langle A_1, \prec_1 \rangle.$$

$$* \qquad *$$
$$*$$

Thus we have also established Theorem 7.3.

We can now stipulate that type $A(\prec)$ denotes an operation whose values for non-wellordered sets are given by the above operation F, and for wellordered sets are given by Definition 8.7. In this way, we obtained an operation that is compatible with similarity. This operation certainly satisfies the condition that, for non-wellordered sets, type$(A, \prec)$ is a set of ordered sets of order type $\langle A, \prec \rangle$. This was needed for the definition of order types in the case of infinitely many arguments; aside from this, we will not need the particular form of the operation type$(A, \prec)$.

The isomorphism types of arbitrary structures can be defined in a similar way. We will, however, not be concerned with this question here.

Fundamental Theorem of Cardinal Arithmetic 10.3. *For an arbitrary cardinal $\kappa \geq \omega$ we have $\kappa^2 = \kappa$.*

Proof. The ordering according to size of the cardinals is a wellordering, and so we can use transfinite induction to prove the assertion. We already know that the assertion is true for $\kappa = \omega$. Assume that $\kappa > \omega$ and the assertion is true for each cardinal λ with $\omega \leq \lambda < \kappa$. Then we have

$$(1) \qquad \lambda_0 + \lambda_1 < \kappa \quad \text{and} \quad \lambda_0 \cdot \lambda_1 < \kappa \quad \text{for} \quad \lambda_0, \lambda_1 < \kappa.$$

Indeed, we may assume that $\lambda_0 \leq \lambda_1$, and as we have $\kappa > \omega$, we may also assume $\lambda_1 \geq \omega$. Thus, by the induction hypothesis we obtain

$$\lambda_0 + \lambda_1 \leq 2\lambda_1 \leq \lambda_1^2 = \lambda_1 < \kappa$$

and

$$\lambda_0 \cdot \lambda_1 \leq \lambda_1^2 < \kappa.$$

We have to show that $\kappa^2 = \kappa$ holds. Let $A = \kappa \times \kappa$ be the set of ordered pairs formed by elements of κ. $|A| = \kappa^2$, and so it is enough to show that $|A| \leq \kappa$. We claim that to this end it is enough to specify a wellordering $\prec$ of $|A|$ such that

$$(2) \qquad \qquad |A \prec x| < \kappa$$

holds for each element of $x \in A$. Indeed, assume that $\prec$ satisfies (2). Let

$$\xi = \text{type } A(\prec).$$

If $\eta < \xi$, then we have $|\eta| < \kappa$ in view of (2), and so $\eta < \kappa$. Thus $\xi \subset \kappa$, and so $\xi \leq \kappa$. Hence

$$|\xi| \leq \xi \leq \kappa.$$

We are now going to describe a wellordering satisfying requirement (2). For an arbitrary $\alpha < \kappa$, write

$$A_\alpha = \{\langle \xi, \eta \rangle : \max(\xi, \eta) = \alpha\}.$$

It is clear that the sets A_α are pairwise disjoint and

$$A = \bigcup_{\alpha < \kappa} A_\alpha.$$

Let $\prec_\alpha$ be the ordering of A_α inherited from the lexicographic product. This is a wellordering according to Theorem 7.6. Let $\prec$ be the ordering of the ordered union of the $\prec_\alpha$'s with respect to the natural ordering $<$ of κ. According to Theorem 7.4, $\prec$ is a wellordering.

Next we show that $\prec$ satisfies (2). Let $x \in A$. Then $x \in A_\alpha$ for some $\alpha < \kappa$. By the definition of ordered union, we have

$$A| \prec x \subset \bigcup_{\beta \leq \alpha} A_\beta.$$

For an arbitrary β clearly

$$A_\beta \subset (\beta \dotplus 1) \times (\beta \dotplus 1)$$

holds. Hence

$$|A| \prec x| \leq \sum_{\beta \leq \alpha} |(\beta \dotplus 1) \times (\beta \dotplus 1)| \leq \sum_{\beta \leq \alpha} |\alpha \dotplus 1|^2 \leq (|\alpha| + 1)^3.$$

Noting that $|\alpha| \leq \alpha < \kappa$, the right-hand side is less than κ in view of (1). Hence

$$|A| \prec \kappa| < \kappa.$$

*   *

*

As a consequence of the theorem just proved, we have

$$\lambda + \kappa = \lambda \cdot \kappa = \max(\lambda, \kappa)$$

for all cardinals λ, κ with $\max(\lambda, \kappa) \geq \omega$; in fact, this is directly shown by the calculations carried out within the proof. In what follows, this relation will be used without any reference.

Corollary 10.4. *For every $\kappa \geq \omega$, the set of all finite sequences formed by elements of κ has cardinality κ.*

Proof. All we need to show is that

$$\sum_{n \in \omega} \kappa^n = \kappa.$$

As $\kappa^{n+1} = \kappa^n \cdot \kappa$, it follows by induction from the preceding theorem that $\kappa^n = \kappa$ for n with $1 \leq n < \omega$. Hence

$$\sum_{n \in \omega} \kappa^n = \kappa \cdot \omega = \kappa.$$

$$* \qquad *$$
$$*$$

Definition 10.2. *The least cardinal greater than the cardinal κ is denoted as κ^+ and is called the successor of κ. That is,*

$$\kappa^+ = \min_{<}\{\lambda : \kappa < \lambda\}.$$

The definition is sound since for every κ we have $2^\kappa > \kappa$. We would like to point out that the notation just introduced will be used for finite cardinals as well. If $\kappa < \omega$, then $\kappa^+ = \kappa + 1 = \kappa \dotplus 1$.

Theorem 10.4. *For each $\kappa \geq \omega$, the cardinality of the set of ordinals of cardinality κ is κ^+, that is,*

$$\kappa^+ = |\{\xi : |\xi| = \kappa\}|.$$

Proof.
$$|\xi| \geq \kappa \iff \xi \geq \kappa$$

and
$$|\xi| < \kappa^+ \iff \xi < \kappa^+.$$

Thus
$$|\xi| = \kappa \iff \kappa \leq \xi < \kappa^+.$$

Therefore,
$$\{\xi : |\xi| = \kappa\} = \{\xi : \kappa \leq \xi < \kappa^+\} = \kappa^+ \setminus \kappa.$$

On the other hand,
$$\kappa^+ = (\kappa^+ \setminus \kappa) \cup \kappa,$$

and so
$$\kappa^+ = |\kappa^+ \setminus \kappa| + \kappa = \max(|\kappa^+ \setminus \kappa|, \kappa).$$

As $\kappa < \kappa^+$, we obtain

$$\kappa^+ = |\kappa^+ \setminus \kappa| = |\{\xi : |\xi| = \kappa\}|.$$

*   *

*

In particular, we have

$$\omega^+ = |\{\xi : |\xi| = \omega\}|,$$

that is, the least uncountable cardinal is the cardinality of the set of all countably infinite ordinals.

The next theorem expresses the fact that the sum of several cardinals can be obtained by forming a maximum or a supremum.

Theorem 10.5. *Let A be a set of cardinals. Then*
1. $\sup A$ *is a cardinal;*
2. *If $\sup A$ is infinite, then*

$$\sum_{\lambda \in A} \lambda = \sup A.$$

Proof. We know that
$$\sup A = \bigcup A.$$

1. Denote the cardinal $|\sup A|$ by κ. Assume $\kappa < \sup A$; then $\kappa \in \lambda$ for some $\lambda \in A$. As $\lambda \subset \sup A$, we then have $|\lambda| \leq \kappa < \lambda$. This is a contradiction, since λ is a cardinal.

Hence
$$|\sup A| = \sup A,$$

and so $\sup A$ is a cardinal.

2. Denote the cardinal $\sup A$ by κ. In view of what we said above, we have $\kappa = \bigcup A$, and so the relation

$$\kappa \leq \sum_{\lambda \in A} |\lambda|$$

is immediate. On the other hand, we have

$$\sum_{\lambda \in A} |\lambda| = \sum_{\lambda \in A} \lambda,$$

since the elements of A are cardinals. Furthermore, for $\lambda \in A$ we have $\lambda \subset \kappa$, and so $\lambda \leq \kappa$; thus $A \subset \kappa \dotplus 1$. Thus, using the assumption $\kappa \geq \omega$, we obtain

$$\sum_{\lambda \in A} \lambda \leq \kappa \cdot |\kappa \dotplus 1| = \kappa^2 = \kappa.$$

$$* \quad *$$
$$*$$

We will now introduce a new notation for infinite cardinals.

Definition 10.3. *Define the* operation ω_α *for an arbitrary ordinal* α *by transfinite recursion on* α:

$$\omega_0 = \omega.$$

Let $\alpha > 0$ *and assume that we have defined* ω_β *for all* $\beta < \alpha$. *Let* ω_α *be the least cardinal that is greater that each* ω_β *defined so far. That is,*

$$\omega_\alpha = \min\{\kappa : \kappa \text{ is a cardinal } \wedge \forall \beta < \alpha \, (\kappa > \omega_\beta)\}.$$

This definition is sound, since such a cardinal always exists according to Theorems 10.4 and 10.5. For historical reasons, we denote ω_α also as $\aleph_\alpha$. This notation is dated to the times when cardinals were not identified with ordinals. Then ω_α denoted the least ordinal of cardinality $\aleph_\alpha$. Today we often write $\aleph_\alpha$ instead of ω_α when we discuss statements involving mainly cardinals.

Theorem 10.6.
1. ω_α *is an infinite cardinal.*
2. $\omega_\alpha < \omega_\beta$ *whenever* $\alpha < \beta$.
3. *For an arbitrary cardinal* $\kappa \geq \omega$, *there is an* α *such that* $\kappa = \omega_\alpha$.
4. $\omega_\alpha = \sup\{\omega_\beta : \beta < \alpha\}$ *for any limit ordinal* α.
5. $\omega_{\alpha \dotplus 1} = (\omega_\alpha)^+$ *for every* α.

Proof. Assertions 1 and 2 follow directly from the definition.

3. Let $\kappa \geq \omega$. In view of Assertion 2, ω_β uniquely determines β. Therefore the set $\{\beta : \omega_\beta < \kappa\}$ of ordinals exists, in view of Convention 11 of Section 1. Using Assertion 2 again, this is a transitive set, so itself is an ordinal. Put

$$\alpha = \{\beta : \omega_\beta < \kappa\}.$$

In view of ω_α being the least cardinal greater than all the ω_β's, we have $\omega_\alpha \leq \kappa$. We cannot have $\omega_\alpha < \kappa$, since that would imply $\alpha \in \alpha$. Hence

$$\kappa = \omega_\alpha.$$

4. If α is a limit ordinal, then we also have $\beta \dot{+} 1 < \alpha$ for each $\beta < \alpha$; furthermore $\alpha > 0$. Hence, for an arbitrary $\beta < \alpha$ we have

$$\sup\{\omega_\gamma : \gamma < \alpha\} \geq \omega_{\beta+1} > \omega_\beta.$$

By virtue of the preceding theorem, $\sup\{\omega_\beta : \beta < \alpha\}$ is a cardinal, and so

$$\sup\{\omega_\beta : \beta < \alpha\}$$

is precisely the least cardinal that is greater than each ω_β.

Assertion 5 is obvious from the definition.

$$* \qquad *$$
$$*$$

Definition 10.4. *Let* $\kappa \geq \omega$. *We call* κ *a* successor cardinal *if* $\kappa = \lambda^+$ *for some* λ; *we call* κ *a* limit cardinal *if it is not a successor cardinal.*

It is clear from the aforesaid that $\kappa = \omega_\alpha$ is a successor cardinal if and only if α is a successor ordinal; further, if κ is a limit cardinal then $\kappa = \sup\{\lambda : \lambda < \kappa\} = \sum_{\lambda < \kappa} \lambda$, where λ runs over cardinals. For example, $\aleph_0$, $\aleph_\omega$ are limit cardinals, and $\aleph_1$, $\aleph_2$, ..., $\aleph_n$, ... are successor cardinals.

Definition 10.5. *Let* $\langle A, \prec \rangle$ *be an ordered set, and let* $B \subset A$. *We say that* B *is* cofinal *in* A *if for every* $x \in A$ *there is* $y \in B$ *for which* $x \preceq y$.

For example, ω is cofinal in $\langle \mathbb{R}, < \rangle$; $\{\omega\}$ is cofinal in $\omega \dot{+} 1$.

Hausdorff Cofinality Theorem 10.7. *Given an arbitrary ordered set* $\langle A, \prec \rangle$, *there is a wellordered subset* B *cofinal in it such that*

$$\text{type } B(\prec) \leq |A|.$$

Proof. Consider a wellordering $\prec_1$ of A for which

$$\text{type } A(\prec_1) = |A|.$$

Define B as the set

$$\{y \in A : \forall z \in A\, (y \prec z \implies y \prec_1 z)\}.$$

We are about to show that B is cofinal in A. Let $x \in A$. We claim that

$$y = \min_{\prec_1}\{z \in A : x \preceq z\}$$

is an element of B. Indeed, if $z \in A$ and $y \prec z$, then $x \prec z$, and so $y \prec_1 z$ also holds in view of the choice of y. Hence $y \in B$.

It is easy to see that the two orderings are identical on the set B. Indeed, let $y, z \in B$ and assume that $y \prec z$. It follows from the definition of B that $y \prec_1 z$. Thus $\prec$ wellorders B, and

$$\text{type } B(\prec) = \text{type } B(\prec_1) \leq \text{type } A(\prec_1) = |A|.$$

*   *
*

Definition 10.6. *Let $\langle A, \prec \rangle$ be an arbitrary ordered set. The least ordinal ξ for which $\langle A, \prec \rangle$ has a wellordered subset B of type ξ cofinal in A is denoted as $\text{cf}(\langle A, \prec \rangle)$, and is called the cofinality of $\langle A, \prec \rangle$.*

The definition is sound according to Theorem 10.7, and

$$\text{cf}(\langle A, \prec \rangle) \leq |A|.$$

If

$$\langle A, \prec \rangle \simeq \langle A', \prec' \rangle,$$

then clearly

$$\text{cf}(\langle A, \prec \rangle) = \text{cf}(\langle A', \prec' \rangle).$$

Thus the following definition is also sound.

Definition 10.7. *For an arbitrary order type Θ, let $\langle A, \prec \rangle$ be any ordered set with $\text{type}(\langle A, \prec \rangle) = \Theta$. Then $\text{cf}(\langle A, \prec \rangle)$ is also denoted as $\text{cf}(\Theta)$, and is called the cofinality of Θ.*

Clearly, $\text{cf}(\langle A, \prec \rangle) = 1$ if and only if $\langle A, \prec \rangle$ has a last element. If $\langle A, \prec \rangle$ has no last element and $A \neq \emptyset$, then

$$\text{cf}(\langle A, \prec \rangle) \geq \omega.$$

Examples: $\text{cf}(n) = 1$ if $n > 0$ and $n \in \omega$;

$\qquad\qquad\text{cf}(\omega) = \omega$;

$\qquad\qquad\text{cf}(\omega_\omega) = \omega$, since $\{\omega_n : n \in \omega\}$ is cofinal in $\langle \omega_\omega, < \rangle$.

In what follows, the following concept will be of fundamental importance.

Definition 10.8. *Let ξ be an arbitrary ordinal. ξ is said to be* regular *if*

$$\text{cf}(\xi) = \xi > 1,$$

and it is said to be singular *if*

$$1 < \text{cf}(\xi) < \xi.$$

Examples: ω is regular;

ω_ω is singular;

$\omega \dot{+} \omega$ is singular;

but n for $n \in \omega$ and $\omega \dot{+} 1$ are neither regular nor singular.

The next theorem exhibits an important property of the possible values for $\mathrm{cf}(\Theta)$.

Theorem 10.8. 1. *For an arbitrary order type* $\Theta \neq 0$, $\mathrm{cf}(\Theta)$ *is either* 1 *or it is a regular cardinal.*

2. *Every regular ordinal is a cardinal.*

Proof. If the ordered set $\langle A, \prec \rangle$ and the sets C, B with $C \subset B \subset A$ are such that C is cofinal in B and B is cofinal in A, then C is clearly cofinal in A. From this, it immediately follows that

$$\mathrm{cf}(\langle B, \prec \rangle) \geq \mathrm{cf}(\langle A, \prec \rangle).$$

Now let A be an ordered set with

$$\mathrm{type}(\langle A, \prec \rangle) = \Theta.$$

Put

$$\beta = \mathrm{cf}(\Theta) \geq 1.$$

Choose a set $B \subset A$ such that type $B(\prec) = \beta$ and B is cofinal in A. According to the remark made just before, we have

$$\mathrm{cf}(\beta) = \mathrm{cf}(\langle B, \prec \rangle) \geq \mathrm{cf}(\langle A, \prec \rangle) = \beta.$$

As $\mathrm{cf}(\beta) \leq \beta$ clearly holds, we obtain

$$\mathrm{cf}(\beta) = \beta > 1,$$

and so $\mathrm{cf}(\Theta) = \beta$ is a regular ordinal.

Assertion 2 follows form the Hausdorff Cofinality Theorem. Indeed, let β be a regular ordinal, that is,

$$\mathrm{cf}(\beta) = \beta > 1.$$

According to Theorem 10.7 we have

$$\mathrm{cf}(\beta) \leq |\beta| \qquad \text{and} \qquad |\beta| \leq \beta.$$

Thus $\beta = |\beta|$, and so β is a cardinal; it cannot be finite, since then $\mathrm{cf}(\beta) \leq 1$ would hold.

*   *
*

It is worth pointing out that the theorem immediately implies that if $\xi < \omega_1$ is a limit ordinal, then $\mathrm{cf}(\xi) = \omega$; that is, every countable limit ordinal has a subset of order type ω that is cofinal in it.

As we have seen, not every infinite cardinal is regular. For example, $\aleph_\omega$ is singular. The following corollary is however true.

Corollary 10.5. *Every successor cardinal is regular.*

Proof. If κ is an infinite cardinal, then it is a limit ordinal. Indeed, $\kappa = \xi \dot{+} 1$ is not possible, since in this case we would have

$$|\kappa| = \max(|\xi|, 1) = |\xi| \leq \xi < \kappa.$$

Furthermore, we can see that if ξ is a limit ordinal then a set $A \subset \xi$ is cofinal in $\langle \xi, < \rangle$ if and only if

$$\bigcup A = \sup A = \xi.$$

Now let $\kappa = \lambda^+$ for some $\lambda \geq \omega$. Then κ is a limit ordinal. Assume that

$$A \subset \kappa \qquad \text{and} \qquad \text{type} \, A(\prec) < \kappa.$$

Then $|A| < \kappa$, thus $|A| \leq \lambda$. and, further, $|\xi| \leq \lambda$ for each $\xi \in A$. Thus

$$|\sup A| \leq \left| \bigcup A \right| \leq \lambda^2 = \lambda < \kappa,$$

and so $\sup A < \kappa$. Thus A is not cofinal in κ.

$$* \qquad *$$
$$*$$

Not every regular cardinal is a successor cardinal; in fact, ω is a regular limit cardinal. The question arises whether there are regular limit cardinals greater than ω. Regular limit cardinals are called *inaccessible cardinals*; occasionally, they are also called *weakly inaccessible cardinals*. A cardinal κ that satisfies the requirement

$$\forall \lambda < \kappa \, (2^\lambda < \kappa)$$

is called a *strong limit cardinal*; there are many strong limit singular cardinals. Strong limit cardinals that are regular (and therefore inaccessible) are called *strongly inaccessible*. In particular, ω is a strongly inaccessible cardinal. The problem of the existence of inaccessible cardinals greater than ω will be discussed in Sections A9 of the Appendix and in Section 15.

Next we give a characterization of $\text{cf}(\kappa)$ for infinite cardinals κ.

Theorem 10.9. *For every infinite cardinal* κ, $\text{cf}(\kappa)$ *is the least ordinal* α *such that there is a sequence* $\{\kappa_\xi < \kappa : \xi < \alpha\}$ *of cardinals with*

$$\sum_{\xi < \alpha} \kappa_\xi = \kappa.$$

Proof. Denote the least α satisfying the above requirement by α_0. First we show that $\alpha_0 \leq \text{cf}(\kappa)$. Indeed, let $A \subset \kappa$ be a set such that type $A(<) = \text{cf}(\kappa)$ and A is cofinal in $\langle \kappa, < \rangle$. As we saw in the proof of Corollary 10.5, we then have $\bigcup A = \kappa$. Thus

$$\kappa \leq \sum_{\zeta \in A} |\zeta| \leq \kappa^2 = \kappa,$$

and so

$$\kappa = \sum_{\zeta \in A} |\zeta|.$$

As $|\zeta| < \kappa$ for $\zeta < \kappa$, we obtain that

$$\alpha_0 \leq \text{type } A(<) = \text{cf}(\kappa).$$

We next show that

$$\text{cf}(\kappa) \leq \alpha_0.$$

As $\text{cf}(\kappa) \leq \kappa$, for the proof we may assume that $\alpha_0 < \kappa$. Choose a sequence $\{\kappa_\xi : \xi < \alpha_0\}$ of ordinals such that $\kappa_\xi < \kappa$ for $\xi < \alpha_0$ and

$$\kappa = \sum_{\xi < \alpha_0} \kappa_\xi.$$

Write

$$A = \{\kappa_\xi : \xi < \alpha_0\}.$$

We claim that A is cofinal in $\langle \kappa, < \rangle$. Indeed, assuming that this is not the case, there is an ordinal $\rho < \kappa$ such that $A \subset \rho$. Therefore

$$\sum_{\xi \in \alpha_0} \kappa_\xi \leq |\rho|\, |\alpha_0| < \kappa$$

in view of the assumption $\alpha_0 < \kappa$. This is a contradiction; therefore A must be cofinal in κ. Then we have

$$\text{cf}(\kappa) \leq \text{cf}(\langle A, < \rangle)$$

according to what we said in the proof of Theorem 10.8. On the other hand, by the Hausdorff Cofinality Theorem, we have

$$\text{cf}(\langle A, < \rangle) \leq |A| \leq |\text{type } A(<)| = |\alpha_0| \leq \alpha_0.$$

Thus $\text{cf}(\kappa) \leq \alpha_0$.

$$* \qquad *$$
$$*$$

Theorem 10.10. *Let $\langle A, \prec \rangle$ be an arbitrary ordered set, let $B, C \subset A$, and assume that B and C are both cofinal in A. Then*

$$\operatorname{cf}(\langle B, \prec \rangle) = \operatorname{cf}(\langle C, \prec \rangle).$$

Proof. Write

$$\operatorname{cf}(\langle B, \prec \rangle) = \beta, \qquad \operatorname{cf}(\langle C, \prec \rangle) = \gamma.$$

For reasons of symmetry, it is sufficient to prove that

$$\gamma \leq \beta.$$

According to the definition of cofinality, we can find sets $B_1 \subset B$ and $C_1 \subset C$ such that

$$\operatorname{type}(\langle B_1, \prec \rangle) = \beta, \qquad \operatorname{type}(\langle C_1, \prec \rangle) = \gamma,$$

and, further, B_1 is cofinal in B and C_1 is cofinal in C. Thus both B_1 and C_1 are cofinal in A. Let f be a function such that $\mathrm{D}(f) = B_1$ and

$$f(x) = \min_{\prec}\{y \in C_1 : x \preceq y\}.$$

As C_1 is cofinal in A, this definition is sound. Now $f``B_1$ is cofinal in C, since for every $z \in A$ there is an $x \in B_1$ such that

$$z \preceq x \preceq f(x) \in f``B_1.$$

Thus

$$\operatorname{type} f``B_1(\prec) \geq \gamma.$$

On the other hand, as $f``B_1 \subset C_1$, we also have $\operatorname{type} f``B_1(\prec) \leq \gamma$; that is

$$\operatorname{type} f``B_1(\prec) = \gamma.$$

Now let g be a function such that $\mathrm{D}(g) = f``B_1$ and

$$g(y) = \min_{\prec}\{x : f(x) = y\}.$$

Then g is a monotonic mapping from $f``B_1$ onto B_1, and so

$$\gamma = \operatorname{type} f``B_1(\prec) \leq \operatorname{type} B_1(\prec) = \beta.$$

* *

*

Corollary 10.6. *If κ is a limit cardinal, then there is a strictly increasing sequence $\{\kappa_\xi : \xi < \mathrm{cf}(\kappa)\}$ of cardinals such that*

$$\sum_{\xi < \mathrm{cf}(\kappa)} \kappa_\xi = \kappa.$$

Proof. If κ is a limit cardinal, then the set

$$A = \{\lambda < \kappa : \lambda \text{ is a cardinal}\}$$

is cofinal in κ, since for every $\xi < \kappa$, we have

$$\xi < |\xi|^+ < \kappa.$$

Hence, according to Theorem 10.10, we have

$$\mathrm{cf}(\langle A, < \rangle) = \mathrm{cf}(\kappa).$$

Let $B \subset A$ be such that B is cofinal in A, and so in κ as well, and type $B(<) = \mathrm{cf}(\kappa)$. Let $B = \{\kappa_\xi : \xi < \mathrm{cf}(\kappa)\}$ be a strictly increasing enumeration of order type $\mathrm{cf}(\kappa)$ of B. Then, as we saw earlier,

$$\kappa = \sum_{\xi < \mathrm{cf}(\kappa)} \kappa_\xi.$$

$$* \qquad *$$
$$*$$

Problems

1. Let $\kappa \geq \omega$ and let $\{A_\alpha : \alpha < \kappa\}$ be a system of sets, each having cardinality κ. Prove the following:

a) There is a one-to-one choice function $\{A_\alpha : \alpha < \kappa\}$.

b) There are pairwise disjoint sets $B_\alpha \subset A_\alpha$ such that $|B_\alpha| = \kappa$ for $\alpha < \kappa$.

2. Prove that if Θ is the order type of an uncountable set, then for every $\alpha < \omega_1$, we have either $\alpha \preceq \Theta$ or $\alpha^* \preceq \Theta$ (the relation $\preceq$ for order types was defined in Problem 7.3).

3. Prove that if $\kappa = \aleph_\alpha$ and α is a limit ordinal, then $\mathrm{cf}(\kappa) = \mathrm{cf}(\alpha)$.

4.[+] For an arbitrary set $X \subset \mathbb{R}$, let X' denote the set of accumulation points of X. For an arbitrary ordinal α, define the sequence $X^{(\alpha)}$ by transfinite recursion as follows:

$$X^{(0)} = X, \qquad X^{(\alpha+1)} = (X^{(\alpha)})' \cap X,$$

$$X^{(\xi)} = \bigcap \{X^{(\beta)} : \beta < \xi\}$$

if ξ is a limit ordinal. Prove that

a) there is an ordinal $\alpha < \omega_1$ such that $X^{(\alpha)} = X^{(\alpha+1)}$;

b) if for the α in Part *a)* we have $X^{(\alpha)} = \emptyset$, then X is countable;

c) if F is closed and $F^{(\alpha)} = F^{(\alpha+1)}$, then $F \setminus F^{(\alpha)}$ is countable, and $F^{(\alpha)}$ is a perfect set.

5. The set system $\mathcal{S}$ is a σ-ring if for $A, B \in \mathcal{S}$ we have $A \setminus B \in \mathcal{S}$, and for every countable set $\mathcal{S}' \subset \mathcal{S}$ we have $\bigcup \mathcal{S}' \in \mathcal{S}$. Prove that if $\mathcal{Z} \subset P(\mathbb{R})$ and $|\mathcal{Z}| \leq 2^{\aleph_0}$, then there is a σ-ring $\mathcal{S} \subset P(\mathbb{R})$ such that $\mathcal{Z} \subset \mathcal{S}$ and $|\mathcal{S}| \leq 2^{\aleph_0}$.

6. Define the cofinality of a partially ordered set $\langle P, \prec \rangle$ as follows:

1) For $A \subset P$, we say that A is cofinal in $\langle P, \prec \rangle$ if for an arbitrary $p \in P$, there is a $q \in A$ for which $p \preceq q$.

2) $\mathrm{cf}\langle P, \prec \rangle = \min\{|A| : A$ is cofinal in $\langle P, \prec \rangle\}$.

a) Verify that with the notion of cofinality so defined, the Hausdorff Cofinality Theorem remains valid for partially ordered sets in the sense that every partially ordered set P includes a *well-founded* subset B that is cofinal in P. Here $\langle B, \prec \rangle$ is said to be well-founded if every nonempty set $X \subset B$ contains a $\prec$-minimal element, i.e, an element u such that $\neg v \prec u$ holds for each $v \in X$.

b) Show that $\mathrm{cf}\langle P, \prec \rangle$ may possibly be a singular cardinal.

*c)** Show that if $\mathrm{cf}\langle P, \prec \rangle$ is a singular cardinal, then P includes an infinite independent set. (Independent sets were defined in Problem 9.1.)

11. PROPERTIES OF THE POWER OPERATION

In the course of the proof of the Fundamental Theorem of Cardinal Arithmetic in the preceding section, we saw that we can always determine the values of the sum $\kappa + \lambda$ and of the product $\kappa \cdot \lambda$ for infinite cardinals. In this section, we will discuss the operation λ^κ. As we shall see, this will lead to much more difficult problems. At this point we only know that $2^\kappa > \kappa$ according to Cantor's Theorem (Theorem 4.6). This we obtained by reductio ad absurdum, using the method of diagonalization, but this method is capable of establishing only that 2^κ is not equal to κ. Thus the question immediately arises: is the equality $2^\kappa = \kappa^+$ true?

The cardinal $2^{\aleph_0}$ has been the object of intensive study. The problem of determining the value of $2^{\aleph_0}$ is called the *Continuum Problem*, and the assumption that $2^{\aleph_0} = \aleph_1$ is called the *Continuum Hypothesis*. The problem of determining the value of 2^κ is called the *Generalized Continuum Problem*, and the assumption that $2^\kappa = \kappa^+$ for every cardinal $\kappa \geq \omega$ is called the *Generalized Continuum Hypothesis*, customarily abbreviated as GCH. These problems and conjectures have already been discussed by Cantor. Since for many years no significant progress had been made on the resolution of these questions, the conviction emerged that the assumptions (axioms) of set theory are not sufficient to answer them. The unsolvability of these fundamental problems was one of the driving forces in the development of methods of axiomatic set theory and mathematical logic.

In 1939, K. Gödel showed that GCH cannot be disproved with the aid of the axioms of set theory. Somewhat more precisely, this means that if the axioms of set theory do not lead to a contradiction, then they do not lead to a contradiction even if we add the assumption $\forall \kappa \geq \omega \, (2^\kappa = \kappa^+)$ to the axioms.

Only much later, in 1963, P. Cohen proved that the assertion $2^{\aleph_0} = \aleph_1$ cannot be proved from the axioms of set theory. Later, using his method, it was shown that even the assumption that the continuum is "arbitrarily large" does not lead to a contradiction. In fact, the only restriction on the value of $2^{\aleph_0}$ is J. König's Theorem to be proved below, according to which $\mathrm{cf}(2^{\aleph_0}) \neq \omega$.

What has been said above indicates that there are only a few results that say something significant about the behavior of the value of λ^κ. We will present the two most important ones among these. As we will show below, these results are always sufficient for determining the value of λ^κ under the assumption of GCH.

König's Theorem 11.1. *Let $\{\lambda_\xi : \xi < \alpha\}$ and $\{\kappa_\xi : \xi < \alpha\}$ be two systems of cardinals such that $\lambda_\xi < \kappa_\xi$ holds for every $\xi < \alpha$. Then*

$$\sum_{\xi<\alpha} \lambda_\xi < \prod_{\xi<\alpha} \kappa_\xi.$$

Before we establish this theorem, we would like to point out that this result is in fact a generalization of Cantor's Theorem. Indeed, putting $\lambda_\xi = 1$ and $\kappa_\xi = 2$, we obtain the inequality $|\alpha| < 2^{|\alpha|}$. The theorem will be proved with the aid of a clever use of the diagonalization method.

Proof. Denote the sum by λ, and the product by κ. Choose pairwise disjoint sets L_ξ such that

$$\bigcup_{\xi<\alpha} L_\xi = \lambda \qquad \text{and} \qquad \lambda_\xi = |L_\xi|$$

holds, and introduce the notation

$$K = \underset{\xi<\alpha}{\bigtimes} \kappa_\xi.$$

Then $|K| = \kappa$. We have to prove that

$$\kappa \not\leq \lambda.$$

Assume, on the contrary, that $\kappa \leq \lambda$, that is, there is an $L \subset \lambda$ and a mapping f for which $L \sim_f K$.

The value $f(x)$ of the function f at a place $x \in L$ is a function in K. Denote the value of this function at ξ by $f(x)(\xi)$. For an arbitrary $\xi < \alpha$, define the set

$$K_\xi = \{f(x)(\xi) : x \in L_\xi \cap L\}.$$

As

$$|L_\xi \cap L| \leq |L_\xi| = \lambda_\xi < \kappa_\xi,$$

we have

$$|K_\xi| < \kappa_\xi,$$

and so

$$\kappa_\xi \setminus K_\xi \neq \emptyset \quad \text{for all} \quad \xi \in \alpha.$$

Consider a choice function ϕ such that $\phi(\xi) \in \kappa_\xi \setminus K_\xi$ holds for each $\xi < \alpha$. Then ϕ is an element of K. We claim that

$$\phi \neq f(x)$$

holds for any $x \in L$. Indeed, if $x \in L$, then, on the one hand, $x \in L_\xi \cap L$ for some $\xi < \alpha$, and so

$$f(x)(\xi) \in K_\xi,$$

on the other hand, $\phi(\xi) \notin K_\xi$, and therefore

$$f(x)(\xi) \neq \phi(\xi);$$

thus $f(x) \neq \phi$ holds. This contradicts the assumption that f maps L onto K. This contradiction shows that our assumption was incorrect; that is, $\lambda < \kappa$ indeed holds.

$$* \qquad *$$
$$*$$

Corollary 11.1. Let $\{\kappa_\xi : \xi < \alpha\}$ be a strictly increasing sequence of cardinals, where α is a limit ordinal and $\kappa_0 > 0$. Then

$$\sum_{\xi < \alpha} \kappa_\xi < \prod_{\xi < \alpha} \kappa_\xi.$$

Proof. We have

$$\sum_{\xi < \alpha} \kappa_\xi < \prod_{\xi < \alpha} \kappa_{\xi+1} \leq \prod_{\xi < \alpha} \kappa_\xi.$$

The first inequality follows from König's Theorem, as $\kappa_\xi < \kappa_{\xi+1}$ for $\kappa < \alpha$. To see the second inequality, notice that if $\xi, \eta < \alpha$ and $\xi \neq \eta$ then $\xi + 1 \neq \eta + 1$ and $\xi + 1 < \alpha$. Thus the product on the left-hand side is obtained from the product on the right-hand side by omitting certain factors, each of which is not less than 1.

$$* \qquad *$$
$$*$$

Corollary 11.2. *For each cardinal $\kappa \geq \omega$, we have*

$$\kappa^{\mathrm{cf}(\kappa)} > \kappa.$$

Proof. If κ is regular, then

$$\kappa^{\mathrm{cf}(\kappa)} = \kappa^\kappa = 2^\kappa > \kappa.$$

Assume now that κ is singular. Then κ is a limit cardinal, and so, by Corollary 10.6, there is a strictly increasing sequence $\{\kappa_\xi : \xi < \mathrm{cf}(\kappa)\}$ of cardinals such that

$$\kappa = \sum_{\xi < \mathrm{cf}(\kappa)} \kappa_\xi.$$

We may assume that $\kappa_0 > 0$. Then

$$\kappa < \prod_{\xi < \mathrm{cf}(\kappa)} \kappa_\xi \leq \kappa^{\mathrm{cf}(\kappa)}.$$

*   *
*

Corollary 11.3. *Let κ and λ be cardinals such that $\kappa \geq \omega$ and $\lambda \geq 2$. Then $\mathrm{cf}(\lambda^\kappa) > \kappa$.*

Proof. In view of the assumptions, λ^κ is an infinite cardinal. We have

$$(\lambda^\kappa)^{\mathrm{cf}(\lambda^\kappa)} > \lambda^\kappa$$

according to Corollary 11.2. On the other hand, if $\tau \leq \kappa$, then

$$(\lambda^\kappa)^\tau = \lambda^{\kappa\tau} = \lambda^\kappa.$$

Hence

$$\mathrm{cf}(\lambda^\kappa) \not\leq \kappa.$$

*   *
*

As a special case, this result implies the inequality $\mathrm{cf}(2^{\aleph_0}) \neq \aleph_0$. This latter implies

$$2^{\aleph_0} \neq \aleph_\omega,$$

for example.

The next theorem shows that the value of κ^λ can, on occasion, be calculated from the values of τ^λ for $\tau < \kappa$.

Bernstein–Hausdorff–Tarski Theorem 11.2. *Let $\kappa \geq \omega$ and λ be cardinals such that $0 < \lambda < \mathrm{cf}(\kappa)$. Then, with τ running over cardinals, we have*

$$\kappa^\lambda = \left(\sum_{\tau < \kappa} \tau^\lambda \right) \cdot \kappa.$$

Proof. As $\lambda \geq 1$, it follows from the monotonicity of cardinal exponentiation that the left-hand side is not smaller than the right-hand side. λ^κ, according to its definition, equals the cardinality of the set $^\lambda\kappa$ of functions. We will show that this set satisfies the equality

$$^\lambda\kappa = \bigcup_{\xi < \kappa} {}^\lambda\xi$$

(here ξ runs over ordinals). The set on the right-hand side is clearly a subset of the set on the left-hand side. Assume that $f \in {}^\lambda\kappa$. The range $R(f)$ of f is a subset of κ of cardinality at most λ. As $\lambda < \mathrm{cf}(\kappa)$, the set $R(f)$ is not cofinal in $\langle\kappa, <\rangle$, according to the definition of $\mathrm{cf}(\kappa)$. This means that there is a $\xi < \kappa$ for which $R(f) \subset \xi$; thus $f \in {}^\lambda\xi$. Therefore, the set on the left-hand side is also a subset of the set on the right-hand side; that is, the two sets are equal. Hence the assertion can be verified via the following calculation:

$$\kappa^\lambda = \sum_{\xi<\kappa}|{}^\lambda\xi| = \sum_{\xi<\kappa}|\xi|^\lambda = \sum_{\substack{\tau<\kappa \\ \tau \text{ is a cardinal}}} \tau^\lambda\big|\{\xi : |\xi| = \tau\}\big| \le \left(\sum_{\tau<\kappa}\tau^\lambda\right)\kappa.$$

$$* \qquad *$$
$$*$$

We give two examples as applications of this theorem.

Example 1. For an arbitrary $n \in \omega$, we have

$$\aleph_n^{\aleph_0} = 2^{\aleph_0} \cdot \aleph_n.$$

This was the original form of Bernstein's Theorem. We will prove the assertion by induction. The assertion is obvious for $n = 0$. Assume that the assertion is known to be true for an integer n. Using Theorem 11.2, we obtain

$$\aleph_{n+1}^{\aleph_0} = \left(\sum_{i=0}^{n}\aleph_i^{\aleph_0} + \sum_{k<\omega}k^{\aleph_0}\right) \cdot \aleph_{n+1} = \left(\sum_{i=0}^{n}\aleph_i^{\aleph_0} + 2^{\aleph_0} \cdot \aleph_0\right) \cdot \aleph_{n+1}.$$

On the right-hand side, the sum in parentheses has only finitely many summands; so the sum equals the largest one of these. Hence

$$\aleph_{n+1}^{\aleph_0} = \aleph_n^{\aleph_0} \cdot \aleph_{n+1}.$$

Using the induction hypothesis, we obtain

$$\aleph_{n+1}^{\aleph_0} = \aleph_n^{\aleph_0} \cdot \aleph_{n+1} = 2^{\aleph_0} \cdot \aleph_n \cdot \aleph_{n+1} = 2^{\aleph_0} \cdot \aleph_{n+1}.$$

If we assume the continuum hypothesis, $\aleph_n^{\aleph_0}$ can be expressed even more simply:

$$\aleph_n^{\aleph_0} = \aleph_n$$

for an arbitrary $1 \le n < \omega$. We cannot use Theorem 11.2 to calculate $\aleph_\omega^{\aleph_0}$, as

$$\mathrm{cf}(\aleph_\omega) = \aleph_0.$$

We certainly know by Theorem 11.1 that $\aleph_\omega^{\aleph_0} > \aleph_\omega$.

Example 2. $\aleph_{\omega_1}^{\aleph_0} = \sum_{\alpha < \omega_1} \aleph_\alpha^{\aleph_0}$.

In order to be able to use Theorem 11.2, we need to show that

$$\aleph_0 < \mathrm{cf}(\aleph_{\omega_1}).$$

We claim that

$$\mathrm{cf}(\aleph_{\omega_1}) = \omega_1.$$

According to Theorem 10.9, we have

$$\mathrm{cf}(\aleph_{\omega_1}) = \mathrm{cf}(\langle \{\aleph_\alpha : \alpha < \omega_1\}, < \rangle).$$

The set $\{\aleph_\alpha : \alpha < \omega_1\}$ has order type ω_1 in the ordering $<$, and so its cofinality is ω_1, since ω_1 is a successor cardinal, and so, according to Corollary 10.5, it is regular.

We can now use Theorem 11.2. According to this, we have

$$\aleph_1^{\aleph_0} = \left(\sum_{\alpha < \omega_1} \aleph_\alpha^{\aleph_0} \right) \cdot \aleph_{\omega_1}.$$

Here we can omit $\aleph_{\omega_1}$, since

$$\aleph_{\omega_1} \leq \sum_{\alpha < \omega_1} \aleph_\alpha \leq \sum_{\alpha < \omega_1} \aleph_\alpha^{\aleph_0}.$$

If one is unable to solve a problem in set theory, it may be worthwhile to examine the problem with the assumption of GCH. According to the result of Gödel mentioned above, in set theory it is not possible to disprove a result proved with the aid of GCH.

Theorem 11.3. *Assuming the Generalized Continuum Hypothesis, for an arbitrary cardinal κ, we have*

$$\kappa^\lambda = \begin{cases} 1 & \text{for} \quad \lambda = 0, \\ \kappa & \text{for} \quad 1 \leq \lambda < \mathrm{cf}(\kappa), \\ \kappa^+ & \text{for} \quad \mathrm{cf}(\kappa) \leq \lambda \leq \kappa, \\ \lambda^+ & \text{for} \quad \kappa < \lambda. \end{cases}$$

Proof. According to our assumption, we have

$$2^\tau = \tau^+$$

for an arbitrary cardinal $\tau \geq \omega$. If $1 \leq \lambda < \mathrm{cf}(\kappa)$, then

$$\kappa \leq \kappa^\lambda = \left(\sum_{\tau < \kappa} \tau^\lambda \right) \cdot \kappa \leq \kappa \left(\sum_{\tau < \kappa} 2^{\tau \cdot \lambda} \right) \leq \kappa \left(\sum_{\tau < \kappa} (\max(\tau, \lambda))^+ \right) \leq \kappa^2 = \kappa$$

holds according to Theorem 11.2. Thus $\kappa^\lambda = \kappa$. If $\mathrm{cf}(\kappa) \leq \lambda \leq \kappa$, then by Corollary 11.2 we have

$$\kappa < \kappa^{\mathrm{cf}(\kappa)} \leq \kappa^\lambda \leq 2^{\kappa\lambda} = 2^\kappa = \kappa^+.$$

Therefore $\kappa < \kappa^\lambda \leq \kappa^+$, and so $\kappa^\lambda = \lambda^+$. Finally, if $\kappa < \lambda$, then

$$2^\lambda \leq \kappa^\lambda \leq 2^{\kappa\lambda} = 2^\lambda = \lambda^+.$$

*   *

*

It is interesting to point out that for about fifty years it had been a generally accepted belief that there is no essential restriction on κ^λ other than the theorems stated and proved in this section. In a surprising turn of events, the following result was discovered in 1974.

Silver's Theorem 11.4. *Let κ be a singular cardinal with $\mathrm{cf}(\kappa) > \omega$, and assume that the Generalized Continuum Hypothesis holds for cardinals less than κ, that is $2^\lambda = \lambda^+$ holds for λ with $\omega \leq \lambda < \kappa$. Then $2^\kappa = \kappa^+$.*

The proof of this theorem is surprisingly simple, but it needs some tools not yet discussed. The proof will be given in the second part of the book, after these tools have been introduced. In Sections 21 and 22, there is a brief survey of generalizations of Silver's Theorem and other results about cardinal exponentiation.

Problems

1. Prove that if $2^{\aleph_n} = \aleph_{\omega+1}$ for every $n \in \omega$, then $2^{\aleph_\omega} = \aleph_{\omega+1}$.

2. For an arbitrary cardinal $\kappa \geq \omega$, put $\prod(\kappa) = \min\{\tau : \kappa^\tau > \kappa\}$. Prove that $\prod(\kappa)$ is a regular cardinal.

3. Prove that if $\kappa \geq \omega$ and $\kappa = \sum_{\xi<\mathrm{cf}(\kappa)} \kappa_\xi$ for some sequence $0 < \kappa_0 < \cdots < \kappa_\xi < \ldots$ of cardinals, then we have

$$\kappa^{\mathrm{cf}(\kappa)} = \prod_{\xi<\mathrm{cf}(\kappa)} \kappa_\xi.$$

4. Prove that we have

$$\aleph_\alpha^{\aleph_1} = 2^{\aleph_1} \cdot \aleph_\alpha^{\aleph_0}$$

for every $\alpha < \omega_1$.

5.* Let κ be an arbitrary cardinal, and let λ_0 be the least cardinal such that $2^{\lambda_0} \geq \kappa$. Show that the operation $f(\lambda) = \kappa^\lambda$ assumes only finitely many values for $\lambda \leq \lambda_0$.

6.* Show that if $2^{\aleph_1} = \aleph_2$, then $\aleph_\omega^{\aleph_0} \neq \aleph_{\omega_1}$.

7. Denote by $\kappa^{<\lambda}$ the cardinal $\sum\limits_{\substack{\tau < \lambda \\ \tau \text{ is a cardinal}}} \kappa^\tau$; this is called κ *to the weak power* λ. Assume that $\kappa \geq 2$ and $\lambda \geq \omega$. Prove that

a) $\kappa^{<\lambda} \geq \lambda$;

b) $\kappa^{<\lambda} = \sum_{\tau_0 \leq \tau < \lambda} \kappa^\tau$ for an arbitrary $\tau_0 < \lambda$;

c_1) either there exists a $\tau < \lambda$ such that $\kappa^{<\lambda} = \kappa^\tau$,

c_2) or there are cardinals $\lambda_\xi < \lambda$ for $\xi < \mathrm{cf}(\lambda)$ such that the set $\{\lambda_\xi : \xi < \mathrm{cf}(\lambda)\}$ is cofinal in λ, the sequence $\langle \kappa^{\lambda_\xi} : \xi < \mathrm{cf}(\lambda) \rangle$ is strictly increasing, and $\kappa^{<\lambda} = \sum_{\xi < \mathrm{cf}(\kappa)} \kappa^{\lambda_\xi}$.

d) Prove that $\mathrm{cf}(\kappa^{<\lambda}) \geq \lambda$ if c_1) holds, and $\mathrm{cf}(\kappa^{<\lambda}) = \mathrm{cf}(\lambda)$ if c_2) holds.

e) Show that

$$(\kappa^{<\lambda})^\rho = \begin{cases} \kappa^{<\lambda} & \text{for } 0 < \rho < \lambda, \\ \kappa^\rho & \text{for } \rho \geq \lambda \end{cases}$$

if c_1) holds, and

$$(\kappa^{<\lambda})^\rho = \begin{cases} \kappa^{<\lambda} & \text{for } 0 < \rho < \mathrm{cf}(\lambda), \\ \kappa^\lambda & \text{for } \mathrm{cf}(\lambda) \leq \rho \leq \lambda, \\ \kappa^\rho & \text{for } \lambda \leq \rho \end{cases}$$

if c_2) holds.

f) Show that GCH implies

$$\kappa^{<\lambda} = \begin{cases} \kappa & \text{for } \lambda \leq \mathrm{cf}(\kappa), \\ \kappa^+ & \text{for } \mathrm{cf}(\kappa) < \lambda \leq \kappa^+, \\ \lambda & \text{for } \kappa^+ \leq \lambda. \end{cases}$$

HINTS FOR SOLVING PROBLEMS
MARKED WITH * IN PART I

Section 7

4. We may assume that $\Theta = \text{type}\, \omega(\prec)$ for an appropriately chosen ordering $\prec$ of ω. Define the values $g(n)$ of the mapping $g : \omega \to \mathbb{Q}$ by recursion as follows. If $\langle g(i) : i < n \rangle$ has been defined in such a way that we have

$$g(i) < g(j) \iff i \prec j$$

for $i, j < n$, then there is a rational number $r \in \mathbb{Q}$ such that

$$g(i) < r \iff i \prec n \qquad \text{and} \qquad r < g(j) \iff n \prec j.$$

Choose such a number r as $g(n)$.

The function g maps the ordered set $\langle \omega, \prec \rangle$ onto a subset of $\langle \mathbb{Q}, < \rangle$ monotonically.

5. The proof is similar to that in Problem 4. Making use of the assumption that the ordering $\langle \omega, \prec \rangle$ is dense, we may also arrange for $\mathrm{R}(g) = \mathbb{Q}$ to hold.

7. Write $\Theta = \text{type}\, A(\prec)$. If the assertion is false, then $\langle A, \prec \rangle$ and $\langle A, \succ \rangle$ are wellordered sets according to the assertion of Problem 1. Let $B = \{x \in A : A \text{ has only finitely many elements smaller than } x\}$. Then B is not empty, as $a_0 = \min A \in B$. Put $b_0 = \min_{\succ} B = \max_{\prec} B$. If A has no element larger than b_0, then A is finite. If A has an element larger than b_0, then let c_0 be the smallest among these. In this case, there are only finitely many elements of A that are smaller than c_0. This is a contradiction.

8. Assume, for example, that $\langle A, \prec \rangle$ is wellordered. For an arbitrary $x \in A$, we choose a number $x^* > x$ such that

$$(x, x^*) \cap A = \emptyset.$$

If $(x, +\infty) \cap A = \emptyset$, then $x^* > x$ can be chosen arbitrarily; if, on the other hand, $\exists y > x \,(y \in A)$, then we choose $x^* = \min_{<}\{y \in A : x < y\}$. The intervals $\{(x, x^*) : x \in A\}$ are pairwise disjoint, and so A is countable.

Section 8

6. As $\beta \times \eta \geq \eta$, there is an η for which $\beta \times \eta > \alpha$. Let $\zeta = \min\{\eta : \beta \times \eta > \alpha\}$. According to the assertion of Problem 4, ζ is a successor ordinal, that is, it has form $\zeta = \xi \dotplus 1$. Then

$$\alpha < \beta \times (\xi \dotplus 1) = \beta \times \xi \dotplus \beta.$$

On the other hand, $\beta \times \xi \subset \alpha$. According to the assertion of Problem 2, there is a ρ with $\alpha = \beta \times \xi \dotplus \rho$. Then $\rho < \beta$ in view of the (weak) monotonicity of the operation $\dotplus$.

7. Let

$$B = \{\langle A', \prec' \rangle \in \mathrm{P}(A) \times \mathrm{P}(A \times A) : A' \subset A \wedge \langle A', \prec' \rangle \text{ is wellordered}\}.$$

Let $C = \{\text{type } A'(\prec') \dotplus 1 : \langle A', \prec' \rangle \in B\}$. C consists of ordinals and it is transitive. Hence $\alpha = C$ satisfies the requirements of the problem.

8. Let ϕ be a mapping such that

$$A \times \alpha \sim_\phi A \cup \alpha.$$

If $\exists a \in A \, \forall \beta < \alpha \, (\phi(\langle a, \beta \rangle) \in A)$, then the function $\psi : \alpha \to A$ defined by the stipulation $\psi(\beta) = \phi(\langle a, \beta \rangle)$ maps α in a one-to-one way onto a subset A' of A. We may therefore assume that $\forall a \in A \, \exists \beta < \alpha \, (\phi(\langle a, \beta \rangle) \in \alpha)$. Define the functions $g : A \to \alpha$ and $\psi : A \to \alpha$ as follows:

$$g(a) = \min_{<}\{\beta : \phi(\langle a, \beta \rangle) \in \alpha\},$$

$$\psi(a) = \phi(\langle a, g(a) \rangle) \quad \text{for} \quad a \in A.$$

If $a \neq a'$, then $\langle a, g(a) \rangle \neq \langle a', g(a') \rangle$, and so $\psi(a) \neq \psi(a')$. ψ maps A in a one-to-one way onto a subset of α.

9. Let A be an arbitrary infinite set, and α be an ordinal such that $a \nsim A'$ for each $A' \subset A$; the existence of such an α is guaranteed by Problem 7. We may assume that $A \cap \alpha = \emptyset$. According to Problem 3 of Section 6, for arbitrary infinite cardinals, we have $b + c = b \cdot c$. Thus we have $A \times \alpha \sim A \cup \alpha$. According to Problem 8, by the choice of α there exist a set $B \subset \alpha$ and a one-to-one function f such that $A \sim_f B$. Hence A can be wellordered. It is also easy to see, and we will prove in the next section, that the Axiom of Choice follows from this assertion.

Section 9

5. Let $k \in H(^\kappa 2)$. Then $\mathrm{D}(k) < \omega$. Thus there is a set $X \subset \kappa$ with $|X| < \omega$ such that for every $\phi, \psi \in \mathrm{D}(k)$ with $\phi \neq \psi$, we have $\phi|X \neq \psi|X$.

Define $g \in \tilde{H}(\kappa)$ as follows: $D(g) = \{\phi | X : \phi \in D(k)\}$ and $g(\phi | X) = k(\phi)$.
It is clear that

$$g \in \bigcap \{A_{\phi, k(\phi)} : \phi \in D(k)\}.$$

6. According to the assertion of Problem 4, it is sufficient to prove that
there are 2^{2^κ} ultrafilters on $\tilde{H}(\kappa)$. For each $\Phi \in {}^{\kappa^2}2$ put $\mathcal{S}_\Phi = \{A_{\phi, \Phi(\phi)} :$
$\phi \in {}^\kappa 2\}$. According to Problem 5, $\mathcal{S}_\Phi$ has the Finite Intersection Property.
By Theorem 9.4, $\mathcal{S}_\Phi$ can be extended to an ultrafilter $\mathcal{U}_\Phi$. If $\Phi_0 \neq \Phi_1$ for
$\Phi_0, \Phi_1 \in {}^{\kappa^2}2$, then $\Phi_0(\phi) \neq \Phi_1(\phi)$ for some $\phi \in {}^\kappa 2$, say $\Phi_0(\phi) = 0$ and
$\Phi_1(\phi) = 1$. Then

$$A_{\phi,0} \cap A_{\phi,1} = \emptyset, \quad \text{and} \quad A_{\phi,i} \in \mathcal{U}_{\phi_i} \quad \text{for} \quad i < 2,$$

and so $\mathcal{U}_{\phi_1} \neq \mathcal{U}_{\phi_2}$. Thus the number of ultrafilters is at least 2^{2^κ}. On the
other hand, it is straightforward that their number is at most 2^{2^κ}.

8. *(Rado's Selection Lemma)* Writing $X = \bigtimes_{\alpha < \kappa} A_\alpha$, for $V \in [\kappa]^{<\omega}$ put

$$F_V = \{f \in X : \exists V' \, (V \subset V' \in [\kappa]^{<\omega} \wedge f | V = f_{V'} | V)\}$$

and

$$\mathcal{F} = \{F_V : V \in [\kappa]^{<\omega}\}.$$

According to the assumptions, $\mathcal{F}$ has the Finite Intersection Property, and
so, by Theorem 9.4, it can be extended to an ultrafilter $\mathcal{U}$. For each $\alpha < \kappa$
there is exactly one $x \in A_\alpha$ such that $\bigtimes_{\beta < \kappa, \beta \neq \alpha} A_\beta \times \{x\} \in \mathcal{U}$. Putting
$f(\alpha) = x$ for this x, we will have $f \in F_V$ for every $V \in [\kappa]^\omega$.

If one wishes to invoke some facts from topology or mathematical logic,
it may be observed that the result easily follows either from Tychonoff's
Theorem about the compactness of topological products of compact spaces
or from the Compactness Theorem of First Order Logic. We will give an
example for the application of the above result among the problems of Section
19.

Section 10

6. *c)* According to Part *a)* of the problem we may assume that $\langle P, \prec \rangle$ is
well-founded (see Definition 16.2), $\mathrm{cf}(\langle P, \prec \rangle) = \lambda = |P| > \mathrm{cf}(\lambda) = \kappa \geq \omega$.
For an arbitrary $X \subset P$, let

$$\hat{X} = \{p \in P : \exists q \in X \, p \leq q\}.$$

We claim that for the set

$$S = \{\hat{X} : X \in [P]^{\leq \kappa}\},$$

we have $|S| > \lambda$, where $[P]^{\leq \kappa} = \{x \in P : |x| \leq \kappa\}$.

Assume, contrary to the assertion, that $|S| \le \lambda$, and let

$$S = \{\hat{X}_\alpha : \alpha < \lambda\}$$

be an enumeration of S (possibly with repetitions). Let $\langle \lambda_\nu : \nu < \kappa \rangle$ be a sequence of cardinals such that $\sup\{\lambda_\nu : \nu < \kappa\} = \lambda$. Using the relations

$$\operatorname{cf}(\langle P, \prec \rangle) = \lambda \qquad \text{and} \qquad |Y_\nu| \le \kappa \cdot \lambda_\nu < \lambda,$$

where

$$Y_\nu = \bigcup\{X_\alpha : \alpha < \lambda_\nu\},$$

for an arbitrary $\nu < \kappa$, we can choose an $x_\nu \in P$ with $x_\nu \notin \hat{Y}_\nu$. Let $X = \{x_\nu : \nu < \kappa\}$. Then $X \in [P]^{\le \kappa}$ and $\hat{X} \ne \hat{X}_\alpha$ for each $\alpha < \lambda$. This is a contradiction, and so $|S| > \lambda$.

Assume now, on the contrary, that every independent subset of P is finite. For an arbitrary $V \in [P]^{<\omega}$, put

$$P(V) = \{p \in P : \forall z \in V \; q \not\le p\}.$$

As $|[P]^{<\omega}| = \lambda$, we have

$$|\{P(V) : V \in [P]^{<\omega}\}| \le \lambda.$$

Thus we will get a contradiction from our assumption if we prove that for each $X \in [P]^{\le \kappa}$ the set $\hat{X}$ equals $P(V)$ for some $V \in [P]^{\le \kappa}$. Let $X \in [P]^{\le \kappa}$. Then $P \setminus \hat{X} \ne 0$, as

$$\operatorname{cf}(\langle P, \prec \rangle) = \lambda > \kappa.$$

Let V be the set of minimal elements of $P \setminus \hat{X}$. This set is not empty, because $\langle P, \prec \rangle$ is well-founded, and V, being an independent set, is finite. We claim $\hat{X} = P(V)$. Clearly, $\hat{X}$ is an initial segment of P, and so no element of $\hat{X}$ can be greater than or equal to any element of V. Hence $\hat{X} \subset P(V)$. On the other hand, given $p \notin \hat{X}$, by virtue of the well-foundedness of $\langle P, \prec \rangle$ there is a $q \in V$ such that $q \preceq p$, and so $p \notin P(V)$.

Section 11

5. Let κ be the least infinite cardinal for which the assertion of the problem is not true. Then there is a strictly increasing sequence $\langle \tau_n : n < \omega \rangle$ of cardinals such that $\langle \kappa^{\tau_n} : n < \omega \rangle$ is strictly increasing and for which

$$2^{\tau_n} < \kappa \qquad \text{for} \qquad n < \omega.$$

We claim that if

$$\forall \rho < \kappa \, (\rho^{\tau_n} < \kappa)$$

holds for some $n < \omega$, then

$$\kappa^{\tau_n} = \kappa \qquad \text{or} \qquad \kappa^{\tau_n} = \kappa^{\text{cf}(\kappa)}$$

according as

$$\tau_n < \text{cf}(\kappa) \qquad \text{or} \qquad \tau_n \geq \text{cf}(\kappa).$$

Indeed, if $\tau_n < \text{cf}(\kappa)$, then Theorem 11.2 shows that

$$\kappa^{\tau_n} = \left(\sum_{\rho < \kappa} \rho^{\tau_n}\right)\kappa = \kappa.$$

If $\text{cf}(\kappa) \leq \tau_n$, then κ is singular, as $\tau_n < \kappa$. Put

$$\kappa = \sum_{\xi < \text{cf}(\kappa)} \kappa_\xi,$$

where $\langle \kappa_\xi : \xi < \text{cf}(\kappa) \rangle$ is a strictly increasing sequence of cardinals. Using the assertion of Problem 3, we can see that

$$\kappa^{\tau_n} = \kappa^{\text{cf}(\kappa)\tau_n} = \prod_{\xi < \text{cf}(\kappa)} \kappa_\xi^{\tau_n} = \prod_{\xi < \text{cf}(\kappa)} \kappa = \kappa^{\text{cf}(\kappa)}.$$

We may therefore assume that

$$(\forall n : n_0 \leq n < \omega)\, \exists \rho < \kappa \quad \rho^{\tau_n} \geq \kappa$$

for some $n_0 < \omega$. Let

$$\rho_n = \min_{<}\{\rho : \rho^{\tau_n} \geq \kappa\} \qquad \text{for} \qquad n_0 \leq n < \omega.$$

As the sequence τ_n is increasing, the sequence ρ_n is weakly decreasing. Hence there is an $n_1 \geq n_0$ such that

$$\omega \leq \rho_n = \rho < \kappa \qquad \text{for} \qquad n_1 \leq n < \omega.$$

In this case, we have

$$\kappa^{\tau_n} = \rho^{\tau_n} \quad \text{and} \quad 2^{\tau_n} < \rho \qquad \text{for} \qquad n_1 \leq n < \omega;$$

this contradicts the minimality of κ.

6. If $2^{\aleph_1} = \aleph_2$ then

$$\aleph_n^{\aleph_1} = \aleph_n \qquad \text{for} \qquad 2 \leq n < \omega$$

holds according to Theorem 11.2. By Problem 3 we have

$$\aleph_n^{\aleph_0} = \prod_{2 \leq n < \omega} \aleph_n.$$

and so

$$(\aleph_\omega^{\aleph_0})^{\aleph_1} = \prod_{2 \leq n < \omega} \aleph_n^{\aleph_1} = \prod_{2 \leq n < \omega} \aleph_n = \aleph_\omega^{\aleph_0}.$$

According to Corollary 11.2 of König's Theorem, we have

$$\aleph_{\omega_1}^{\aleph_1} > \aleph_{\omega_1},$$

and so

$$\aleph_\omega^{\aleph_0} \neq \aleph_{\omega_1}.$$

APPENDIX
AN AXIOMATIC DEVELOPMENT
OF SET THEORY

INTRODUCTION

We are going to sketch how the half-way intuitive and half-way axiomatic development given in the first eleven sections can be transformed into a rigorously axiomatic development. To follow this sketch, an elementary acquaintance with the basics of mathematical logic is needed. As this book considers only set theory, it is beyond its scope to provide the necessary background in mathematical logic. To facilitate matters, we will, however, explain the notation and clarify what is meant by an axiomatic development of set theory.

In what follows, we will denote by L_0 a first-order language that, in addition to the variable symbols contains two two-place predicates: $=$ (equality) and $\in$ (being and element of). L_0 is called the language of the Zermelo–Fraenkel axiom system (see Section A1). In a strictly formal presentation, variable symbols would be specified as, say, $\mathfrak{v}_0, \mathfrak{v}_1, \mathfrak{v}_2, \ldots$. For easier readability, we will use a variety of letters to denote variables. In particular, unless otherwise indicated, the letters x, y, z, u, v, w, A, B, C (possibly with subscripts) will always denote variables.

Later we will consider languages L that are extensions of the language L_0. Such languages may contain other predicate and function symbols. We assume that the reader knows how to define terms and well-formed formulas in a language. The set of terms and well-formed formulas in a language L are denoted by $\mathrm{Term}(L)$ and $\mathrm{Wff}(L)$, respectively. We will use the symbols $\neg$ (not), $\wedge$ (and), $\vee$ (or), $\implies$ (implies), $\iff$ (if and only if) to denote the logical connectives, and $\exists$ (there exists), $\forall$ (for all) to denote the quantifiers of the language.

A rigorous axiomatic foundation of set theory would mean that one enumerates the well-formed formulas of the language L representing the axioms in an effective manner, and from these axioms, by using the axioms and the rules of inference of the first-order functional calculus one derives the well-formed formulas representing the known theorems of set theory.

In practice, this program is rarely, if ever, carried out; formal proofs are very tedious, if not impossible, to read; it is more enlightening to indicate instead how such formal proofs could be obtained if one really took the time. Therefore, in what follows, if there is no danger of misunderstanding, instead of the well-formed formulas and terms of the language, we will use their customary, more readable, abbreviated forms. The conventions for such abbreviated forms will not be listed here, since such abbreviations have already been used in the first part of the book. Such a convention is, for example, that instead of the formula $(\phi \wedge \psi) \iff \theta$ we will write $\phi \wedge \psi \iff \theta$ in the same way as instead of $(3 \cdot 5) + 7$ it is customary to write $3 \cdot 5 + 7$.

Another, more important deviation from formal derivations is that during the discussion these formulas will be considered as statements invested with mathematical content, and from these we derive other formulas by using the usual arguments of mathematical practice. It is known by Gödel's Completeness Theorem that, in this case, a derivation based on the rules of functional calculus can also be given.

A1. THE ZERMELO–FRAENKEL
AXIOM SYSTEM OF SET THEORY

Before we list the axioms of set theory, we make one more important remark. The fact that in the language we have the only predicate symbols $=$ and $\in$, and we do not introduce a one-place predicate symbol to say that something is a set implicitly means that we only discuss sets; other objects are of no concern to us. The attentive reader may have noticed that this practice was tacitly followed even in the first part of the book, after the introduction of good sets. That this will not impose undue restrictions on us will be clear exactly from the fact that the axiomatic development can be carried out in this way.

The today generally accepted Zermelo–Fraenkel axiom system of set theory contains the following axioms.

Axiom of Empty Set:

$$\mathbf{A}_0 \qquad\qquad \exists x \forall u \, (u \notin x).$$

Here $u \notin x$ abbreviates $\neg u \in x$.

Axiom of Extensionality:

$$\mathbf{A}_1 \qquad \forall x \forall y \, \big(\forall u \, (u \in x \iff u \in y) \implies x = y\big).$$

We actually have $\iff$ in this formula in place of $\implies$. This is because the symbol $=$ is considered as a symbol of logic rather than of set theory. This means, according to the rules of logic, that equal objects behave the same way; formally,

$$(x = y) \implies (\phi(x) \iff \phi(y))$$

for any formula ϕ.

Axiom of Pairing:

$$\mathbf{A}_2 \qquad\qquad \forall x \forall y \exists z \forall u \, (u \in x \iff u = x \lor u = y)$$

Axiom of Power Set:

$$\mathbf{A}_3 \qquad \forall x \exists y \forall u \, \big(u \in y \iff \forall v (v \in u \implies v \in x)\big).$$

Axiom of Union:

A₄ $\forall x \exists y \forall u \left(u \in y \iff \exists v \left(u \in v \land v \in x \right) \right).$

Axiom of Infinity:

A₅
$$\exists x \Big(\exists u \big(u \in x \land \forall v (v \notin u) \big) \land$$
$$\forall u \Big(u \in x \implies \exists v \big(v \in x \land \forall w \left(w \in v \iff w \in u \lor w = u \right) \big) \Big) \Big).$$

Axiom of Replacement:
For an arbitrary well-formed formula $\phi(u, v, x_1, \ldots, x_n)$ of L_0 all the free variables of which are among the pairwise distinct variables u, v, $x_1, \ldots, x_n$, and for variables x and y different from those listed and from each other, we have

A₆(ϕ)
$$\forall x_1 \ldots \forall x_n \Big(\forall u \exists! v \, \phi(u, v, x_1, \ldots, x_n) \implies$$
$$\forall x \exists y \forall v \Big(v \in y \iff \exists u \big(u \in x \land \phi(u, v, x_1, \ldots, x_n) \big) \Big) \Big).$$

Here and in what follows $\exists! v \, \phi(v)$ abbreviates the assertion "there is exactly one v for which $\phi(v)$," which stands for the formula

$$\exists v \Big(\phi(v) \land \forall v \forall z \big(\phi(v) = \phi(z) \implies v = z \big) \Big).$$

The formula $\phi(v)$ here may contain free variables other than v, not explicitly indicated here. It is important that, in expanding the abbreviation just discussed, conflicts should be avoided. That is, the variable z must be different from the free variables occurring in $\phi(v)$.

Note that the Axiom of Replacement is not a single axiom: **A₆(ϕ)** is a separate axiom for each choice of the formula ϕ. Such a group of axioms, described by specifying the way each axiom in the group can be obtained, is called an *Axiom Scheme*. For this reason, the Axiom of Replacement is more accurately referred to as the Axiom Scheme of Replacement. Each axiom **A₆(ϕ)** is called an instance of this scheme.

Axiom of Regularity:

A₇ $\forall x \left(\exists u \left(u \in x \right) \implies \exists v \left(v \in x \land \forall w \left(\neg(w \in x \land w \in v) \right) \right) \right).$

The system of these axioms is denoted by ZF.
These axioms correspond to the stipulations made in Section 1 as follows: **A₀** corresponds to Convention 4, **A₁** to Convention 3, **A₂** to Convention 5,

A_3 to Convention 7, A_4 to Convention 8. A_5 is equivalent to assuming the existence of ω. Axiom $A_6(\phi)$ is implicitly stated in Convention 11. Finally, the Axiom of Regularity A_7 has not been mentioned or used before. This is a simplifying assumption, invented by J. von Neumann, that is usually included among the Zermelo–Fraenkel axioms; we did not want to use it in the intuitive development given in Part I of the book. We will discuss its role in simplifying the development separately. It is worth noting that among the axioms given above, axioms A_0 and A_5 ensure that sets exist at all. A_0 is in fact superfluous, since it follows from A_5 immediately; the rationale for the inclusion of A_0 is partly historical, and partly the fact that an important part of set theory can be developed without axiom A_5. The reason axiom A_5 was given in this form was to simplify the technical difficulties involved in the definition of ω (see Definition A5.2 below).

The rest of the axiom are conditional existence axioms, axioms by which the existence of some sets enables one to conclude the existence of certain other sets.

Our aim below is to show that the other stipulations discussed in Section 1 are meaningful and derivable in the axiom system presented, and to give helpful hints as to what order the theorems proven so far should be discussed in the axiomatic development.

A2. DEFINITION OF CONCEPTS;
EXTENSION OF THE LANGUAGE

In Part I of this book, we did not only prove theorems (that is, derived formulas) from ZF; we also introduced new concepts. For example, we defined the property $\subset$, the operation $\{x, y\}$, and the concept of empty set. Our first objective is to clarify the role of this procedure in the axiomatic development. When, for example, we introduce the symbol $\subset$ and use it in the formulation of further assertions, we in fact extend the language L_0 used so far to a language L which also contains the new two-place predicate symbol $\subset$ in addition to the symbols $=$ and $\in$. The property $x \subset y$ is defined as

$$\forall u \, (u \in x \implies u \in y);$$

that is, we add the formula

$$\forall x \forall y \, (x \subset y \iff \forall u \, (u \in x \implies u \in y))$$

to ZF, and from then on we work in the extended system ZF$'$.

In general, a new *property* will be described in an extension ZF$'$ of ZF with the aid of an extension L' of the language L_0 as follows. We add a new n-place relation symbol B to the current language L'; in this way, we obtain an extended language L''. Then we consider a suitable formula $\phi \in \text{Wff}(L')$ with free variables $x_1, \ldots, x_n$, and we add the formula

$$\forall x_1 \ldots \forall x_n \, (\phi(x_1, \ldots, x_n) \iff B(x_1, \ldots, x_n))$$

(defining the property B) to the current system ZF$'$ to obtain the extended axiom system ZF$''$.

In defining the operation $\{x, y\}$, we proceed as follows. We consider the formula

$$\phi(x, y, z) \equiv \forall u \, (u \in z \iff u = x \lor u = y),$$

which is called the *defining postulate* of the operation $\{,\}$. Then, with the aid of axioms $\mathbf{A}_1$ and $\mathbf{A}_2$, we prove that

$$\forall x \forall y \exists ! z \, \phi(x, y, z)$$

holds, and we add the formula

$$\forall x \forall y \, \phi(x, y, \{x, y\})$$

to the axioms; this formula is called *the definition* of the operation $\{,\}$.

In general, if L' is an extension of the language, and we are given the axiom system ZF' with ZF $\subset$ ZF' $\subset$ Wff(L'), then in defining an *n-place operation* $\mathcal{F}$ we add the new n-place function symbol $\mathcal{F}$ to the language L'; consider an appropriate formula

$$\phi(x_1, \ldots, x_n, u) \in \text{Wff}(L')$$

(the *defining postulate* of $\mathcal{F}$) all of whose free variables are among $u, x_1, \ldots, x_n$. We next prove that

$$\text{ZF}' \vdash \forall x_1 \ldots \forall x_n \exists! u \, \phi(x_1, \ldots, x_n, u),$$

and then we add the formula

$$\forall x_1 \ldots \forall x_n \, \phi\big(x_1, \ldots, x_n, \mathcal{F}(x_1, \ldots, x_n)\big)$$

(the *definition* of $\mathcal{F}$) to ZF' to obtain the extended axiom system ZF''.

A special case of defining new operations was the definition of new sets, e.g., the definition of the empty set. In this case, we add a 0-place function symbol, that is, a constant symbol, in our case the symbol $\emptyset$, to the language, and consider the formula

$$\phi(u) \iff \forall v \, (\neg v \in u);$$

using axioms $\mathbf{A}_0$ and $\mathbf{A}_1$, we prove that $\exists! u \, \phi(u)$. Then we add the formula $\phi(\emptyset)$ to the axiom system.

It is to be observed that in the course of introducing new operations, we need to prove a theorem concerning the operation being well defined, whereas the introduction of new properties, such as e.g. $\subset$, can be carried out without any reference to the axioms given in the language L_0.

If $\mathcal{F}$ and $\mathcal{G}$ have already been added to the language, their various compositions are always defined, e.g., if $\mathcal{F}$ is a two-place function symbol and $\mathcal{G}$ is a one-place function symbol, the composition $\mathcal{F}(x, \mathcal{G}(y))$ is, of course, a term of the language. Since the formula

$$\forall x \forall y \exists! z \, (z = \mathcal{F}(x, \mathcal{G}(y)))$$

is provable in the extended axiom system, we can further extend the axiom system and give a name, say $\mathcal{H}$, to this composition, by adding the axiom

$$\mathcal{H}(x, y) = \mathcal{F}(x, \mathcal{G}(y)).$$

It is important to point out that we in fact "stay in the original axiom system ZF" when we use this procedure of adding definitions. The relevant concept is described by the following.

Definition A2.1. *Let $L \subset L'$ be two arbitrary languages, and assume*

$$\Sigma \subset \Sigma', \quad \Sigma \subset \mathrm{Wff}(L), \quad \Sigma' \subset \mathrm{Wff}(L').$$

We say that Σ' is a conservative extension of Σ if for every $\phi \in \mathrm{Wff}(L)$, if $\Sigma' \vdash \phi$, then also $\Sigma \vdash \phi$.

We say that Σ' is an effective conservative extension if the derivation of ϕ in Σ can be calculated from the derivation of ϕ in Σ'.

Informally, an extension being conservative means that if a formula is provable in the extended system, then it is already provable in the original system. Effectiveness means that the proof of the formula in the extended system can "systematically" be converted to a proof in the original system. In fact, we will find it useful to work with a more restrictive concept of conservative extension:

Definition A2.2. *A conservative extension $\Sigma' \subset \mathrm{Wff}(L')$ of $\Sigma \subset \mathrm{Wff}(L)$ is called a strict conservative extension if for every $\phi' \in \mathrm{Wff}(L')$ there is a formula $\phi \in \mathrm{Wff}(L)$ such that $\Sigma' \vdash \phi' \iff \phi$.*

We say that Σ' is an effective strict conservative extension of Σ if it is an effective conservative extension, and if for every $\phi' \in \mathrm{Wff}(L')$ one can calculate a formula $\phi \in \mathrm{Wff}(L)$ such that $\Sigma' \vdash \phi' \iff \phi$, and the proof of $\phi' \iff \phi$ in Σ' can be calculated.

In other words, a conservative extension is a strict conservative extension if every "new" formula is provably equivalent to an "old" formula; a strict conservative extension that is also an effective conservative extension is an effective strict conservative extension if for every new formula an equivalent old formula can be found in a "systematic" way, and a proof of this equivalence can also be found in a "systematic" way.

Theorem A2.1. *The extensions used in the definition of concepts (properties and operations) are effective strict conservative extensions.*

We will omit the proof of Theorem A2.1. This theorem certainly shows that when we develop set theory by repeatedly introducing new concepts we in effect stay within the framework of the original axiom system ZF.

A3. A SKETCH OF THE
DEVELOPMENT. METATHEOREMS

We have already introduced the property $x \subset y$, the operation $\{x, y\}$, and the constant $\emptyset$. The operation $\langle x, y \rangle$ described in Definition 1.2 can be given by the defining postulate

$$\phi(x, y, z) \equiv \{\{x\}, \{x, y\}\},$$

and Theorem 1.3 can be proved about it in the same way as was done in Section 1, without the need for additional remarks.

Next we are going to sketch how the rest of the definitions given in Section 1 can be handled.

Using axioms A_1, A_2, and A_4, we can obtain the operations $P(x)$ and $\bigcup x$ with the aid of the defining postulates

$$\phi(x, y) \equiv \forall u \, (u \in y \iff u \subset x)$$

and

$$\phi(x, y) \equiv \forall u \, (u \in y \iff \exists v \, (u \in v \wedge v \in x).$$

As we mentioned, the Axiom of Replacement $A_6(\phi) \in \text{Wff}(L_0)$ corresponds to Convention 11 in Section 1. As an illustration, we will show that this axiom can be used to establish the assertion

$$\forall x \exists y \forall u \, (u \in y \iff \exists v \, (u = \{v\} \wedge v \in x));$$

that is, for every set x there is a set y whose elements are the values that the operation $\{u\}$ assumes for elements of x. In Part I this set was denoted as $y = \{\{u\} : u \in x\}$. A convenient way to obtain this set would appear to be as follows. Write

$$\phi_0 \equiv \phi(u, v) \equiv (v = \{u\}).$$

Then, clearly,

$$\forall u \exists! v \phi_0(u, v)$$

holds, and so the assertion to be proved is apparently identical to the axiom $A_6(\phi_0)$. This argument, however, cannot be used, since ϕ_0 is not a formula of the language L_0; it is a formula of a language L' obtained by extending the language L_0 in a way described in Section A2. Thus $A_6(\phi_0)$ is not an axiom of ZF.

To make the above argument complete, we need the following theorem.

Theorem A3.1. *Assume $L_0 \subset L'$, let ZF' be a strict conservative extension of ZF, and let $\phi' \subset \text{Wff}(L')$. Then $ZF' \vdash \mathbf{A_6}(\phi')$.*

Proof. Write

$$\phi' \equiv \phi'(u, v, x_1, \ldots, x_n),$$

where all the free variables of ϕ' are among $u, v, x_1, \ldots, x_n$, and x and y are distinct variables different from them. According to Theorem A2.1, there is a formula $\phi(u, v, x_1, \ldots, x_n) \in \text{Wff}(L_0)$ such that

$$ZF' \vdash \phi'(u, v, x_1, \ldots, x_n) \iff \phi(u, v, x_1, \ldots, x_n).$$

Then clearly

$$ZF' \vdash \forall u \exists! v \, \phi'(u, v, x_1, \ldots, x_n) \iff \forall u \exists! v \, \phi(u, v, x_1, \ldots, x_n).$$

Thus we have

$$ZF' \vdash \mathbf{A_6}(\phi) \implies \mathbf{A_6}(\phi') \quad \text{and} \quad \mathbf{A_6} \in (\phi) \, ZF';$$

Hence

$$ZF' \vdash \mathbf{A_6}(\phi').$$

$$* \quad *$$
$$*$$

Theorem A3.1 says that the Axiom of Replacement can be used also for formulas of the extended language. Theorem A3.1 and Theorem A2.1 are not theorems of set theory, but statements that are outside set theory; they are statements about set theory. These types of statements are usually called *metatheorems*. A theorem of set theory must be given as a formula of the language L_0, or possibly as a formula of the extended language. A proof of a theorem of set theory is a formal derivation, according to the rules of first-order functional calculus, of the formula representing the theorem, from the formulas representing the axioms.

Theorem A3.1 does not claim that a single formula is derivable in ZF', and for this reason alone it is not a theorem of the formal system. It in fact claims the derivability of an infinite number of formulas. Its proof is not a derivation in the formal sense; it is, rather, a description of a method, i.e., an *algorithm*, of how one can obtain a formal derivation from the axioms of ZF' of the statement $\mathbf{A_6}(\phi')$. For each particular choice of ϕ' one can follow this method to construct a derivation of $\mathbf{A_6}(\phi')$ in ZF' and to convince oneself that $\mathbf{A_6}(\phi')$ is indeed a theorem of ZF'. For this reason, Theorem A3.1 is a theorem *about*, and not *within*, the formal system ZF'.

As we mentioned above, our purpose here is not to give formal derivations of key results of axiomatic set theory; such a project would be unwieldy, and the result totally unreadable. Our purpose is, rather, to show that it is possible to give such formal derivations, and, indeed, how these derivations can be obtained. Metatheorems, such as Theorem A3.1 above, are a great help in this respect, since they describe in a systematic way how a large number of similar formal derivations, useful in this project, can be obtained. In what follows, we will present and use a number of such metatheorems. We will not burden this outline by explicitly indicating on every occasion whether a result presented is a theorem within the formal system or whether it is a metatheorem about the formal system. The reader will, however, be well served by trying to make this distinction in each case on her (or his) own.

Theorem A3.2. *Given an operation* $\mathcal{F}(u, x_1, \ldots, x_n)$ *all the free variables of which are among the variables* $u, x_1, \ldots, x_n$ *and* x *is a variable different from these, it is possible to introduce an operation* $\mathcal{G}(x, x_1, \ldots, x_n)$ *such that*

$$\forall v \left(v \in \mathcal{G}(x, x_1, \ldots, x_n) \iff \exists u \left(u \in x \wedge \mathcal{F}(u, x_1, \ldots, x_n) = v \right) \right).$$

The operation $\mathcal{G}(x, x_1, \ldots, x_n)$ described in this theorem will be denoted as $\{\mathcal{F}(u, x_1, \ldots, x_n) : u \in x\}$. The theorem shows that in the axiomatic development we are justified to use the stipulations stated in Convention 11 of Section 1.

Proof. The operation $\mathcal{F}(u, x_1, \ldots, x_n)$ is defined in some strict conservative extension ZF$'$ in a language $L' \supset L_0$. Let $\phi(u, v, x_1, \ldots, x_n)$ be the formula

$$v = \mathcal{F}(u, x_1, \ldots, x_n)$$

of the language L'. Clearly, ZF$' \vdash \forall u \exists ! v\, \phi(u, v, x_1, \ldots, x_n)$. Thus, according to Theorem A3.1, we have

$$\text{ZF}' \vdash \forall x \exists y \forall v \left(v \in y \iff \phi(u, v, x_1, \ldots, x_n) \wedge u \in x \right).$$

Let $\psi(x, y, x_1, \ldots, x_n)$ be the formula

$$\forall v \left(v \in y \iff \phi(u, v, x_1, \ldots, x_n) \wedge u \in x \right).$$

According to the Axiom of Extensionality, we can then also derive the formula

$$\forall x \exists ! y\, \psi(x, y, x_1, \ldots, x_n).$$

We now extend the language L' by adding the $(n+1)$-place function symbol

$$\mathcal{G}(x, x_1, \ldots, x_n),$$

with the formula

$$\forall x \forall x_1 \ldots \forall x_n \, \phi\big(x, \mathcal{G}(x, x_1, \ldots, x_n), x_1, \ldots, x_n\big)$$

being its definition.

*　　*
*

Our next problem is that there are no axioms in ZF that correspond to Convention 6, which had basic importance for the intuitive development. We now turn to dealing with this problem. The following theorem corresponds to Convention 6.

Theorem A3.3. *Given a formula* $\phi(u, x_1, \ldots, x_n)$ *and an operation* $\mathcal{F}(x_1, \ldots, x_n)$ *which do not have free variables other than those indicated, we can define an operation* $\mathcal{G}(x_1, \ldots, x_n)$ *such that*

$$\forall u \, \big(u \in \mathcal{G}(x_1, \ldots, x_n) \iff u \in \mathcal{F}(x_1, \ldots, x_n) \wedge \phi(u, x_1, \ldots, x_n)\big)$$

holds in an appropriate extension of ZF.

This operation $\mathcal{G}$ will usually be denoted as

$$\{u \in \mathcal{F}(x_1, \ldots, x_n) : \phi(u, x_1, \ldots, x_n)\}$$

Proof. Let L' be the extension of L_0 in which ϕ and $\mathcal{F}$ are defined; let ZF$'$ be the corresponding extension of ZF. We may assume that the operations $\{u\}$ and $\bigcup$ and the constant $\emptyset$ are also defined in L' in ZF$'$.
Let $\psi(u, v, x_1, \ldots, x_n)$ be the an following formula of L':

$$\big(\phi(u, x_1, \ldots, x_n) \wedge v = \{u\}\big) \vee \big(\neg\phi(u, x_1, \ldots, x_n) \wedge v = \emptyset\big).$$

Clearly,

$$\forall x_1 \ldots \forall x_n \forall u \exists! v \psi(u, v, x_1, \ldots, x_n).$$

In view of this formula, we can introduce an operation $H(u, x_1, x_2, \ldots, x_n)$ such that

$$\forall x_1 \ldots \forall x_n \forall u \, \psi\big(u, H(u, x_1, \ldots, x_n), x_1, \ldots, x_n\big)$$

holds (i.e., the latter formula is the definition of H, while ψ above is its defining postulate).
Using Theorem A3.2 now, we can introduce the operation

$$K(x, x_1, \ldots, x_n) = \{H(u, x_1, \ldots, x_n) : u \in x\}.$$

Hence we can also introduce the operation

$$\mathcal{G}(x_1, \ldots, x_n) = \bigcup K\big(\mathcal{F}(x_1, \ldots, x_n), x_1, \ldots, x_n\big).$$

We claim that $\mathcal{G}$ satisfies the requirements of the theorem. Let $x_1, \ldots, x_n$ be arbitrary sets. Assume $u \in \mathcal{G}(x_1, \ldots, x_n)$. Then by the definition of the operation $\bigcup$ there is a v such that $u \in v$ and

$$v \in K\big(\mathcal{F}(x_1, \ldots, x_n), x_1, \ldots, x_n\big).$$

Then $v = H(w, x_1, \ldots, x_n)$ for some $w \in \mathcal{F}(x_1, \ldots, x_n)$, and $v \neq \emptyset$ (as $u \in v$). According to the definition of H,

$$v = \{w\}$$

holds, and so

$$u = w$$

in view of $u \in v$. Thus

$$u \in \mathcal{F}(x_1, \ldots, x_n) \qquad \text{and} \qquad \phi(u, x_1, \ldots, x_n).$$

Conversely, if $u \in \mathcal{F}(x_1, \ldots, x_n)$ and $\phi(u, x_1, \ldots, x_n)$, then

$$H(u, x_1, \ldots, x_n) = \{u\},$$

$$\{u\} \in K\big(\mathcal{F}(x_1, \ldots, x_n), x_1, \ldots, x_n\big),$$

$$u \in \bigcup K\big(\mathcal{F}(x_1, \ldots, x_n), x_1, \ldots, x_n\big).$$

$$* \qquad *$$
$$*$$

Theorem A3.3 is often called the *Axiom Scheme of Comprehension*. Occasionally, this scheme is also listed as part of the axiom system, even though, as we just saw, each of its instances can be proved in ZF. We would like to point out that the proof we presented above is not "unnecessarily complicated." It is known that if we omit the Axiom of Union $\mathbf{A}_4$, then the Axiom Scheme of Comprehension, i.e. Theorem A3.3, is no longer provable.

Theorem A3.3 means that the possibilities opened up by Convention 6 are also available in the axiomatic development if property is taken to mean a property described by a first-order formula.

It is our opinion that, for an axiomatic development, it is necessary to deal with the conceptual difficulties described in Sections A2 and A3. Since this would have been premature in the first part of the book, we decided to at first give a semi-axiomatic introduction to set theory.

A4. A SKETCH OF THE DEVELOPMENT.
DEFINITIONS OF SIMPLE OPERATIONS
AND PROPERTIES (CONTINUED)

In what follows, we will no longer point out repeatedly that after the introduction of each new concept we always work in an extended axiom system. One of the difficulties in following the development described in the first part of the book is that for the defined properties we often did not introduce symbols. For example, we said that a set is a relation if it consists of ordered pairs. This means that we introduced a one-place predicate (which may be denoted, e.g., by $\text{Rel}(x)$) with the aid of the following definition:

$$\text{Rel}(x) \iff \forall u \in x\, \exists v \exists w\, (u = \langle v, w \rangle).$$

(Here and in what follows we use $\forall u \in x\, \phi$ and $\exists u \in x\, \phi$ to abbreviate the formulas $\forall u\, (u \in x \implies \phi)$ and $\exists u\, (u \in x \land \phi)$, respectively.) To make up for this lapse, here we introduce notation for the concepts introduced in Section 1, but for definitions given later, this task will mostly be left to the reader.

We introduce the property $\text{Singlevalued}(x)$ as follows:

$$\text{Singlevalued}(x) \iff \forall u \forall v \forall w (\langle u, v \rangle \in x \land \langle u, w \rangle \in x \implies v = w).$$

The property $\text{Function}(x)$ can be given as

$$\text{Function}(x) \iff \text{Rel}(x) \land \text{Singlevalued}(x).$$

Then we need the operations defining domain and range:

$$D(x) = \{ u \in \bigcup\bigcup x : \exists v\, (\langle u, v \rangle \in x) \},$$

$$R(x) = \{ v \in \bigcup\bigcup x : \exists u\, (\langle u, v \rangle \in x) \},$$

According to Theorem A3.3, these definitions are sound, and it is easy to show that their definitions imply the formulas

$$\forall u\, \big(u \in D(x) \iff \exists v\, (\langle u, v \rangle \in x) \big),$$

$$\forall v \left(v \in \mathrm{R}(x) \iff \exists u \left(\langle u, v \rangle \in x \right) \right).$$

The *inverse operation* can be described as

$$x^{-1} = \{ u \in \mathrm{PP} \bigcup \bigcup x : \exists v \exists w \left(u = \langle v, w \rangle \wedge \langle w, v \rangle \in x \right) \};$$

here $\mathrm{P}x$, more usually written as $\mathrm{P}(x)$, denotes the power set of x; the formal definition was given in the preceding section. The definition is sound again by Theorem A3.3. With this definition, it is easy to show that

$$\forall x \forall u \left(u \in x^{-1} \iff \exists v \exists w \left(u = \langle v, w \rangle \wedge \langle w, v \rangle \in x \right) \},$$

that is x^{-1} is obtained by reversing the pairs in x. In describing the operation $\circ$ used in defining the *composition of functions*, it is expedient to give the definition in more generality as follows:

$$x \circ y = \{ z \in \mathrm{PP} \bigcup \bigcup (x \cup y) :$$
$$\exists u \exists v \exists w \left(\langle u, v \rangle \in y \wedge \langle v, w \rangle \in x \wedge z = \langle u, w \rangle \right) \}.$$

This definition is again sound, according to Theorem A3.3. With its aid, we can verify that $x \circ y$ consists of all ordered pairs for which there is a v such that $\langle u, v \rangle \in y$ and $\langle v, w \rangle \in x$.

The *value of the function x at the place y* is a two-place operation that can be introduced, e.g., as follows: Consider the formula

$$\phi(x, y, z) \equiv \left(\exists ! z \left(\langle y, z \rangle \in x \right) \wedge \langle y, z \rangle \in x \right) \vee \left(\neg \exists ! z \left(\langle y, z \rangle \in x \right) \wedge z = \emptyset \right).$$

We show that $\forall x \forall y \exists ! z \, \phi(x, y, z)$, and then we take $\phi(x, y, z)$ to be the defining postulate of $x(y)$, that is, we stipulate

$$\forall x \forall y \, \phi \left(x, y, x(y) \right).$$

It is easy to see that $x(y)$ is the unique z such that $\langle y, z \rangle \in x$ if there is exactly one such z, and it is $\emptyset$ otherwise.

In Section 2, we defined by Id_A the identity function on the set A. This is a one-place operation that can be defined as follows:

$$\mathrm{Id}_x = \{ \langle u, u \rangle : u \in x \}.$$

Rewriting Definition 2.1 presents no problems:

$$\text{One-one}(x) \iff \text{Singlevalued}(x) \wedge \text{Singlevalued}(x^{-1}),$$
$$x \sim_z y \iff \text{Function}(z) \wedge \text{One-one}(z) \wedge \mathrm{D}(z) = x \wedge \mathrm{R}(z) = y,$$
$$x \sim y \iff \exists z \, (x \sim_z y).$$

Having done this, we can proceed to prove Theorem 2.1, according to which $\sim$ is an equivalence property.

At this point, we arrived at the hardest problem in presenting an axiomatic development in set theory. We defined the concept of "being the same size" for infinite sets, but we have no examples for infinite sets. In the semi-axiomatic development above, the role of Sections 2–7 was to overcome this difficulty, and the development of set theory was continued in Section 8.

A5. A SKETCH OF THE DEVELOPMENT. BASIC THEOREMS, THE INTRODUCTION OF ω AND $\mathbb{R}$ (CONTINUED)

Proceeding with the development, we next describe the notion of ordered set $\langle A, \prec \rangle$ following Definition 7.3, and then the notion of monotone mapping similarly as given in Definition 7.6, without again giving examples for these notions. We define the concept of wellordered set as in Definition 7.9. Then we show that similarity $\simeq$ is an equivalence property (Theorem 7.1) and that it preserves the property of being wellordered. Having proved Theorems 8.1–8.3, we then give the definition of $\in_A = \in \restriction A$ as follows:

$$\in \restriction A = \in_A = \{\langle x, y \rangle \in \mathrm{PP}(A) : x \in A \land y \in A \land x \in y\};$$

the definition can be justified by Theorem A3.3.

We define the notions of transitive set and ordinal as

$$\mathrm{Trans}(x) \iff \forall u \, (u \in x \implies u \subset x),$$

$$\mathrm{Ordinal}(x) \iff \mathrm{Trans}(x) \land \langle x, \in_x \rangle \text{ is wellordered};$$

then we can prove Theorems 8.4, 8.5 and 8.6. We remark that the procedure adopted in Section 8 of setting aside new variable symbols $\alpha, \beta, \ldots$ to run over ordinals can easily be rephrased formally as carrying out a conservative extension as follows.

Given $L_0 \subset L'$ and $\mathrm{ZF} \subset \mathrm{ZF}' \subset \mathrm{Wff}(L')$, consider the extension L'' of L' by adding new variable symbols $\alpha, \beta, \gamma, \ldots$. Having proved in ZF' that $\exists x \, \mathrm{Ordinal}(x)$, for each new variable symbol α and for each formula $\phi(x) \in \mathrm{Wff}(x)$ (with possibly other free variables, not indicated), we add the formulas

$$\forall \alpha \, \phi(\alpha) \iff \forall x \, (\mathrm{Ordinal}(x) \implies \phi(x))$$

and

$$\exists \alpha \, \phi(\alpha) \iff \exists x \, (\mathrm{Ordinal}(x) \land \phi(x))$$

to ZF'. One can prove that this is a strict conservative extension.

Define the ordering of ordinals as

$$\alpha < \beta \iff \alpha \in \beta;$$

then Theorems 8.7, 8.8, and Corollaries 8.1, 8.2, and 8.3 can be established in turn. The definition of the type $A(\prec)$ operation given in Definition 8.7 can be reproduced for wellordered sets by the following defining postulate:

$$\phi(x, y) \iff (\exists A \exists \prec (x = \langle A, \prec \rangle \text{ is wellordered})$$
$$\wedge \, \text{Ordinal}(y) \wedge \langle y, \in_y \rangle \simeq \langle A, \prec \rangle)$$
$$\vee (\neg \exists A \exists \prec (x = \langle A, \prec \rangle \text{ is wellordered}) \wedge y = \emptyset).$$

Naturally, at this point we disregard the remark about finite sets made after Definition 8.7, since finite sets have not yet been defined in the axiomatic approach.

We next define the operation $x \dotplus 1$ as $x \cup \{x\}$, and we prove Theorem 8.10 in the form that for an arbitrary α the set $\alpha \dotplus 1$ is the least ordinal greater than α, and then prove Theorem 8.11.

Thereafter we prove Theorems 9.1 and 9.2. It is important to point out that both of these are metatheorems, and they are only theorems of the axiom system ZF$'$ for a given formula $\phi(x)$ or a given operation $\mathcal{G}(x)$. Furthermore, parameters are also allowed in ϕ and $\mathcal{G}$. The case of Theorem 9.2 is somewhat subtle, so we will formulate it again and comment on its proof. The formal proof of Theorem 9.1 is more closely an imitation of the informal proof given above, and we will not comment on it further.

Transfinite Recursion Theorem A5.1. *Given an arbitrary operation* $\mathcal{G}(x, x_1, \ldots, x_n)$, *there is an operation* $\mathcal{F}(x, x_1, \ldots, x_n)$ *such that*

$$\forall x_1 \ldots \forall x_n \forall \alpha \left(\mathcal{F}(\alpha, x_1, \ldots, x_n) \right.$$
(*)
$$\left. = \mathcal{G}\big(\mathcal{F}(\alpha, x_1, \ldots, x_n) | \alpha, x_1, \ldots, x_n\big)\right).$$

If $\mathcal{F}$ *and* $\mathcal{F}'$ *both satisfy* (*), *then*

$$\forall x_1 \ldots \forall x_n \forall \alpha \left(\mathcal{F}(\alpha, x_1, \ldots, x_n) = \mathcal{F}'(\alpha, x_1, \ldots, x_n)\right).$$

Sketch of Proof. Assume the symbol $\mathcal{G}$ is contained in the language L' with the axiom system ZF$'$. Consider the following formula $\phi \in L'$:

$$\phi(u, \alpha, x_1, \ldots, x_n) \equiv \exists f \left(\text{Function}(f) \wedge \text{D}(f) = \alpha \dotplus 1 \right.$$
$$\left. \wedge \, \forall \beta \leq \alpha \left(f(\beta) = \mathcal{G}(f|\beta, x_1, \ldots, x_n)\right) \wedge u = f(\alpha)\right).$$

Prove by transfinite induction on α that

$$\forall x_1 \ldots \forall x_n \forall \alpha \exists! u \, \phi(u, \alpha, x_1, \ldots, x_n).$$

The details of the proof are essentially the same as in the proof of Theorem 9.2, but now we can make a rigorous appeal to the Axiom of Replacement in ZF' to justify the existence of f on the right-hand side of the symbol $\equiv$ in the formula defining ϕ. Then we can introduce the new operation $\mathcal{F}$ in the usual way, by adding the new symbol $\mathcal{F}$ to L', and then adding the formula

$$\phi(\mathcal{F}(\alpha, x_1, \ldots, x_n), \alpha, x_1, \ldots, x_n)$$

to ZF'.

Next we consider the Axiom of Infinity $\mathbf{A}_5$. With the notation introduced so far, this axiom can be written as (i.e., can be proven equivalent to, in the appropriate $\mathrm{ZF}' \backslash \{\mathbf{A}_5\}$)

$$\exists x \left(\emptyset \in x \wedge \forall y \in x \, (y \overset{.}{+} 1 \in x) \right).$$

Definition A5.1.

1. $$\mathrm{NL}(x) \iff x = \emptyset \vee \exists y \, (x = y \overset{.}{+} 1).$$

$\mathrm{NL}(x)$ *says that if x is an ordinal then it is not a limit ordinal.*

2. $$\mathrm{Int}(x) \iff \mathrm{Ordinal}(x) \wedge \mathrm{NL}(x) \wedge \forall y \in x \, \mathrm{NL}(y).$$

$\mathrm{Int}(x)$ *intends to say that x is a finite ordinal (i.e., a nonnegative integer).*

Lemma A5.1. *If $\mathrm{Int}(x)$ and $y \in x$, then $\mathrm{Int}(y)$.*

Proof. Assume $\mathrm{Int}(x)$ and $y \in x$. x is an ordinal, and so it is transitive. Thus if $z \in y$, then $z \in x$, and so $\mathrm{NL}(z)$. We also have $\mathrm{NL}(y)$, as $y \in x$. Thus $\mathrm{Int}(y)$.

$$* \qquad *$$
$$*$$

Theorem A5.2. *There exists the set of the finite ordinals, that is, there is a set A such that*
$$\forall u \left(u \in A \iff \mathrm{Int}(u) \right).$$

Proof. Let x be a set satisfying the Axiom of Infinity. By transfinite induction, we prove that for all α we have

$$\mathrm{Int}(\alpha) \implies \alpha \in x.$$

If $\alpha = 0$ then $\alpha \in x$. If $\alpha > 0$ and the assertion holds for every $\beta < \alpha$, then in case α is not finite, the assertion holds vacuously. If α is finite then it has form $\alpha = \beta \overset{.}{+} 1$ (according to Definition A5.1), and so by

Lemma A5.1, we also have $\text{Int}(\beta)$. Thus $\beta \in x$ by the induction hypothesis, and so $\alpha = \beta \dotplus 1 \in x$, as x satisfies Axiom $\mathbf{A_5}$. Hence, the sought after set

$$A = \{\alpha \in x : \text{Int}(x)\}$$

exists according to Theorem A3.3. By Lemma A5.1, it is a transitive set that consists of ordinals. From now on this set will be denoted as ω.

* *

*

Definition A5.2. $\omega = \{\alpha : \text{Int}(\alpha)\}$.

As ω is an ordinal, we have $\omega \notin \omega$, and so ω is not a finite ordinal. ω is the least ordinal that is not finite, since for $\beta \in \omega$ we have $\text{Int}(\beta)$.

Next we state the theorem on induction (mathematical induction) and recursion on ω.

Theorem A5.3. 1. If $\emptyset \in x$ and $\forall n \in \omega \, (n \in x \implies n \dotplus 1 \in x)$ then $\omega \subset x$.

2. If $\mathcal{G}(x, x_1, \ldots, x_n)$ is an arbitrary operation, then there is an operation $\mathcal{F}(y, x_1, \ldots, x_n)$ such that, for arbitrary fixed sets $x_1, \ldots, x_n$, the equation $f(y) = \mathcal{F}(y, x_1, \ldots, x_n)$ defines a function f with $\text{D}(f) = \omega$ such that

$$\forall n < \omega \, \big(f(n) = \mathcal{G}(f|n, x_1, \ldots, x_n)\big).$$

These assertions are immediate consequences of the Transfinite Induction and Transfinite Recursion Theorems.

From now on we agree that the letters $n, m, \ldots$ when used as variables run over elements of ω (these letters can also be used outside the language, when discussing questions of syntax).

At this point, we can define the operations addition, multiplication, and exponentiation for nonnegative integers by recursion as follows.

Definition A5.3.
1. $n + 0 = n$ and $n + (m \dotplus 1) = (n + m) \dotplus 1$ for $m, n < \omega$.
2. $n \cdot 0 = 0$ and $n(m \dotplus 1) = n \cdot m + n$ for $m, n < \omega$.
3. $n^0 = 1$ and $n^{m+1} = n^m \cdot n$ for $m, n < \omega$.

One can prove the usual algebraic properties of these operations by induction.

Then the set $\mathbb{Z}$ of integers can be defined in the usual way as the difference ring of ω, the set of rational numbers, as the quotient field of $\mathbb{Z}$, and the set of real numbers $\mathbb{R}$, as the Dedekind-completion of $\mathbb{Q}$. One can easily check that the set-theoretical constructions needed to do this can be described with the aid of the results presented so far.

A6. THE ZFC AXIOM SYSTEM. A WEAKENING OF THE AXIOM OF CHOICE. REMARKS ON THE THEOREMS OF SECTIONS 2–7

Axiom of Choice:

$$\mathbf{A_8} \quad \forall x \exists f \left(\mathrm{Function}(f) \wedge \mathrm{D}(f) = x \right.$$
$$\left. \wedge\, \forall u \in x \left(u \neq \emptyset \implies f(u) \in u \right) \right).$$

We did not formulate the Axiom of Choice earlier, since it would have been quite lengthy to present it without defined concepts.

In the literature, the axiom system $\mathrm{ZF} \cup \{\mathbf{A_8}\}$ is denoted by ZFC, where C refers to the word Choice. In ZFC we can prove the wellordering theorem, and with the aid of the latter we can give Definition 10.1 of the cardinality operation. There is nothing that we can add here to the discussion in Sections 10 and 11; the discussion there can be considered satisfactory from an axiomatic point of view.

There is one more point we would like to mention in connection with the Axiom of Choice. On several occasions we mentioned that the Axiom of Choice is used in the proofs of many classical theorems of mathematics. This does not mean, however, that the full force of the Axiom of Choice is needed for these proofs.

A frequently discussed weakening of the Axiom of Choice is the Axiom of Dependent Choice (DC):

$$\forall X \forall R \left(\mathrm{Rel}(R) \wedge \forall u \in X \exists v \in X\, uRv \implies \forall a \in X \exists f \left(\mathrm{Function}(f) \right. \right.$$
$$\left. \left. \wedge\, \mathrm{D}(f) = \omega \wedge \mathrm{R}(f) \subset X \wedge f(0) = a \wedge \forall n \in \omega \left(f(n)Rf(n+1) \right) \right) \right).$$

In words: if every element u of X is in relation R with some element v of X, then for an arbitrary $a \in X$ there is a sequence $\{a_0, a_1, \ldots, a_n, \ldots\} \subset X$ such that $a_0 = a$ and $a_n R a_{n+1}$ for every n. DC can be easily proved in ZFC, but it is known that $\mathrm{ZF} \cup \{\mathrm{DC}\}$ does not imply the Axiom of Choice. The ambitious reader can check that the classical results of analysis and point set theory can be proved in $\mathrm{ZF} \cup \{\mathrm{DC}\}$.

In this section we would like to say a few more remarks about the theorems proved in Sections 2–7.

The attentive reader might ask what precisely "assuming the existence of ω" means. To dissolve any concerns on this point, we would remark that Theorem 2.2 can easily be proved from Definition A5.1 by induction, and Theorem 3.1 is an easy consequence of Theorem 2.2 and Definition A5.2. In the proof of Theorem 3.2, in order to avoid any reference to the cardinality operation defined with the aid of the Axiom of Choice, it is necessary to show that if $n \in \omega$ and $A \subset n$, then it is meaningful to talk about the "number of elements" of A; that is, there is an $m \leq n$ such that $A \sim m$. This can be proved by induction on n. It is worthwhile to check that the proofs of Theorems 3.4 and 3.6 can also be carried out in $ZF \cup \{DC\}$. Theorem 3.8, of course, served only as an illustration, and the concepts discussed in it can only be defined after carrying out the procedure described at the end of Section A5. Sections 4, 5, and 6 discuss the ideas of Cantor's classical set theory, and the proofs can be carried out essentially without change in the axiomatic framework. An exception is Theorem 6.1, but that result only served as an illustration. In its proof, it is preferable to avoid a reference to the addition of cardinals, since in order to justify the definition of cardinal addition, it is necessary to use the properties of the cardinality operation. The assertion can be deduced as a consequence of Theorem 8.11, as

$$\sup\{a_\gamma : \gamma \in \Gamma\} = \alpha$$

exists, and

$$a_\gamma \leq |\alpha| < 2^{|\alpha|} \qquad \text{for} \qquad \gamma \in \Gamma.$$

A7. THE ROLE OF THE AXIOM OF REGULARITY

As we mentioned above, we did not introduce the Axiom of Regularity during the discussion of intuitive set theory since it is somewhat unnatural. At this point we will establish some of its consequences and some statements equivalent to it, and we will indicate its role in simplifying the discussion. First we will prove some results in the axiom system $\mathrm{ZF}^* = \mathrm{ZF} \setminus \{\mathbf{A_7}\}$.

Definition A7.1 *(The Cumulative Hierarchy).* For an arbitrary ordinal α, define R_α by transfinite recursion as follows: $R_0 = \emptyset$, $R_\alpha = \mathrm{P}(R_\beta)$ for $\alpha = \beta \dotplus 1$, and

$$R_\alpha = \bigcup_{\beta < \alpha} R_\beta$$

if α is a limit ordinal. We say that x has rank or x is ranked if there is an α such that $x \in R_\alpha$.

To indicate that x has rank, we will say $R(x)$. The operation $\mathrm{rk}(x)$ on ranked sets is defined as follows:

$$\mathrm{rk}(x) = \min_{<}\{\alpha : x \in R_\alpha\};$$

$\mathrm{rk}(x)$ is often called the *rank* of x.

Theorem A7.1.
1. R_α is transitive.
2. $R_\beta \subset R_\alpha$ for $\beta < \alpha$.
3. $\mathrm{rk}(x)$ is a successor ordinal for an arbitrary ranked set x.
4. If $R(x)$ and $y \in x$, then $\mathrm{rk}(y) < \mathrm{rk}(x)$.
5. If $\forall y \in x\, R(y)$, then $R(x)$.

Proof. 1. We prove the assertion by transfinite induction. For $\alpha = 0$ the assertion is obvious. If α is a limit ordinal, then R_α is transitive, since, according to the induction hypothesis, it is a union of transitive sets. Assume next that $\alpha = \beta \dotplus 1$ for some β and $u \in R_\alpha = \mathrm{P}(R_\beta)$. Then $u \subset R_\beta$. If $v \in u$, then $v \in R_\beta$, and so $v \subset R_\beta$ by the induction hypothesis; thus $v \in R_\alpha$. Hence $u \subset R_\alpha$, and so R_α is transitive.

2. We prove the assertion $\forall \beta < \alpha\, (R_\beta \subset R_\alpha)$ by transfinite induction on α. If α is not a successor ordinal, then the assertion is straightforward. If

$\alpha = \beta \dotplus 1$, then $R_\alpha = P(R_\beta)$. Hence $R_\beta \in R_\alpha$, and so, by Assertion 1, $R_\beta \subset R_\alpha$. If $\gamma < \beta$, then $R_\gamma \subset R_\beta \subset R_\alpha$ according to the induction hypothesis.

3. If $x \in R_\alpha$ and α is not a successor ordinal, then there is a $\beta < \alpha$ such that $x \in R_\beta$.

4. Assume $\alpha = \mathrm{rk}(x)$ and $y \in x$. Then $x \in R_\alpha$ and $\alpha = \beta \dotplus 1$ for some β; thus $x \subset R_\beta$, and so $y \in R_\beta$. Therefore, $\mathrm{rk}(y) \leq \beta < \alpha = \mathrm{rk}(x)$.

5. Assume each element of x has rank. Put

$$\alpha = \sup\{\mathrm{rk}(y) : y \in x\}.$$

Then for every $y \in x$, there is a $\beta \leq \alpha$ for which $y \in R_\beta$. By Assertion 2 we then have $x \subset R_\alpha$, and so $x \in R_{\alpha \dotplus 1}$.

*　　*
*

Next we prove the following in ZF*.

Theorem A7.2. $A_7 \iff \forall x \, R(x)$.

Proof. Assume that $\forall x \, R(x)$, and let $x \neq \emptyset$ be arbitrary. Write

$$\alpha = \min\{\mathrm{rk}(y) : y \in x\},$$

and let $y \in x$ be a set such that $\mathrm{rk}(y) = \alpha$. If $z \in y$, then we have

$$\mathrm{rk}(z) < \mathrm{rk}(y) = \alpha$$

according to Assertion 4 of Theorem A7.1, and so $y \cap x = \emptyset$.

To prove the converse, for an arbitrary set x define the transitive closure $T(x)$ of a set as follows. First define $T_n(x)$ for $n < \omega$ by recursion as follows:

$$T_0(x) = \{x\} \quad \text{and} \quad T_{n+1}(x) = \bigcup T_n(x) \quad \text{for} \quad n < \omega.$$

Put

$$T(x) = \bigcup_{n<\omega} T_n(x).$$

Then $x \in T(x)$ and $T(x)$ is transitive, since if $u \in T(x)$, then $u \in T_n(x)$ for some $n \in \omega$, and then we also have

$$u \subset T_{n+1}(x) \subset T(x).$$

It is easy to see that if T is a transitive set and $x \in T$, then $T(x) \subset T$; thus the term transitive closure is indeed justified. Assume now, contrary to the assertion to be proved, that $\mathbf{A}_7$ holds and yet there is a set, say, y, that has no rank. Let $T = T(y)$. Then the set

$$x = \{z \in T : \neg R(z)\}$$

is not empty, since $y \in x$. According to $\mathbf{A}_7$ there is a $z \in x$ with $z \cap x = \emptyset$. Then, T being transitive, we have $z \subset T$, and so every element of z has rank. Hence Assertion 5 of Theorem A7.1 implies that z has rank, and this contradicts the definition of x.

*   *

*

In the first part of the book, we said that the cardinality operation cannot be defined from the usual axioms of set theory without using the Axiom of Choice. When saying this, however, we did not include the Axiom of Regularity among the usual axioms. The following theorem shows that it is possible to define the cardinality operation in ZF.

Theorem A7.3. *Put*

$$\mathcal{F}(x) = \left\{y \in R_{\mathrm{rk}(x)} : y \sim x \wedge \forall z \left(z \sim y \implies \mathrm{rk}(y) \leq \mathrm{rk}(z)\right)\right\}.$$

The set $\mathcal{F}(x)$ consists of the sets of minimal rank that are equivalent to x. The operation $\mathcal{F}(x)$ is compatible with the property $\sim$.

Proof. The definition of $\mathcal{F}$ is sound according to Theorem A3.3. $\mathcal{F}(x)$ is not empty, since there is a $y \in R_{\mathrm{rk}(x)}$ such that $y \sim x$; for example, $y = x$ is such. If $\mathcal{F}(x) = \mathcal{F}(y)$, then $x \sim y$, since for an arbitrary $z \in \mathcal{F}(x)$ we have $z \sim x$ and $z \sim y$. If $x \sim y$, then, again by the transitivity of the equivalence, for an arbitrary z we have $z \sim x \iff z \sim y$. Hence $z \in \mathcal{F}(x) \iff z \in \mathcal{F}(y)$.

*   *

*

Finally we show that in ZF (with the Axiom of Regularity) the concept of ordinal has a simpler definition than the one given earlier.

Theorem A7.4. $\mathrm{Ordinal}(x) \iff \mathrm{Trans}(x) \wedge \forall y \in x \ \mathrm{Trans}(y).$

Proof. As each element of an ordinal is also an ordinal, it is clear that the left-hand side implies the right-hand side.

The reverse implication is proved by transfinite induction on the rank of x. Assume that the assertion is true for every set that has rank less than that of x, and, further, that $\mathrm{Trans}(x) \wedge \forall y \in x \ \mathrm{Trans}(y)$ holds. Then for each $y \in x$ we have $\mathrm{Trans}(y) \wedge \forall z \in y \ \mathrm{Trans}(z)$; thus y is an ordinal by the induction hypothesis. Hence x is transitive and consists of ordinals, that is, it is an ordinal itself.

A8. PROOFS OF RELATIVE CONSISTENCY.
THE METHOD OF INTERPRETATION

In the first part of the book, there were several assertions about which we said that they cannot be proved without the use of certain axioms. We now describe how a result concerning the unprovability of an assertion can be formulated precisely, and we will present two general methods that can be used to establish such results. According to the well-known Gödel Incompleteness Theorem, the *consistency* (i.e., of the state of being contradiction free) of ZF cannot be proved inside ZF. On the other hand, it is known that, in an axiom system with sufficient expressive power, it is possible to formulate its own consistency. This formula is denoted by $\text{Con}(\Sigma)$ for the axiom system Σ. As it is not possible to prove $\text{Con}(\text{ZF})$, it is possible to imagine a state of affairs such that one can prove, e.g., the formula $0 = 1$ in ZF. For this reason we need the following definition.

Definition A8.1. *Let* $L_0 \subset L$, $\Sigma \subset \text{Wff}(L)$, *and let* $\phi \in \text{Wff}(L)$. *We say that the formula is* relatively consistent *with* Σ *if the formula* $\text{Con}(\Sigma)$ *implies* $\text{Con}(\Sigma \cup \{\phi\})$. *The proof of such an implication is called a* relative consistency proof.

Most relative consistency proofs can be carried out in Peano Arithmetic, but we will be satisfied with proving the implication

$$\text{Con}(\Sigma) \implies \text{Con}(\Sigma \cup \{\phi\}).$$

in Σ itself. The simplest method used to prove relative consistency is the *method of interpretation*. We will illustrate this with the proof of

$$\text{Con}(\text{ZF}^*) \implies \text{Con}(\text{ZF}),$$

due to John von Neumann.

Definition A8.2. *Let* $H(x)$ *be a property formulated in* ZF^* *such that*

$$\text{ZF}^* \vdash \exists x \, H(x),$$

and let $E(x, y)$ *be a property with two variables. The pair* $H(x)$, $E(x, y)$, *which will be denoted by* Δ, *will be an interpretation of the language* L_0

of set theory in which the sets with property H(x) will be called sets in the model, and the property E(x, y) will be used to "interpret" the relation $\in$. To clarify, for an arbitrary $\phi \in \mathrm{Wff}(L_0)$ we define the interpretation ϕ^Δ of ϕ as follows: We introduce the variables $\bar{x}, \bar{y}, \ldots$, to run over sets x with property H(x). For arbitrary variables $x, y, \ldots$ in ϕ, the subformulas $x \in y$ and $x = y$ in ϕ will be replaced with $E(\bar{x}, \bar{y})$ and $\bar{x} = \bar{y}$, and the quantifiers $\forall x, \exists x$ in ϕ will be replaced with $\forall \bar{x}$ and $\exists \bar{x}$. We say that Δ interprets the axiom system $\Sigma \subset \mathrm{Wff}(L_0)$ if for each $\phi \in \Sigma$ the formula ϕ^Δ is provable in ZF*.

As we can see, the concept of interpretation is defined outside the formal system. We are now ready to establish the following.

Theorem A8.1. *Assume Δ interprets the system*

$$\mathrm{ZF} \subset \mathrm{Wff}(L_0)$$

of formulas in ZF*. *Then the implication*

$$\mathrm{Con}(\mathrm{ZF}^*) \implies \mathrm{Con}(\mathrm{ZF})$$

is provable in ZF*.

The proof, easily given by using standard ideas in mathematical logic, will not be presented here. It should be intuitively clear that if one can derive a contradiction from ZF, then this contradiction can also be derived from the formulas $\{\phi^\Delta : \phi \in \mathrm{ZF}\}$. As these formulas can be proved in ZF*, this means that a contradiction can be derived from ZF* alone. Furthermore, if we have an algorithm for proving the formulas $\{\phi^\Delta : \phi \in \mathrm{ZF}\}$ in ZF*, then this algorithm also gives an algorithm to prove the relative consistency of ZF. Finally, if Δ interprets ZF, and certain notions (such as properties, operations, and constants) can be defined in ZF, then each of these concepts can also be interpreted, and its interpretation will satisfy the interpretation of its definition.

Theorem A8.2. $\mathrm{Con}(\mathrm{ZF}^*) \implies \mathrm{Con}(\mathrm{ZF})$.

Proof. Let $H(x) = R(x)$ ($R(x)$ was defined in Definition A7.1), and $E(x, y) = x \in y$. According to Theorem A8.1, we have to prove the formulas $\mathbf{A}_0^\Delta, \ldots, \mathbf{A}_5^\Delta, \mathbf{A}_7^\Delta, \mathbf{A}_6^\Delta(\phi)$ in ZF*. In what follows, it will be convenient to introduce the notion of *class*.

Given an arbitrary property $B(x)$ with x among its variables, instead of $B(x)$, we will often write $x \in B$ and we call B a *class*. Operations whose natural definitions designate only sets as elements of its values will from now on be considered as defined also for classes. For example, if the class B consists of elements x with property $B(x)$, then $\bigcup B$ consists of elements x with property $\exists y (x \in y \wedge B(y))$. Some operations cannot be defined in this way; for example, the operation $\{B\}$ cannot be defined in this way since this would require that B be an element of the class $\{B\}$, whereas only sets can be elements of classes, according to the way we introduced classes above. We will also use the self-explanatory symbols

$$B \cap B', \qquad B \cup B', \qquad B \subset B', \qquad \bigcup B$$

Accordingly, we will write $x \in R$ instead of $R(x)$. In Theorem A7.1 we proved that $x \subset R$ holds for each $x \in R$. We may call R a *transitive class*. In what follows, a number of assertions remain valid for any interpretation Δ in which H is a transitive class and $x \in y = E(x, y)$. Interpretations of this kind are called *transitive interpretations*.

Next we study the interpretation of each of the axioms.

$$\mathbf{A}_0^\Delta \equiv \exists \bar{x} \forall \bar{u} \, (\bar{u} \notin \bar{x}).$$

We have $\emptyset \in R$, $\forall u \, (u \notin \emptyset)$ and so, *a fortiori*,

$$\forall \bar{u} \, (\bar{u} \notin \emptyset),$$

that is, $\emptyset$ satisfies $\mathbf{A}_0^\Delta$.

$$\mathbf{A}_1^\Delta \equiv \forall \bar{x} \forall \bar{y} \, \big(\forall \bar{u} \, (\bar{u} \in \bar{x} \iff \bar{u} \in \bar{y}) \iff \bar{x} = \bar{y} \big).$$

Let $\bar{x}, \bar{y} \in R$. Then $\bar{x}, \bar{y} \subset R$, and so

$$\forall \bar{u} \, (\bar{u} \in \bar{x} \iff \bar{u} \in \bar{y}) \iff \forall u \, (u \in \bar{x} \iff u \in \bar{y}).$$

Hence $\mathbf{A}_1^\Delta$ follows from $\mathbf{A}_1$.

$$\mathbf{A}_2^\Delta \equiv \forall \bar{x} \forall \bar{y} \exists \bar{z} \forall \bar{u} \, (\bar{u} \in \bar{z} \iff \bar{u} = \bar{x} \vee \bar{u} = \bar{y}).$$

Let $\bar{x}, \bar{y} \in R$. Then $\{\bar{x}, \bar{y}\} \in R$ according to Assertion 5 of Theorem A7.1. For $\bar{z} = \{\bar{x}, \bar{y}\}$, we have

$$\forall u \, (u \in \bar{z} \iff u = \bar{x} \vee u = \bar{y}).$$

A fortiori, we have

$$\forall \bar{u} \, (\bar{u} \in \bar{z} \iff \bar{u} = \bar{x} \vee \bar{u} = \bar{y}).$$

$$\mathbf{A}_3^\Delta \equiv \forall \bar{x} \exists \bar{y} \forall \bar{u} \, (\bar{u} \in \bar{y} \iff \bar{u} \subset^\Delta \bar{x}).$$

For $\bar{x}, \bar{u} \in R$ we have $\bar{x}, \bar{u} \subset R$; hence

$$\bar{u} \subset^\Delta \bar{x} \equiv \forall \bar{v} \, (\bar{v} \in \bar{u} \implies \bar{v} \in \bar{x}) \equiv \forall v \, (v \in \bar{u} \implies v \in \bar{x}) \equiv \bar{u} \subset \bar{x}.$$

According to the definition of R, we have $\mathrm{P}(\bar{x}) \in R$. Write $\bar{y} = \mathrm{P}(\bar{x})$. We have

$$\forall u \, (u \in \bar{y} \iff u \in \bar{x}),$$

and so, *a fortiori*,

$$\forall \bar{u} \, (\bar{u} \in \bar{y} \iff \bar{u} \in \bar{x}).$$

$$\mathbf{A}_4^\Delta \equiv \forall \bar{x} \exists \bar{y} \forall \bar{u} \, (\bar{u} \in \bar{y} \iff \exists \bar{v} \, (\bar{u} \in \bar{v} \wedge \bar{v} \in \bar{x})).$$

Let $\bar{x} \in R$; then $\bigcup \bar{x} \subset R$ according to Assertion 1 of Theorem A7.1, and so $\bigcup \bar{x} \in R$ according to Assertion 5 of the same theorem. Write $\bar{y} = \bigcup \bar{x}$. We have

$$\forall u \, (u \in \bar{y} \iff \exists v \, (u \in v \wedge v \in \bar{x})).$$

Hence, *a fortiori*,

$$\forall \bar{u} \, (\bar{u} \in \bar{y} \iff \exists v \, (\bar{u} \in v \wedge v \in \bar{x})).$$

But

$$\exists v \, (\bar{u} \in v \wedge v \in \bar{x})) \iff \exists \bar{v} \, (\bar{u} \in \bar{v} \wedge \bar{v} \in \bar{x})$$

as $\bar{x} \subset R$.

$$\mathbf{A}_5^\Delta \equiv \exists \bar{x} \, (\emptyset^\Delta \in \bar{x} \wedge \forall \bar{u} \, (\bar{u} \in \bar{x} \implies \bar{u} \cup^\Delta \{\bar{u}\}^\Delta \in \bar{x})).$$

It is easy to show by transfinite recursion that $\alpha \in R$ holds for every ordinal α. Thus $\omega \in R$. We are going to show that $\bar{x} = \omega$ satisfies $\mathbf{A}_5^\Delta$.

We saw in the proof of $\mathbf{A}_1^\Delta$ that $\emptyset \in R$ and $\emptyset^\Delta = \emptyset$. If $\bar{u} \in R$, then we have

$$\bar{u} \cup \{\bar{u}\} \in R$$

by Assertion 5 of Theorem A7.1. By a calculation similar to the calculations above, it is easy to show that

$$\bar{u} \cup^\Delta \{\bar{u}\}^\Delta = \bar{u} \cup \{\bar{u}\};$$

in particular, the left-hand side here is well defined. Thus ω indeed satisfies $\mathbf{A}_5^\Delta$.

$$\mathbf{A}_7^\Delta \equiv \forall \bar{x} \, (\bar{x} \neq \emptyset^\Delta \implies \exists \bar{y}(\bar{y} \in \bar{x} \wedge \bar{y} \cap^\Delta \bar{x} = \emptyset^\Delta)).$$

Let $\bar{x} \in R$, and assume that $\bar{x} \neq \emptyset^\Delta$. Then, clearly, $\bar{x} \neq \emptyset$, since $\emptyset^\Delta = \emptyset$. Let y be an element of $\bar{x}$ of minimal rank. Then, by Assertion 4 of Theorem A7.1 we have $y \cap \bar{x} = \emptyset$. For this y, we have $y \in R$; hence $y = \bar{y}$, and

$$\bar{y} \cap^\Delta \bar{x} = y \cap \bar{x} = \emptyset.$$

$\mathbf{A_6^\Delta}$: Let $\phi \in \mathrm{Wff}(L_0)$, assume all variables of ϕ are among $u, v, x_1, \ldots, x_n$, and let x, y be variables different from these.

$$\mathbf{A_6^\Delta}(\phi) \equiv \forall \bar{x}_1 \ldots \forall \bar{x}_n \left(\forall \bar{u} \exists! \bar{v} \, \phi(\bar{u}, \bar{v}, \bar{x}_1, \ldots, \bar{x}_n) \implies \right.$$

$$\left. \forall \bar{x} \exists! \bar{y} \forall \bar{v} \left(\bar{v} \in \bar{y} \iff \exists \bar{u} \left(\bar{u} \in \bar{x} \wedge \phi(\bar{u}, \bar{v}, \bar{x}_1, \ldots, \bar{x}_n) \right) \right) \right).$$

Assume $\bar{x}_1, \ldots, \bar{x}_n, \bar{x} \in R$ are given, and assume that

$$\forall \bar{u} \exists! \bar{v} \, \phi^\Delta(\bar{u}, \bar{v}, \bar{x}_1, \ldots, \bar{x}_n)$$

holds. Then, according to $\mathbf{A_6}(\phi^\Delta)$, there is a y such that

$$y = \{ \bar{v} : \bar{u} \in \bar{x} \wedge \phi^\Delta(\bar{u}, \bar{v}, \bar{x}_1, \ldots, \bar{x}_n) \}.$$

According to Assertion 5 of Theorem A7.1, we have $y \in R$, and so $\bar{y} = y$ satisfies the axiom.

*   *
*

The theorem just proved shows that the Axiom of Regularity does not lead to a contradiction if there was no contradiction in ZF* itself. On the other hand, the "natural" model R of the Axiom of Regularity admits sufficiently many sets, so that its adoption does not represent a significant restriction. For this reason, in the current literature, and especially in the part of literature concerned with axiomatic questions, the Axiom of Regularity is almost always assumed.

A9. PROOFS OF RELATIVE CONSISTENCY. THE METHOD OF MODELS

Let $L_0 \subset L'$, $\mathrm{ZF} \subset \mathrm{ZF}' \subset \mathrm{Wff}(L')$. It may turn out that we can prove the consistency of a system Σ in the axiom system ZF' by specifying a set model satisfying Σ. We will illustrate this method by presenting a sketch of the proof of the following theorem; this theorem was known in its essentials even to Zermelo.

Theorem A9.1. *The following is provable in ZF: If $\kappa > \omega$ is a strongly inaccessible cardinal, then the axioms of ZF hold on the structure*

$$\mathfrak{A} = \langle R_\kappa, \in \restriction R_\kappa \rangle$$

(written, somewhat informally, also as $\langle R_\kappa, \in \rangle$). In ZFC, it can be proved that the Axiom of Choice also holds in $\langle R_\kappa, \in \rangle$.

According to Gödel's Completeness Theorem, this means that the consistency of ZF is provable in the axiom system

$$\mathrm{ZF}_1 = \mathrm{ZF} \cup \{\exists \kappa > \omega \, (\kappa \text{ is strongly inaccessible})\}.$$

Therefore, in view of Gödel's Second Incompleteness Theorem, in ZF one cannot prove the existence of strongly inaccessible cardinals greater than ω.

Thus we obtained a relative consistency proof; namely we established that

$$\mathrm{Con}(\mathrm{ZF}) \implies \mathrm{Con}\big(\mathrm{ZF} \cup \{\neg \exists \kappa > \omega \, (\kappa \text{ is strongly inaccessible}\}\big).$$

A more precise formulation of Theorem A9.1 can be given as follows. In the axiom system ZF, we redefine the syntactical concepts of language, formula, term, etc., for a language $\tilde{L}_0$ that contains the only relation symbol $\tilde{\in}$ in addition to the symbol for equality. We denote by $\widetilde{\mathrm{ZF}}$ the set of formulas in this language corresponding to the formulas $\mathbf{A}_0, \ldots, \mathbf{A}_6(\phi), \mathbf{A}_7$. Then we define what it means for a formula to be true or false in a structure in the usual way, and the theorem then says that

$$\langle R_\kappa, \in \rangle \models \phi \qquad \text{for any} \qquad \phi \in \widetilde{\mathrm{ZF}}.$$

In this form the result is not a metatheorem, since the property

$$\langle R_\kappa, \in \rangle \models \phi$$

is definable in ZF.

Proof of Theorem A9.1. We can proceed in close similarity to the proof of Theorem A8.2. Denoting by $\bar{x}, \bar{y}, \ldots$ the variables running over elements of R_κ; noting that R_κ is a transitive set; and noting that, κ being a limit ordinal, for each $x \in R_\kappa$ we also have $P(x) \in R_\kappa$, the proofs of the assertions $R_\kappa \models A_i$ for $i = 0, 1, 2, 3, 4, 5, 7$ can be almost word for word copied from the proof of Theorem A8.2.

Next, by using transfinite induction on α, we prove that we have $|R_\alpha| < \kappa$ for $\alpha < \kappa$. Indeed, if $\alpha = \beta + 1$, then

$$|R_{\beta+1}| \le 2^{|R_\beta|} < \kappa,$$

by the induction hypothesis and by κ being strongly inaccessible. If α is a limit ordinal, then

$$|R_\alpha| \le \sum_{\beta < \alpha} |R_\beta| < \kappa,$$

as κ is a regular cardinal. Consider

$$A_6(\phi) \in \text{ZF} \quad \text{for} \quad \phi = \phi(u, v, x_1, \ldots, x_n),$$

and assume that for arbitrary $\bar{x}_1, \ldots, \bar{x}_n$ we have

$$\langle R_\kappa, \in \rangle \models \forall u \exists! v \, \phi(u, v, \bar{x}_1, \ldots, \bar{x}_n).$$

Then there is a function f with $D(f) = R_\kappa$, $R(f) \subset R(\kappa)$ such that

$$\langle R_\kappa, \in \rangle \models \phi\big(\bar{u}, f(\bar{u}), \bar{x}_1, \ldots, \bar{x}_n\big)$$

holds for each $\bar{u} \in R_\kappa$. Then

$$\text{rk}\big(f(\bar{u})\big) < \kappa \quad \text{for each} \quad \bar{u} \in R_\kappa.$$

Now let $\bar{x} \in R_\kappa$ be arbitrary; then $\bar{x} \in R_\gamma$ for some $\gamma < \kappa$, and so $|\bar{x}| < \kappa$ according to what we said above. Hence

$$\alpha \overset{def}{=} \sup\{\text{rk}\big(f(\bar{u})\big) : \bar{u} \in \bar{x}\} < \kappa,$$

as κ is regular.

Write $y = \{f(\bar{u}) : \bar{u} \in \bar{x}\}$. Then $y \subset R_\alpha$, and so

$$y \in R_{\alpha+1} \subset R_\kappa.$$

Thus $\bar{y} = y$ satisfies the axiom.

Given $\bar{x} \in R_\kappa$ and a choice function f on $\bar{x}$, we have $f \in R_\kappa$, and so we can also prove the assertion

$$\langle R_\kappa, \in \rangle \models \mathbf{A_8}$$

in ZFC.

$$* \qquad *$$
$$*$$

We would like to point out that the assumption $\kappa > \omega$ was used only in order to prove that $\omega \in R_\kappa$, and so, that $\langle R_\kappa, \in \rangle \models \mathbf{A_5}$. That is, the proof above shows that $\langle R_\omega, \in \rangle \models \mathrm{ZF} \setminus \{\mathbf{A_5}\}$.

We would like to observe that the relative consistency of the nonexistence of strongly inaccessible cardinals can be proved by using the above ideas without any reference to the Incompleteness Theorems of Gödel. To this end, we only have to establish the following lemma.

Lemma A9.1. *If κ is the smallest strongly inaccessible cardinal greater than ω, then*

$$\langle R_\kappa, \in \rangle \models \text{There is no strongly inaccessible cardinal greater than } \omega.$$

The proof will be left to the reader.

The desire of proving the relative consistency of the existence of strongly inaccessible cardinals seems legitimate. It is, however, impossible to fulfill this desire.

Theorem A9.2. *If $\mathrm{ZFC} \cup \{\mathrm{Con}(\mathrm{ZFC})\}$ is consistent then*

$$\mathrm{Con}(\mathrm{ZFC}) \implies \mathrm{Con}\big(\mathrm{ZFC} \cup \{\exists \kappa > \omega \,(\kappa \text{ is strongly inaccessible})\}\big)$$

cannot be proved in ZFC.

Proof. Assume

$$\mathrm{ZFC} \vdash \mathrm{Con}(\mathrm{ZFC}) \implies$$
$$\mathrm{Con}\big(\mathrm{ZFC} \cup \{\exists \kappa > \omega \,(\kappa \text{ is strongly inaccessible})\}\big).$$

Then, according to Theorem A9.1, we have

$$\mathrm{ZFC} \cup \{\mathrm{Con}(\mathrm{ZFC})\} \vdash \mathrm{Con}\big(\mathrm{ZFC} \cup \{\mathrm{Con}(\mathrm{ZFC})\}\big).$$

This is, however, impossible in view of Gödel's Second Incompleteness Theorem.

$$* \qquad *$$
$$*$$

We point out that the existence of strongly inaccessible cardinals greater than ω is equiconsistent with the existence of inaccessible cardinals greater than ω.

Theorem A9.3. *If there is a structure $\langle M, \in \rangle$ such that $\langle M, \in \rangle \models$ ZFC, and for some element κ of M we have $\langle M, \in \rangle \models (\kappa$ is inaccessible), then there is an $M' \subset M$ such that*

$$\langle M', \in \rangle \models \text{ZFC} \qquad \text{and} \qquad \langle M', \in \rangle \models \kappa \text{ is strongly inaccessible.}$$

This theorem is an easy corollary of Gödel's results about *constructible sets*. A discussion of these results is beyond the scope of this book.

The assumption of the existence of strongly inaccessible cardinals is usually regarded as a natural strengthening of set theory. It follows from what was said above that it is not only impossible to prove their existence, it is also impossible to prove the consistency of their existence in a satisfactory way, at least by the present means of mathematics. This situation, however, does not differ very much from the situation that the proof of the consistency of the axiom system of set theory appears hopeless.

In the second part of the book, we will see that the assumption of the existence of inaccessible cardinals, or even much larger cardinals, plays a significant role in modern set theory. The role of large cardinals is even more important in certain other axiomatic investigations in set theory that are not the subject of the present book. In what follows, it will turn out that the answers to a number of questions whose formulations have nothing to do with large cardinals *per se* depend on the existence of large cardinals. Thus the study of these questions would not be possible without a thorough investigation of large cardinals.

PART II
TOPICS IN COMBINATORIAL
SET THEORY

12. STATIONARY SETS

In Definition 9.1 in the first part of the book, we discussed the concepts of filter and ultrafilter. The purpose of the present section is to describe a naturally arising filter on an ordinal ξ (as the underlying set) with $\mathrm{cf}(\xi) > \omega$.

Before we can indicate the significance of this filter, we need to introduce a few simple concepts.

Definition 12.1. Let X be a nonempty set, called the underlying set. For an arbitrary set $\mathcal{A} \subset \mathrm{P}(X)$, the set $\{X \setminus A : A \in \mathcal{A}\}$ is denoted as $\mathrm{co}(\mathcal{A})$, and is called the dual of $\mathcal{A}$.

The dual of a filter on an underlying set X is called an ideal on X. We will also give a direct definition of ideals.

Definition 12.2. Let X be a nonempty set. A set system $\mathcal{I} \subset \mathrm{P}(X)$ is called an ideal on X if it is not empty and satisfies the following requirements.
1. $X \notin \mathcal{I}$.
2. If $A, B \in \mathcal{I}$ then $A \cup B \in \mathcal{I}$.
3. If $A \in \mathcal{I}$ and $B \subset A$ then $B \in \mathcal{I}$.

It is clear that $\mathcal{F} \subset \mathrm{P}(X)$ is a filter on X if and only if $\mathrm{co}(\mathcal{F})$ is an ideal.

Our definition above slightly deviates from the one used in the literature. In the definition of ideal, it is not customary to include Condition 1. Set systems satisfying Conditions 1, 2, 3 are usually called *proper ideals*. We find our definition more useful.

Definition 12.3. The set system $\mathcal{I} \subset \mathrm{P}(X)$ is called a prime ideal on X if it is an ideal and there is no ideal $\mathcal{I}' \subset \mathrm{P}(X)$ on X for which $\mathcal{I} \subsetneq \mathcal{I}'$.

It is easy to verify that $\mathcal{I} \subset \mathrm{P}(X)$ is a prime ideal on X if and only if $\mathrm{co}(\mathcal{I})$ is an ultrafilter; indeed, it is clear from the proof of Theorem 9.4 that an ultrafilter is a maximal filter.

Definition 12.4. The ideal $\mathcal{I} \subset \mathrm{P}(X)$ is called a principal ideal on X if $\bigcup \mathcal{I} \in \mathcal{I}$.

Definition 12.5. Let $\kappa \geq \omega$ be a cardinal. A filter $\mathcal{F} \subset \mathrm{P}(X)$ is called a κ-complete filter if for every set system $\mathcal{F}' \subset \mathcal{F}$ with $|\mathcal{F}'| < \kappa$ we have $\bigcap \mathcal{F}' \in \mathcal{F}$.

Definition 12.6. *Let $\kappa \geq \omega$ be a cardinal. An ideal $\mathcal{I} \subset P(X)$ is called a κ-complete ideal if for every set system $\mathcal{I}' \subset \mathcal{I}$ with $|\mathcal{I}'| < \kappa$ we have $\bigcup \mathcal{I}' \in \mathcal{I}$.*

It is clear that $\mathcal{F} \in P(X)$ is a κ-complete filter if and only if $\text{CO}(\mathcal{F})$ is a κ-complete ideal on X.

Filters and ideals that are ω_1-complete are also called *σ-complete* by tradition. A σ-complete ideal is also called a *σ-ideal*.

Example: If $X = \mathbb{R}$, then the sets of Lebesgue measure zero form a σ-complete ideal; similarly for the sets of first category.

As can be seen from these examples, the elements of an ideal $\mathcal{I}$ can be considered "small" subsets of the underlying set; the elements of the dual filter $\text{CO}(\mathcal{I})$, "large" subsets; and those subsets that are not elements of the ideal, "not small" subsets.

The successful applications of the Lebesgue measure and the Baire category show that the σ-ideals obtained in a natural way have great significance. In spite of this, only at a relatively late stage in the evolution of set theory was the theory of stationary sets developed, in its definitive form in 1956, as a result of the theorem of G. Fodor. This theory is of fundamental importance in set theory today, and it significantly simplifies the proofs of a number of results originally obtained prior to the existence of this theory.

Definition 12.7. *Given an arbitrary ordinal ξ, we call a set $A \subset \xi$ closed in ξ if for each $\eta < \xi$ and for each nonempty $B \subset A \cap \eta$ we have $\sup B = \bigcup B \in A$.*

This is exactly the same as saying that A is closed in the ordering topology of the set $\langle \xi, \in \rangle$. For this reason, if there is no danger of misunderstanding, we may say shortly that A is a closed subset of (or in) ξ, or, shortly, A is ξ-closed, or, simply, closed.

Definition 12.8. *Given an arbitrary ordinal ξ, the set $A \subset \xi$ is called a closed unbounded set in ξ or, shortly, a club in ξ, or a ξ-club if A is closed in ξ and it is cofinal in ξ. The set system*

$$\{A \subset \xi : \exists B \subset A \, (B \text{ is a club in } \xi)\}$$

is denoted by $\mathcal{C}(\xi)$. The elements of $\mathcal{C}(\xi)$ are called ξ-large sets.

As one can surmise, Definition 12.8 is not of much use unless we impose some restrictions on the ordinal ξ; for example, every cofinal subset of ω is a club in ω.

It is easy to see that $\mathcal{C}(\xi)$ is not empty for an arbitrary $\xi > 0$; in fact, if $\eta < \xi$, then

$$\xi \setminus \eta \in \mathcal{C}(\xi).$$

In what follows we will denote by $\text{Lim}(\xi)$ the *set of limit ordinals* less than ξ. It is easily seen that, for example, $\text{Lim}(\omega_1) \in \mathcal{C}(\omega_1)$. This example shows that there is a club in ω_1 the complement of which has cardinality $\aleph_1$; in fact, the set $\{\eta + 1 : \eta < \omega_1\}$ clearly has cardinality $\aleph_1$.

The following theorem underlines the importance of our definition for $\mathrm{cf}(\xi) > \omega$.

Theorem 12.1. *Assume ξ is a limit ordinal with $\mathrm{cf}(\xi) > \omega$. Then $C(\xi)$ is a $\mathrm{cf}(\xi)$-complete filter.*

Proof. It is sufficient to show that, given a system $\mathcal{A} \subset \mathrm{P}(\xi)$ of clubs in ξ with $|\mathcal{A}| < \mathrm{cf}(\xi)$, the set $\bigcap \mathcal{A}$ is a club in ξ. The assertion that $\bigcap A$ is a closed set is true for the intersection of closed sets in any topological space; even without any reference to topology, the assertion can easily be verified directly.

We claim that $\bigcap \mathcal{A}$ is cofinal in ξ. Let $\eta < \xi$. By recursion on n, we are going to define a sequence $\{\eta_n : n \in \omega\}$ of ordinals.

Put $\eta_0 = \eta$. Let $n \in \omega$, and assume that we have defined η_n is such a way that $\eta_n < \xi$. Pick a set

$$A_n \subset \xi \setminus \eta_n, \qquad A_n \neq \emptyset$$

such that we have

$$|A_n| < \mathrm{cf}(\xi) \qquad \text{and} \qquad A \cap A_n \neq \emptyset \quad \text{for every} \quad A \subset \mathcal{A}.$$

This is clearly possible, since $A \setminus \eta_n \neq \emptyset$ for $A \in \mathcal{A}$, and $|\mathcal{A}| < \mathrm{cf}(\xi)$. Put

$$\eta_{n+1} = \sup A_n + 1.$$

As $|A_n| < \mathrm{cf}(\xi)$, we have $\eta_{n+1} < \xi$. This completes the definition of the sequence

$$\{\eta_n : n < \omega\}.$$

Clearly

$$\eta = \eta_0 < \cdots < \eta_n < \cdots < \xi.$$

Put

$$\tau = \sup\{\eta_n : n \in \omega\}.$$

Then $\eta \leq \tau < \omega$; the latter inequality holds since $\mathrm{cf}(\xi) > \omega$.

Let $A \in \mathcal{A}$. According to the construction, we have

$$A \cap (\eta_{n+1} \setminus \eta_n) \neq \emptyset \qquad \text{for} \qquad n \in \omega,$$

and so

$$\sup(A \cap \tau) = \tau.$$

As A, being an element of $\mathcal{A}$, is a club in ξ, this implies that $\tau \in A$; thus $\tau \in \bigcap \mathcal{A}$.

*   *
*

Definition 12.9. *The set $A \subset \xi$ is called ξ-stationary if $\xi \setminus A \notin \mathcal{C}(\xi)$, that is, if A intersects every ξ-club.*

For historical reasons, we will introduce the following, somewhat redundant notation. The system of ξ-stationary sets is denoted by $\mathrm{Stat}(\xi)$, the set system $\mathrm{CO}(\mathcal{C}(\xi))$, by $\mathrm{NS}(\xi)$. The elements of $\mathrm{NS}(\xi)$ are called ξ-nonstationary sets. Instead of ξ-stationary set and ξ-nonstationary set, we will often say stationary set in ξ and nonstationary set in ξ, respectively, or, somewhat less precisely, simply stationary set and nonstationary set.

Note that, as $\xi \setminus \eta$ is a ξ-club for each $\xi < \eta$, every ξ-stationary subset of an ordinal ξ is cofinal in ξ. As we will see below, the concept of ξ-stationary sets is useful mainly if $\mathrm{cf}(\xi) > \omega$, although on at least one occasion (in Definition 20.5 below) the concept will be useful even if $\mathrm{cf}(\xi) = \omega$. We contemplate no use for stationary sets when $\mathrm{cf}(\xi) < \omega$. Anyway, it is easy to see that, in the case when $0 < \mathrm{cf}(\xi) \leq \omega$, a subset of ξ is ξ-stationary if and only if its complement is bounded (i.e., not cofinal) in ξ.

Theorem 12.1 leads to the following.

Corollary 12.1. *If ξ is a limit ordinal with $\mathrm{cf}(\xi) > \omega$, then $\mathrm{NS}(\xi)$ is a $\mathrm{cf}(\xi)$-complete ideal in ξ.*

As we mentioned already, $\mathrm{Lim}(\omega_1)$ is an ω_1-club. Hence $B = \{\eta + 1 : \eta < \omega_1\}$ is a set of cardinality $\aleph_1$ that is not stationary in ω_1. In what follows, we would like to give a characterization of stationary sets. To this end, we need the following definitions.

Definition 12.10. *Let ξ be an ordinal, and let $\langle B_\alpha : \alpha < \xi \rangle$ be a sequence of order type ξ of subsets of ξ. The set*

$$\left\{ \alpha < \xi : 0 < \alpha \in \bigcap_{\eta < \alpha} B_\eta \right\}$$

is denoted as $\Delta_{\alpha < \xi} B_\alpha$, and is called the diagonal intersection of the sequence *$\langle B_\alpha : \alpha < \xi \rangle$.*

Definition 12.11. *The filter $\mathcal{F} \subset \mathrm{P}(\xi)$ is called a normal filter if $\mathcal{F}$ is closed with respect to diagonal intersection.*

The requirement in the definition says that if each element of a sequence of length ξ belongs to $\mathcal{F}$, then the diagonal intersection of this sequence also belongs to $\mathcal{F}$. For a regular cardinal $\kappa > \omega$ replacing ξ, Theorem 12.1 can be sharpened as follows.

Theorem 12.2. *Let $\kappa > \omega$ be a regular cardinal, and let $\{B_\alpha : \alpha \in \kappa\} \subset \mathrm{P}(\kappa)$ be a sequence of order type κ of κ-clubs. Then $\Delta_{\alpha<\kappa} B_\alpha$ is also a κ-club.*

Proof. We are first going to show that

$$B = \underset{\alpha<\kappa}{\Delta} B_\alpha$$

is κ-closed (i.e., closed in κ). Assume that we have

$$\emptyset \neq B' \subset B, \qquad B' \subset \eta$$

for some $\eta < \kappa$. Denote the union $\bigcup B' = \sup B'$ by ξ. We have to show that $\xi \in B$. If ξ is not a limit ordinal, then $\xi \in B' \subset B$. For this reason, we may assume that ξ is a limit ordinal. Thus we have $\alpha + 1 < \xi$ for $\alpha < \xi$; consider a fixed $\alpha < \xi$. Then

$$\sup\big(B' \setminus (\alpha + 1)\big) = \sup\big((\xi \setminus (\alpha + 1)) \cap B'\big) = \xi.$$

On the other hand, by the definition of diagonal intersection, we have

$$B' \setminus (\alpha + 1) \subset B_\alpha \cap \eta.$$

B_α is closed, and so

$$\sup\big(B' \setminus (\alpha + 1)\big) = \xi \in B_\alpha.$$

This holds for every $\alpha < \xi$, and so

$$\xi \in \bigcap_{\alpha<\xi} B_\alpha.$$

Hence $\xi \in B$ as we wanted to show.

Next we claim that B is cofinal in κ. We proceed similarly as in the proof of Theorem 12.1.

Let $\eta < \kappa$. By recursion on n we define a sequence $\{\eta_n : n \in \omega\}$ of ordinals. Let $\eta_0 = \eta$. Assume that we have already defined $\eta_n < \kappa$ for some n. Let

$$A_n \subset \kappa \setminus \eta_n$$

be a set such that

$$A_n \cap B_\alpha \neq 0$$

for each $\alpha < \eta_n$, and $|A_n| < \kappa$. It is possible to choose A_n in such a way, since $\eta_n < \kappa$ and

$$B_\alpha \cap (\kappa \setminus \eta_n) \neq \emptyset \qquad \text{for} \qquad \alpha < \eta_n.$$

Let $\eta_{n+1} = \sup A_n + 1$. Then $\eta_{n+1} < \kappa$, as κ is a regular cardinal. This completes the definition of the sequence $\{\eta_n : n < \omega\}$.

It is clear that

$$\eta = \eta_0 < \cdots < \eta_n < \cdots < \kappa.$$

Write $\tau = \sup\{\eta_n : n \in \omega\}$. As $\omega < \kappa = \mathrm{cf}(\kappa)$, we can see that

$$\eta \leq \tau < \kappa.$$

We claim that $\tau \in B$. To show this, let $\alpha < \tau$. Then there is an $n \in \omega$ such that $\alpha < \eta_n$. According to the construction, we have

$$B_\alpha \cap (\eta_{m+1} \setminus \eta_m) \neq \emptyset \qquad \text{whenever} \qquad n \leq m < \omega.$$

Thus

$$\sup B_\alpha \cap \tau = \tau \in B_\alpha,$$

as B_α is closed in κ. Since $\alpha < \tau$ was arbitrary, $\tau \in B$ also follows, as claimed.

*　　*
*

The following definitions and Lemma 12.1 below make it possible to apply Theorem 12.2 to the ideal $\mathrm{NS}(\kappa)$.

Definition 12.12. *Let A be a set of ordinals. Let f be an ordinal-valued function with $A \subset \mathrm{D}(f)$. We say that f is a regressive function (or, a pressing-down function) on the set A if we have $f(\alpha) < \alpha$ for each $\alpha \in A$ with $\alpha \neq 0$, and we have $f(0) = 0$ if $0 \in A$.*

Definition 12.13. *Let ξ be an arbitrary ordinal. The ideal $\mathcal{I} \subset \mathrm{P}(\xi)$ on ξ is called a normal ideal on ξ (or in $\mathrm{P}(\xi)$) if, for each subset $A \notin \mathcal{I}$ of ξ and for every regressive function f on A, there is an ordinal $\rho < \xi$ such that $f^{-1}(\{\rho\}) \notin \mathcal{I}$.*

Lemma 12.1. *Let ξ be an arbitrary ordinal and $\mathcal{F} \subset \mathrm{P}(\xi)$, a filter. The filter $\mathcal{F}$ is a normal filter if and only if $\mathcal{I} = \mathrm{co}(\mathcal{F})$ is a normal ideal.*

Proof. First we are going to show that if the filter $\mathcal{F}$ is a normal filter, then $\mathcal{I} = \mathrm{co}(\mathcal{F})$ is a normal ideal.

To this end, let $A \notin \mathcal{I} = \mathrm{co}(\mathcal{F})$ and let f be a regressive function on A. Then we will show that there is $\alpha \in \xi$ such that $f^{-1}(\{\alpha\}) \notin \mathcal{I}$.

Assume the contrary. Then, for every $\alpha < \xi$, we have

$$\xi \setminus f^{-1}(\{\alpha\}) = B_\alpha \in \mathcal{F}.$$

As $\mathcal{F}$ is closed with respect to diagonal intersection, we have

$$B = \mathop{\Delta}_{\alpha < \xi} B_\alpha \in \mathcal{F}.$$

We claim that
$$B \cap A = \emptyset.$$

We have $0 \notin B$, since B is a diagonal intersection. Given $0 < \eta \in A$, we have $f(\eta) < \eta$ and $\eta \notin B_{f(\eta)}$. Hence $\eta \notin B$, verifying our claim. We thus obtained that
$$A \subset \xi \setminus B,$$

and so $A \in \mathcal{I}$. This contradicts the assumption that $A \notin \mathcal{I}$.

Next we are going to show that if $\mathcal{F}$ is a filter such that $\mathcal{I} = \mathrm{co}(\mathcal{F})$ is a normal ideal, then $\mathcal{F}$ is closed with respect to diagonal intersection. Let $B_\alpha \in \mathcal{F}$, $\alpha < \xi$. Assume, on the contrary, that

$$B = \mathop{\Delta}_{\alpha < \xi} B_\alpha \notin \mathcal{F}, \qquad \text{that is} \qquad A \overset{def}{=} \xi \setminus B \notin \mathcal{I}.$$

We define a regressive function on A. According to the definition of B, for each $\eta \in A$ there is an $\alpha < \eta$ for which $\eta \notin B_\alpha$. Let

$$f(\eta) = \min\{\alpha : \alpha < \eta \wedge \eta \notin B_\alpha\} \qquad \text{for} \qquad \eta \in A.$$

Then f is a regressive function on A. As $\mathcal{I}$ is a normal ideal and $A \notin \mathcal{I}$, there exists an $\alpha < \xi$ such that

$$f^{-1}(\{\alpha\}) \notin \mathcal{I}.$$

Let $C = f^{-1}(\{\alpha\})$. As we just saw, $C \notin \mathcal{I}$. If $\eta \in C$, then $f(\eta) = \alpha$. Thus $B_\alpha \cap C = \emptyset$. This contradicts the assumption $B_\alpha \in \mathcal{F}$.

*   *
*

Fodor's Theorem 12.3. *If $\kappa > \omega$ is a regular cardinal, then* $\mathrm{NS}(\kappa)$ *is a normal ideal.*

The result was established by G. Fodor [F] in 1956. Several authors call this result the Pressing-Down Lemma; they also prefer to call pressing-down functions what we called regressive functions.

Proof. According to Theorem 12.2, the filter $\mathcal{C}(\kappa)$ is closed with respect to diagonal intersection. So, according to Lemma 12.1, $\mathrm{NS}(\kappa) = \mathrm{co}(\mathcal{C}(\kappa))$ is a normal ideal.

*   *
*

We would like to note that it is not difficult to give a direct proof, one not relying on the concept of diagonal intersection, of Fodor's Theorem. We followed the approach described above, since in this way Theorems 12.2 and

12.3 can be proved similarly to the simpler Theorem 12.1. Fodor described his theorem in 1956 in terms of stationary sets and regressive functions. The concept of normal ideal was introduced by D. Scott in 1961.

In what follows, we summarize the properties discussed so far of the set systems $\mathcal{C}(\kappa)$, $\text{Stat}(\kappa)$, and $\text{NS}(\kappa)$.

If $\kappa > \omega$ is a regular cardinal, then, according to Theorems 12.1 and 12.2, $\mathcal{C}(\kappa)$ is a κ-complete filter closed with respect to diagonal intersection, i.e., it is a normal filter, and, according to Theorem 12.3, $\text{NS}(\kappa) = \text{CO}(\mathcal{C}(\kappa))$ is a normal ideal that, clearly, includes all subsets of cardinality less than κ of κ.

Furthermore, the following assertions are straightforward:

1. If $A \in \text{Stat}(\kappa)$, then $|A| = \kappa$.
2. If $B \in \mathcal{C}(\kappa)$, then $B \in \text{Stat}(\kappa)$.
3. If $B \in \mathcal{C}(\kappa)$ and $A \in \text{Stat}(\kappa)$, then $A \cap B \in \text{Stat}(\kappa)$.

The first assertion easily follows from the remark made just before, and the second assertion is obvious. As for the third assertion, it can be seen as follows. We know that

$$A = (A \cap B) \cup (A \setminus B) \qquad \text{and} \qquad A \setminus B \in \text{NS}(\kappa).$$

Thus

$$A \cap B \notin \text{NS}(\kappa).$$

Notation: If X is a set and κ is a cardinal, then $[X]^{<\kappa}$ denotes the set $\{A \subset X : |A| < \kappa\}$, and $[X]^{\kappa}$, the set $\{A \subset X : |A| = \kappa\}$. Similarly, $[X]^{\leq \kappa}$ denotes the set $\{A \subset X : |A| \leq \kappa\}$.

In what follows, the next lemma will often be used, so we include a proof here.

Lemma 12.2. *If κ is a regular cardinal, $\mathcal{I}$ is a normal ideal on κ, and $[\kappa]^{<\kappa} \subset \mathcal{I}$, then $\mathcal{I}$ is also κ-complete.*

Proof. Let $\lambda < \kappa$, and let $\{A_\alpha : \alpha < \lambda\}$ be a set system such that

$$A_\alpha \in \mathcal{I} \qquad \text{and} \qquad A = \bigcup \{A_\alpha : \alpha < \lambda\}.$$

We use reduction ad absurdum. Assume that $A \notin \mathcal{I}$. As $\lambda < \kappa$, we have $\lambda \in \mathcal{I}$, and so from $A \notin \mathcal{I}$ we can conclude that $A \setminus \lambda \notin \mathcal{I}$. Define the function f on the set $A \setminus \lambda$ as follows: Let

$$f(\eta) = \min\{\alpha < \lambda : \eta \in A_\alpha\}$$

for each $\eta \in A \setminus \lambda$. Then f is a regressive function on the set $A \setminus \lambda$. As $A \setminus \lambda \notin \mathcal{I}$, by the normality of the ideal $\mathcal{I}$, there is an $\alpha < \lambda$ such that

$$f^{-1}(\{\alpha\}) \notin \mathcal{I}.$$

Then $A_\alpha \setminus \lambda \notin \mathcal{I}$ holds, which contradicts the assumption $A_\alpha \in \mathcal{I}$.

$$* \qquad *$$
$$*$$

In what follows, we will often use the following characterization of stationary sets.

Theorem 12.4 *(W. Neumer). Let $\kappa > \omega$ be a regular cardinal, and let $A \subset \kappa$. The set A is κ-stationary if and only if for every regressive function f defined on A, there is an $\alpha < \kappa$ such that*

$$|f^{-1}(\{\alpha\}) \cap A| = \kappa.$$

Proof. The "only if" part of the theorem follows from Fodor's Theorem, since every set stationary in κ has cardinality κ. To prove the converse, assume that A is not stationary in κ, and let B be a κ-club such that $A \cap B = \emptyset$. Define a function f on the set A as follows. For each $\xi \in A$ let

$$f(\xi) = \bigcup(B \cap \xi).$$

We show that this function fails to satisfy the requirements in the theorem. It is clear that $f(\xi) \leq \xi$; in fact, f is even regressive. To see this, note that if $B \cap \xi = \emptyset$, then $f(\xi) = 0$. Further, if $B \cap \xi \neq \emptyset$ and $\alpha = \sup(B \cap \xi)$, then we have $\alpha \in B$, as B is closed in κ. Thus $\alpha \notin A$, since $A \cap B = \emptyset$. Hence

$$\alpha = f(\xi) < \xi.$$

Let $\rho < \kappa$ be an arbitrary ordinal. As B is cofinal in κ, there is an $\eta \in B$ for which $\rho < \eta$. Then, for each $\xi > \eta$ we have $f(\xi) \geq \eta$. Hence

$$f^{-1}(\{\rho\}) \subset \eta + 1;$$

that is,

$$|f^{-1}(\{\rho\})| < \kappa.$$

Corollary 12.2. *If $\kappa > \omega$ is a regular cardinal, and $\mathcal{I} \subset \mathrm{P}(\kappa)$ is a normal ideal on κ with $[\kappa]^{<\kappa} \subset \mathcal{I}$, then $\mathrm{NS}(\kappa) \subset \mathcal{I}$; that is, $\mathrm{NS}(\kappa)$ is the smallest extension of $[\kappa]^{<\kappa}$ to a normal ideal.*

Proof. According to Lemma 12.2, the ideal $\mathcal{I}$ is κ-complete. If $A \in \mathrm{NS}(\kappa)$ then, by Neumer's Theorem, there is a regressive function f on A such that

$$f^{-1}(\{\rho\}) \in [\kappa]^{<\kappa} \qquad \text{for any} \qquad \rho < \kappa.$$

Then, however, we have $f^{-1}(\{\rho\}) \in \mathcal{I}$, and so $A \in \mathcal{I}$.

$$* \qquad *$$
$$*$$

It is worth pointing out that the following special case for $\kappa = \omega_1$ of the "only if" part of the above theorem was already proved in 1924 by P. S. Aleksandrov and P. S. Uryson: If f is a regressive function on ω_1, then there is an ordinal $\alpha < \omega_1$ such that $|f^{-1}(\{\alpha\})| = \omega_1$. They used this to prove that ω_1 with the ordering topology is a non-metrizable topological space.

From the theorems proved so far, one can see only that for $\kappa = \mathrm{cf}(\kappa) > \omega$ the elements of the set $\mathcal{C}(\kappa)$ are "large," and the nonstationary sets are "small." The next result indicates that among the stationary sets there are some that are neither small nor large.

Theorem 12.5 *(R. Solovay). Let $\kappa > \omega$ be a regular cardinal, and $A \subset \kappa$, a κ-stationary set. Then A can be split as the union of κ many, pairwise disjoint, κ-stationary sets.*

Even the case $\kappa = \omega_1$ of this theorem is not easy; in fact, it is not obvious that ω_1 can be split as the union of two disjoint ω_1 stationary sets. In this section, we prove only the particular case, stated as Theorem 12.5A, involving successor cardinals of this theorem. The proof we describe here gives a stronger statement, which will be used in the solutions of the exercises below. It also helps one to understand Solovay's proof given in Section 17. For another proof of Theorem 12.5A, see Theorem 17.2.

Theorem 12.5A. *Let $\kappa \geq \omega$, and let $A \subset \kappa^+$ be a κ^+-stationary set. Then A can be represented as the union of κ many, pairwise disjoint, κ-stationary sets.*

Before the proof, we give another definition.

Definition 12.14. *Let $\kappa > \omega$ be a regular cardinal, and let $A \subset \kappa$. We say that the function f is essentially bounded on the set A if there is an ordinal $\rho < \kappa$ such that the set $\{\xi \in A : f(\xi) \geq \rho\}$ is not stationary in κ.*

In what follows, we will denote the set $\{\xi \in A : f(\xi) \geq \rho\}$ by the symbol $A(f \geq \rho)$.

Proof of Theorem 12.5A. We may assume that for the stationary set $A \subset \kappa^+$ we also have

$$A \subset \left(\kappa^+ \setminus (\kappa \dotplus 1)\right) \cap \operatorname{Lim}(\kappa^+).$$

Define the function g on the set A by the stipulation $g(\xi) = \operatorname{cf}(\xi)$ for $\xi \in A$. As

$$\operatorname{cf}(\xi) \leq \kappa \qquad \text{for} \qquad \xi < \kappa^+,$$

g is a regressive function on A. Thus, by Fodor's Theorem, there is a stationary subset B of A and a cardinal $\lambda < \kappa^+$ such that

$$g(\xi) = \operatorname{cf}(\xi) = \lambda$$

holds for $\xi \in B$. Now, for each $\xi \in B$, choose a strictly increasing sequence

$$\langle \nu_\xi(\eta) : \eta < \lambda \rangle$$

of ordinals tending to ξ, i.e., a sequence such that

$$\sup\{\nu_\xi(\eta) : \eta < \lambda\} = \xi.$$

For an arbitrary ordinal $\eta < \lambda$, we define a function f_η on B by writing

$$f_\eta(\xi) = \nu_\xi(\eta) \qquad \text{for each} \qquad \xi \in B.$$

Then, for each $\eta < \lambda$, f_η is a regressive function on B. We claim that there is an $\eta < \lambda$ for which the function f_η is not essentially bounded on B. Indeed, assume, on the contrary, that the function f_η is essentially bounded on B for each $\eta < \lambda$. This means that for each ordinal $\eta < \lambda$, there is an ordinal $\rho_\eta < \kappa^+$ for which

$$B_\eta \stackrel{def}{=} B(f_\eta \geq \rho_\eta) \in \mathrm{NS}(\kappa^+).$$

As κ^+ is a regular cardinal, we have

$$\rho \stackrel{def}{=} \sup\{\rho_\eta : \eta < \lambda\} < \kappa^+.$$

The set system $\mathrm{NS}(\kappa^+)$ being a κ^+-complete ideal, we also have

$$B \setminus \bigcup_{\eta < \lambda} B_\eta \in \mathrm{Stat}(\kappa^+).$$

Hence there is an ordinal α such that

$$\alpha \in B \setminus \bigcup_{\eta < \lambda} B_\eta \quad \text{and} \quad \rho < \alpha.$$

Then, by the definition of B_η we have

$$\nu_\alpha(\eta) = f_\eta(\alpha) < \rho_\eta,$$

and so

$$\alpha = \sup\{\nu_\alpha(\eta) : \eta < \lambda\} \leq \sup\{\rho_\eta : \eta < \lambda\} = \rho.$$

This contradicts the choice of α. Hence there is an ordinal $\eta < \lambda$ such that the function f_η is not essentially bounded on B. Denote such a function f_η by f. For each $\alpha < \kappa$, let

$$B_\alpha = f^{-1}(\{\alpha\}) \cap B$$

and

$$M = \{\alpha : B_\alpha \in \mathrm{Stat}(\kappa^+)\}.$$

The sets B_α for $\alpha \in M$ are pairwise disjoint stationary sets, so to complete the proof of the theorem it is sufficient to show that $|M| = \kappa^+$. Let $\beta < \kappa^+$. As the function f is not essentially bounded on B, we have

$$B(f \geq \beta) = \{\xi \in B : f(\xi) \geq \beta\} \in \mathrm{Stat}(\kappa^+).$$

Thus, again by Fodor's Theorem, there is a value α assumed by the function f on a stationary subset of the set $B(f \geq \beta)$. Clearly, $\alpha \geq \beta$. This α is an element of M. Thus we obtained that M is cofinal in κ^+. Thus $|M| = \kappa^+$.

$$* \qquad *$$
$$*$$

We will prove Solovay's Theorem in its full generality only in Section 17 (cf. Theorem 17.3). The proof will only have to be carried out for regular cardinals $\kappa > \omega$ that are limit cardinals. In Section 10, such cardinals were called inaccessible cardinals. (In the Appendix above we also gave some information concerning the existence problems connected with such cardinals.)

We find it useful to reflect at this point on what it would mean if Solovay's Theorem were not true for a cardinal $\kappa > \omega$ and for a stationary set κ. For this, we need some preliminary remarks.

Given an ideal $\mathcal{I}$ on the underlying set X and $A \notin \mathcal{I}$, we denote by $\mathcal{I} + (X \setminus A)$ the set

$$\{Y \subset X : Y \cap A \in \mathcal{I}\}.$$

This is usually called the ideal *generated by* $\mathcal{I} \cup \{X \setminus A\}$.

Lemma 12.3. *Let $\mathcal{I}$ be an ideal on the underlying set X, and assume $A \notin \mathcal{I}$. Then*

1. $\mathcal{I} + (X \setminus A)$ *is an ideal on X.*
2. *If $\mathcal{I}$ is a λ-complete ideal, then $\mathcal{I} + (X \setminus A)$ is λ-complete.*
3. *If $X = \xi$ for some ordinal ξ and $\mathcal{I}$ is a normal ideal, then $\mathcal{I} + (\xi \setminus A)$ is normal.*

The proof is left to the reader.

$$* \qquad *$$
$$*$$

It is now easy to see that the assertion that, for some regular cardinal $\kappa > \omega$ and for some κ-stationary set A, Solovay's Theorem is not true means the following: for the ideal $\mathcal{I}^* = \mathrm{NS}(\kappa) + (\kappa \setminus A)$ there are no κ many pairwise disjoint subsets of κ that do not belong to $\mathcal{I}^*$. This property of ideals is worthy of being a separate subject of study.

Definition 12.15. *Let $\mathcal{I} \subset \mathrm{P}(\kappa)$ be an ideal, and let λ be an arbitrary cardinal greater than 1 but not greater than κ. We say that $\mathcal{I}$ is a λ-saturated ideal if for every system $\mathcal{G} \subset \mathrm{P}(\kappa) \setminus \mathcal{I}$ of pairwise disjoint sets we have $|\mathcal{G}| < \lambda$.*

It is easy to show that $\mathcal{I}$ is a 2-saturated ideal if and only if it is a prime ideal.

We mention that if there is an X with $\emptyset \neq X \subset \kappa$ and $|X| < \lambda$ such that $\kappa \setminus X \in \mathcal{I}$, then $\mathcal{I}$ is clearly λ-saturated. If we want to study nontrivial λ-saturated ideals, then we need to assume that

$$[\kappa]^{<\lambda} \subset \mathcal{I}.$$

The concept of λ-saturated ideals is often defined also for $\lambda > \kappa$.

Definition 12.16. *Let $\mathcal{I} \subset \mathrm{P}(\kappa)$ be an ideal, and let λ be a cardinal greater than κ. We say that $\mathcal{I}$ is a λ-saturated ideal if for every set system $\mathcal{G} \subset \mathrm{P}(\kappa) \setminus \mathcal{I}$ such that $A \cap B \in \mathcal{I}$ for any two $A, B \in \mathcal{G}$ with $A \neq B$ we have $|\mathcal{G}| < \lambda$.*

This is a natural generalization of Definition 12.15. If, for $\lambda \leq \kappa$, we call the ideals satisfying the requirements described in Definition 12.16 *strongly λ-saturated*, then it is easy to see that, for κ-complete ideals, the property of λ-saturatedness and strong λ-saturatedness coincide.

Definition 12.17. *The ideal $\mathcal{I} \subset \mathrm{P}(\kappa)$ is called τ-dense if there is a sequence $\{A_\alpha : \alpha < \tau\} \subset \mathrm{P}(\kappa) \setminus \mathcal{I}$ of sets such that for every $B \subset \kappa$ with $B \notin \mathcal{I}$, there is an $\alpha < \tau$ such that $A_\alpha \setminus B \in \mathcal{I}$.*

It is clear that if an ideal $\mathcal{I} \subset \mathrm{P}(\kappa)$ is τ-dense then it is also τ^+-saturated. We will further clarify these concepts through the problems below and the attached hints at the end of the book.

Problems

1. For arbitrary cardinals κ, λ with $\kappa = \mathrm{cf}(\kappa) > \lambda = \mathrm{cf}(\lambda) \geq \omega$, denote the set

$$\{\alpha < \kappa : \mathrm{cf}(\alpha) = \lambda\}$$

by $S_{\kappa,\lambda}$. Prove that $S_{\kappa,\lambda} \in \mathrm{Stat}(\kappa)$.

2. Show that

a) if ξ is a limit ordinal and B is a ξ-club, then there is a strictly increasing *continuous sequence*

$$\{\alpha_\nu : \nu < \mathrm{cf}(\xi)\} \subset B$$

of ordinals that is cofinal in B; being continuous here means that we have $\alpha_\nu = \sup\{\alpha_\mu : \mu < \nu\}$ for each limit ordinal ν;

b) if we assume in addition that ξ is regular, then we can write B in the form

$$B = \{\alpha_\nu : \nu < \xi\},$$

where $\{\alpha_\nu : \nu < \xi\}$ is a strictly increasing continuous sequence.

3. Let κ be a cardinal with $\kappa > \mathrm{cf}(\kappa) > \omega$.

a) Prove that the set $A \subset \kappa$ is κ-stationary if and only if for every regressive function defined on α there is an ordinal $\alpha < \kappa$ such that

$$f^{-1}(\alpha) \in \mathrm{Stat}(\kappa).$$

b) Show that it is not in general possible to replace $f^{-1}(\alpha)$ with $f^{-1}(\{\alpha\})$ in Part a) of this problem.

c) Show that the assertion in Part a) of this problem remains valid if we replace the condition "$f^{-1}(\alpha)$ is stationary" with "$f^{-1}(\alpha)$ is cofinal in κ."

4. Let κ be a regular cardinal, $A \subset \kappa$, $|A| = \kappa$. Prove that the set

$$\{\alpha < \kappa : A \cap \alpha \text{ is cofinal in } \alpha \quad \text{and} \quad \text{type}(A \cap \alpha) = \alpha\}$$

is a κ-club.

5. Assume the set $A \subset \omega_1 \times \omega_1$ has order type ω_1^2 in the anti-lexicographic product of the sets $\langle \omega_1, < \rangle$, $\langle \omega_1, < \rangle$. (See Definition 7.13.) Prove that there is an ordinal $\alpha < \omega_1$ such that the set

$$B = \big\{ \xi < \alpha : \{\eta < \alpha : \langle \eta, \xi \rangle \in A\} \text{ is cofinal in } \alpha \big\}$$

is cofinal in α.

6. Let $\kappa = \text{cf}(\kappa) > \omega$. Let $\{A_\alpha : \alpha < \kappa\} \subset P(\kappa)$ be a sequence consisting of pairwise disjoint sets. Prove that the set $A = \bigcup_{\alpha < \kappa} A_\alpha$ is stationary if and only if either $B = \{\min_< A_\alpha : \alpha < \kappa\} \in \text{Stat}(\kappa)$ or there is an $\alpha < \omega$ for which $A_\alpha \in \text{Stat}(\kappa)$.

Definition. *Let* $\kappa = \text{cf}(\kappa) > \omega$, $A \in \text{Stat}(\kappa)$. *Denote by* $\mathcal{F}_0(\kappa, A)$ *the following property. If* $\{A_\alpha : \alpha < \kappa\} \subset \text{Stat}(\kappa) \cap P(A)$, *then the sequence* $\{A_\alpha : \alpha < \kappa\}$ *has a disjoint refinement consisting of stationary sets, that is, there are pairwise disjoint sets* $A'_\alpha \subset A_\alpha$ *with* $A'_\alpha \in \text{Stat}(\kappa)$.

The assertion that $\mathcal{F}_0(\kappa, A)$ always holds would be a joint generalization of Theorem 12.5 and of the assertion of Problem 10.1*b)*. Property $\mathcal{F}_0(\kappa, A)$ was first formulated by G. Fodor. It turns out that its validity is independent of the axiom system ZFC. In what follows, we formulate a few problems on this topic.

Definition. *We say that the sets A and B are almost disjoint if* $|A \cap B| < |A| \cap |B|$.

7. Prove that if $\kappa = \text{cf}(\kappa) > \omega$, then for each $A \subset \text{Stat}(\kappa)$ the ideal $\text{NS}(\kappa)$ is κ^+-saturated on A (i.e., the ideal $\text{NS}(\kappa) + (\kappa \setminus A)$ is κ^+-saturated) if and only if there is no sequence $\{B_\alpha : \alpha < \kappa^+\} \subset P(A)$ of pairwise almost disjoint stationary sets.

8.* Prove that if $\kappa = \text{cf}(\kappa) > \omega$, $A \subset \text{Stat}(\kappa)$, and for each $B \subset A$ the ideal $\text{NS}(\kappa)$ is not κ^+-saturated on B, then $\mathcal{F}_0(\kappa, A)$ holds.

9.* Prove that $\mathcal{F}_0(\omega_1, A)$ is false for a set $A \in \text{Stat}(\omega_1)$ if and only if there is a set $B \subset A$ with $B \in \text{Stat}(\omega_1)$ such that the nonstationary ideal on B, i.e., $\text{NS}(\omega_1) + (\omega_1 \setminus B)$, is ω_1-dense.

10.* Assume that κ is a regular cardinal such that $2^\lambda \le \kappa$ for $\lambda < \kappa$. Let $\{A_\alpha : \alpha < \kappa^+\}$ be a sequence of order type κ^+ such that $A_\alpha \in \text{NS}(\kappa)$ for each $\alpha < \kappa$. Prove that there is a set $D \subset \kappa^+$ with

$$|D| = \kappa \quad \text{and} \quad \bigcup\{A_\alpha : \alpha \in D\} \in \text{NS}(\kappa).$$

13. Δ-SYSTEMS

In the next two sections, we will discuss a number of results, interesting in their own right and also applicable in several branches of mathematics; the proofs of many of these results use the theory of stationary sets.

Let $\mathcal{F}$ be a system of sets. In trying to solve various problems in set theory, a question of the following type often arises: Is it true that if $\mathcal{F}$ has many elements, then there is a system $\mathcal{F}' \subset \mathcal{F}$ having relatively many elements such that any two elements of $\mathcal{F}'$ are "very different" from each other? There are numerous quantitative formulations of this problem. Among the Problems below, we will mention a line of investigation that radically differs from the one discussed below.

In this section, the concept of "very different" is made precise in the following definition.

Definition 13.1. *The set system $\mathcal{F}$ is called a Δ-system if there is a set D such that $F_1 \cap F_2 = D$ for any two distinct elements F_1, F_2 of $\mathcal{F}$.*

Paul Erdős and Richard Rado systematically studied the question that, given a set system of a certain cardinality consisting of sets of a specified size, how large a Δ-system it must include.

In what follows, we will present only a sufficient condition due to them to ensure that $\mathcal{F}$ should include a Δ-system of the same cardinality as $\mathcal{F}$ itself. Theorem 13.1 giving this condition is one of the most frequently used results in set theory nowadays. The proof will rely on results of the preceding section.

Theorem 13.1. *Let $\kappa \geq \omega$ be a regular cardinal, λ, an arbitrary cardinal less than κ. Let $\mathcal{F}$ be a set system of cardinality κ each of whose elements has cardinality λ. If λ, κ are such that $\forall \tau < \kappa \, (\tau^\lambda < \kappa)$, then there is a Δ-system $\mathcal{F}' \subset \mathcal{F}$ with $|\mathcal{F}'| = \kappa$.*

Remark: The condition assumed on the cardinals κ, λ in the theorem is often expressed by saying that κ is *inaccessible from λ.*

Proof. We will first carry the proof out for the case $\kappa = \omega$, since in this case the general method used in the second part of the proof is not applicable.

According to the assumptions, we have $\lambda = n$ for some integer n with $1 \leq n < \omega$. If $n = 1$, then the assertion is obvious. Assume that $n > 1$ and that for $\lambda = n - 1$ the assertion has already been established. By our assumptions, $\mathcal{F}$ consists of countably infinitely many sets of n elements each. We distinguish two cases. In the first case, we assume that there is an element x that belongs to infinitely many $F \in \mathcal{F}$. Let

$$\mathcal{F}^{\mathrm{I}} = \{F \in \mathcal{F} : x \in F\}.$$

Then

$$\mathcal{F}^{\mathrm{II}} = \{F \setminus \{x\} : F \in F^{\mathrm{I}}\}$$

is an infinite set system consisting of sets of $n-1$ elements each. According to the induction hypothesis, there is a set $\mathcal{F}^{\mathrm{III}} \subset \mathcal{F}^{\mathrm{II}}$ such that $|\mathcal{F}^{\mathrm{III}}| = \omega$ and $\mathcal{F}^{\mathrm{III}}$ is a Δ-system. Then the set

$$\mathcal{F}' = \{F \cup \{x\} : F \in \mathcal{F}^{\mathrm{III}}\} \subset \mathcal{F}$$

is an infinite Δ-system.

We may therefore assume that there is no element x that belongs to infinitely many sets in $\mathcal{F}$. In this second case, by recursion we define a set

$$\{F_n : n \in \omega\} \subset \mathcal{F}$$

consisting of pairwise disjoint sets. Assume that we have already defined the sets

$$F_0, F_1, \ldots, F_{k-1} \in \mathcal{F}.$$

As

$$\left| \bigcup \{F_i : i < k\} \right| < \omega \qquad \text{and} \qquad |\mathcal{F}| = \omega,$$

there is a set $F \in \mathcal{F}$ such that

$$F \cap \bigcup \{F_i : i < k\} = \emptyset.$$

Let F_k be such a set. Then $\{F_n : n < \omega\}$ is an infinite Δ-system; in fact, it consists of pairwise disjoint sets.

For the remainder of the proof, we may therefore assume that $\kappa > \omega$. According to the assumptions, we have

$$\left| \bigcup \mathcal{F} \right| \leq \kappa, \qquad \lambda < \kappa.$$

Therefore, we may also assume that

$$\mathcal{F} \subset \mathrm{P}(\kappa).$$

Let $\langle F_\alpha : \alpha < \kappa \rangle$ be an enumeration of order type κ of the system $\mathcal{F}$, each element listed exactly once. Denote by $\tilde{\lambda}$ the smallest infinite cardinal greater than λ. That is, $\tilde{\lambda} = \lambda^+$ if $\lambda \geq \omega$, and $\tilde{\lambda} = \omega$ if λ is finite. According to the assumptions on the cardinals λ, κ, we have $\lambda^+ \leq 2^\lambda < \kappa$ if $\lambda \geq \omega$, and so it is certainly true that $\tilde{\lambda}$ is a regular cardinal that is greater than λ but less than κ. Put

$$A = \{\alpha < \kappa : \mathrm{cf}(\alpha) = \tilde{\lambda}\}.$$

As $\tilde{\lambda}$ is a regular cardinal, we know according to Problem 12.1 that $A \in \mathrm{Stat}(\kappa)$. Define the function g on the set A by putting

$$g(\alpha) = \sup(F_\alpha \cap \alpha)$$

for each $\alpha \in A$. As $|F_\alpha| < \tilde{\lambda}$ and $\mathrm{cf}(\alpha) = \tilde{\lambda}$ is regular, we have

$$g(\alpha) = \sup(F_\alpha \cap \alpha) < \alpha.$$

Thus g is a regressive function on A. According to Fodor's Theorem, there is an ordinal $\rho < \kappa$ and a set $B \subset A$ with $B \in \mathrm{Stat}(\kappa)$ such that

$$g(\alpha) = \rho \qquad \text{for} \qquad \alpha \in B.$$

As $\rho < \kappa$, we have

$$|[\rho]^{\leq \lambda}| \leq |\max(\rho, \omega)|^\lambda < \kappa,$$

according to our assumptions. As $|B| = \kappa$ and κ is regular, there is a set $C \subset B$ with $|C| = \kappa$ and there is a set $D \subset \rho$ with $|D| \leq \lambda$ such that

$$F_\alpha \cap \alpha = D \qquad \text{for} \qquad \alpha \in C.$$

By transfinite recursion we now define a sequence $\langle \alpha_\mu : \mu < \kappa \rangle$ such that the set system

$$\{F_{\alpha_\mu} : \mu < \kappa\} = \mathcal{F}' \subset \mathcal{F}$$

is a Δ-system. Assume that $\langle \alpha_\mu : \mu < \nu \rangle$ has already been defined for some ordinal $\nu < \kappa$. Then

$$\left| \{F_{\alpha_\mu} : \mu < \nu\} \right| \leq |\nu| \cdot \lambda < \kappa,$$

and so

$$\sup\{F_{\alpha_\mu} : \mu < \nu\} < \kappa.$$

Let

$$\alpha_\nu = \min\{\alpha \in C : \alpha > \sup \bigcup \{F_{\alpha_\mu} \cup \alpha_\mu : \mu < \nu\} \wedge \alpha > \sup D\}.$$

This defines α_ν, completing the definition of the sequence $\langle \alpha_\mu : \mu < \kappa \rangle$. We are going to show that $\mathcal{F}'$ is a Δ-system. Let $\mu < \nu < \kappa$. Then $\alpha_\mu < \alpha_\nu$. As $\alpha_\nu \in C$, we have

$$F_{\alpha_\nu} \cap \alpha_\nu = D \subset F_{\alpha_\mu}.$$

Furthermore, by the definition of α_ν,

$$F_{\alpha_\mu} \subset \alpha_\nu$$

holds, and so

$$F_{\alpha_\mu} \cap F_{\alpha_\nu} = F_{\alpha_\mu} \cap \alpha_\nu \cap F_{\alpha_\nu} = F_{\alpha_\mu} \cap D = D.$$

*   *

*

Problems

1. Show that none of the conditions of Theorem 13.1 can be omitted.

2. Prove that for all nonnegative integers $n, k < \omega$, there is $m < \omega$ such that every system of cardinality m of sets of n elements each includes a Δ-system of k elements. Denote by $d(n, k)$ the smallest such integer m.

3. Prove that

$$d(n, 3) \leq 2^{n-1}n! + 1.$$

4. Prove that there is a set system $\mathcal{F} \subset [\omega]^\omega$ with $|\mathcal{F}| = 2^{\aleph_0}$ that consists of pairwise almost disjoint sets. (See the definition before Problem 12.7.)

5.* Prove that if $\lambda \geq \kappa \geq \omega$ and $\rho = \lambda^{<\kappa}$, then there is a system $\mathcal{F} \subset [\rho]^\kappa$ consisting of pairwise almost disjoint sets such that $|\mathcal{F}| = \lambda^\kappa$.

6.* Assume GCH, and let $\lambda \geq \kappa \geq \omega$ with $\mathrm{cf}(\lambda) = \mathrm{cf}(\kappa)$. Prove that then there exists a system $\mathcal{F} \subset [\rho]^\kappa$ consisting of pairwise almost disjoint sets such that $|\mathcal{F}| > \lambda$.

7.* Prove that if $\lambda \geq \kappa \geq \omega$, $\mathrm{cf}(\lambda) \neq \mathrm{cf}(\kappa)$, and $X \in [\lambda]^\kappa$, then there is a

$$Y \subset X, \qquad Y \in [\lambda]^\kappa$$

such that Y is not cofinal in λ.

8.* Assume that GCH holds, and let $\lambda \geq \kappa \geq \omega$ with $\mathrm{cf}(\lambda) \neq \mathrm{cf}(\kappa)$. Prove then that for an arbitrary set system $\mathcal{F} \subset [\lambda]^\kappa$ consisting of almost disjoint sets, we have

$$|\mathcal{F}| \leq \lambda.$$

9. Verify that if $\kappa \geq \omega$ is a regular cardinal and $\mathcal{F} \subset [\kappa]^\kappa$ with $|\mathcal{F}| = \kappa$ is a set system consisting of pairwise almost disjoint sets, then there is a set $A \subset \kappa$ for which $A \notin \mathcal{F}$ and

$$|A \cap F| < \kappa$$

holds for every set $F \in \mathcal{F}$.

10. Show that if $\kappa \geq \omega$ and $\mathcal{F} \subset [\kappa]^{\mathrm{cf}(\kappa)}$ with $|\mathcal{F}| = \kappa$ is a system consisting of pairwise almost disjoint sets, then there is an

$$X \in [\kappa]^{\mathrm{cf}(\kappa)} \qquad (X \notin \mathcal{F})$$

such that $\mathcal{F} \cup \{X\}$ is a set system consisting of pairwise almost disjoint sets.

11. Show that there is a set $\mathcal{F} \subset \mathrm{P}(\omega)$ for any two distinct elements A, B of which we have neither $A \subset B$ nor $B \subset A$, and for which $|\mathcal{F}| = 2^{\aleph_0}$.

12. Prove that if

$$\mathcal{F} \subset [\omega]^\omega, \qquad |\mathcal{F}| = \omega,$$

then there is a set $B \in [\omega]^\omega$ that intersects every element of $\mathcal{F}$ but does not include any of them.

13.* Prove that there is a set system $\mathcal{F} \subset \mathrm{P}(\omega)$ of cardinality $2^{\aleph_0}$ consisting of pairwise almost disjoint sets for which there is no set B satisfying the conditions of Problem 13.12.

14. RAMSEY'S THEOREM AND ITS
GENERALIZATIONS. PARTITION CALCULUS

In 1930, the British mathematical logician F. P. Ramsey proved the result, today known as Ramsey's Theorem, that we will discuss at the beginning of this section. Ramsey's set-theoretical theorem has surprising and deep applications in several branches of mathematics. We will discuss only one application to model theory, but we would like to mention that this theorem also has applications in geometry and analysis.

The possibility of studying various generalizations of the theorem is even more important. The so-called Ramsey theory is now an important subject of finite combinatorics, while the study of its transfinite generalizations brought about an important branch of set theory, now called partition calculus, in the the wake of investigations by Paul Erdős and Richard Rado. The calculus of partitions has important applications outside set theory mainly in set-theoretic topology and universal algebra. In this section, we will give a glimpse of some results of partition calculus, without in any way aiming at completeness. For this reason, right at the outset, we formulate the problem studied in Ramsey's Theorem in somewhat general terms.

Definition 14.1. *Let X be an arbitrary set, λ, a cardinal, and γ, an ordinal. Given a mapping f of the set $[X]^\lambda$ into γ, we call f a λ-partition of (or with) γ colors of the set X. If $Y \subset X$, $\nu < \gamma$, and $f``[Y]^\lambda = \{\nu\}$, then we say that Y is* homogeneous *of (or in) color ν with respect to f. The set $Y \subset X$ is said to be* homogeneous *with respect to f if it is homogeneous in some color $\nu < \gamma$.*

The aforementioned theorem of Ramsey says that if $1 \leq r, k < \omega$ and f is a k-partition with r colors of an infinite set X, then there exists an infinite subset Y of X that is homogeneous with respect to f.

The generalizations of this result investigate the question of how large a homogeneous subset can be guaranteed if the cardinality of the set X is large. In order to study this question quantitatively, we will introduce a general symbol.

Definition 14.2. *Let γ be an ordinal, and let κ, λ, κ_ν $(\nu < \gamma)$ be cardinals. The symbol*

$$\kappa \to (\kappa_\nu)^\lambda_{\nu < \gamma}$$

is used to indicate that the following assertion is true:

For an arbitrary set X of cardinality κ and for every λ-partition f of X with γ colors, there is a set $Y \subset X$ and an ordinal $\nu < \gamma$ such that Y is homogeneous in color ν with respect to f and $|Y| = \kappa_\nu$.

To indicate that this assertion is not true, it will be denoted as

$$\kappa \not\to (\kappa_\nu)^\lambda_{\nu < \gamma}.$$

Occasionally, in the problems but not in the main text, partition relations involving order types rather than cardinals are mentioned. The meanings of these can be described in a similar way. For example, given order types Θ, Θ_ν, in the relation

$$\Theta \to (\Theta_\nu)^\lambda_{\nu < \gamma},$$

the λ-partition of a set of order type Θ is considered, and the existence of a homogeneous subset of order type Θ_ν is claimed for some $\nu < \gamma$.

We observe that here we in effect defined a property

$$R(\kappa, \lambda, \gamma, \langle \kappa_\nu : \nu < \gamma \rangle)$$

depending on the cardinals κ, λ, the ordinal γ, and the sequence $\langle \kappa_\nu : \nu < \gamma \rangle$. The above, today generally accepted, graphic symbol, introduced by Richard Rado, is so devised that if we increase the cardinals on the left-hand side of the arrow ($\to$) in a true assertion, the resulting assertion will again be true, while if we decrease the cardinals or ordinals on the right-hand side of the arrow in a true assertion, the resulting assertion will again be true. If all cardinals κ_ν are equal to τ then it is natural to write

$$\kappa \to (\tau)^\lambda_\gamma \qquad \text{instead of} \qquad \kappa \to (\kappa_\nu)^\lambda_{\nu < \gamma},$$

$$\kappa \not\to (\tau)^\lambda_\gamma \qquad \text{instead of} \qquad \kappa \not\to (\kappa_\nu)^\lambda_{\nu < \gamma}.$$

This symbol describes a nontrivial assertion only if $\lambda \geq 1$, $\gamma > 2$, and $\kappa_\nu \geq \lambda$; we will not bother the reader with discussing its truth value in the trivial cases. The combinatorial phenomenon discussed in Ramsey's Theorem appears only for $\lambda \geq 2$; we nevertheless want to mention the following assertion concerning the case $\lambda = 1$.

For each cardinal κ and every sequence $\langle \kappa_\nu : \nu < \gamma \rangle$ of cardinals with $0 < \kappa_\nu \leq \kappa$ for $\nu < \gamma$, the assertion

$$\kappa \to (\kappa_\nu)^1_{\nu < \gamma}$$

is true if and only if, for every sequence $\langle \lambda_\nu : \nu < \gamma \rangle$ of cardinals with $\lambda_\nu < \kappa_\nu$ for $\nu < \gamma$, the relation

$$\sum_{\nu < \gamma} \lambda_\nu < \kappa$$

holds. Thus, for example, for $\kappa \geq \omega$ the relation $\kappa \to (\kappa)^1_\gamma$ holds if and only if $\gamma < \mathrm{cf}(\kappa)$.

Next we present one among the several well-known proofs of Ramsey's Theorem; familiarity with this proof will be useful for us in later sections.

Ramsey's Theorem 14.1. *Assume* $1 \leq r < \omega$ *and* $1 \leq k < \omega$. *Then* $\omega \to (\omega)_k^r$.

Proof. We will prove the theorem by induction on r. For $r = 1$ the assertion is straightforward.

Let $r \geq 1$ and assume that the assertion is true for r; we will show that it is true for $r + 1$ as well. Let $f : [\omega]^{r+1} \to \kappa$ be an $(r+1)$-partition of ω with k colors. Let $\mathcal{U} \subset \mathrm{P}(\omega)$ be an ultrafilter that includes the set

$$\{\omega \setminus A : A \in [\omega]^{<\omega}\};$$

the existence of such an ultrafilter is guaranteed by Corollary 9.1.

We are going to define an r-partition of ω with k colors as follows. Let $V \in [\omega]^r$. For an arbitrary $i < k$, denote by $F_i(V)$ the set

$$\{x \in \omega \setminus V : f(V \cup \{x\}) = i\}.$$

We then clearly have

$$\omega = V \cup \bigcup_{i<k} F_i(V),$$

and the summands on the right-hand side are pairwise disjoint. Hence, by $\mathcal{U}$ being an ultrafilter, for an arbitrary $V \in [\omega]^r$ there is exactly one $i < k$ such that

$$F_i(V) \in \mathcal{U}.$$

Let $f'(V) = i$ for this unique i. The mapping f' so defined is an r-partition with k colors of ω. We denote the set $F_{f(V)}(V)$ by $G(V)$. Then

$$G(V) \in \mathcal{U}$$

holds for every element $V \in [\omega]^r$.

We now define a sequence $\{x_n : n < \omega\}$ by recursion on n. If $n < \omega$ and the sequence $\{x_m : m < n\}$ has already been defined, then put

$$x_n = \min\Big\{\bigcap\{G(V) : V \in [\{x_m : m < n\}]^r\} \setminus \max\{x_m + 1 : m < n\}\Big\}.$$

If $n < r$, then the set system $[\{x_m : m < n\}]^r$ is empty; the intersection of the empty set system in the above formula is considered to be equal to ω. The definition is sound, since the intersection of finitely many elements of $\mathcal{U}$ is an infinite subset of ω. Thus we have defined an infinite subset

$$A = \{x_n : n \in \omega\}$$

such that $x_0 < x_1 < \cdots < x_n < \ldots$ holds. The function $f'|[A]^r$ is an r-partition of the set A with k colors, and so, by the induction hypothesis, there is a set $Y \subset A$ with $|Y| = \omega$ that is homogeneous with respect to f'. We claim that Y is homogeneous with respect to f. Let

$$W_0, W_1 \in [Y]^{r+1}, \qquad z_0 = \max W_0, \qquad z_1 = \max W_1,$$
$$W_0 = V_0 \cup \{z_0\}, \qquad W_1 = V_1 \cup \{z_1\}, \qquad V_0, V_1 \in [Y]^r.$$

Then, by our construction,

$$z_j \in G(V_j) = F_{f'(V_j)}(V_j),$$

and so

$$f(W_j) = f'(V_j)$$

for $j < 2$. On the other hand,

$$f'(V_0) = f'(V_1),$$

as Y is homogeneous with respect to f'. Thus

$$f(W_0) = f(W_1),$$

and so Y is homogeneous with respect to f as well.

$$* \qquad *$$
$$*$$

Next we present an application of Ramsey's Theorem to model theory. For this, the reader needs to be familiar with the fundamentals of the theory of first-order languages and structures. While we will include a definition of most of the required concepts, we would still recommend that the reader not conversant with the basics of mathematical logic skips these pages, at least on the first reading of this book.

Definition 14.3. Prerequisites. *Let L be a first-order structure, $\mathfrak{A} = \langle A, \mathcal{J} \rangle$ a structure of a signature corresponding to L. In our notation $\mathcal{J}$ is a function defined on the relation and function symbols of the first-order language L.*

If p is an n-place relation symbol for some $n < \omega$, then $\mathcal{J}(p)$ is an n-place relation on the set A.

If g is an n-place function symbol ($n < \omega$), then $\mathcal{J}(g)$ is a function of n variables on the set A.

If in particular $n = 0$, and C is a function symbol of 0 variables, called a constant symbol, then $\mathcal{J}(C)$ is an element of A.

The set of well-formed formulas of the language L is denoted by $\mathrm{Wff}(L)$. If $\phi \in \mathrm{Wff}(L)$, then $V(\phi)$ denotes the set of free variables of the formula ϕ.

If $\phi \in \mathrm{Wff}(L)$, $V(\phi) = \{u_0, \ldots, u_{n-1}\}$, $x_0, \ldots, x_{n-1} \in A$, and $\mathbf{x} = \langle x_0, \ldots, x_{n-1} \rangle$, then $\mathfrak{A} \models \phi(\mathbf{x})$ indicates that the formula ϕ is satisfied by $\mathfrak{A}$ when we give the value x_i to the variable u_i for each $i < n$. We assume familiarity with the precise definition of this concept.

For the rest of the prerequisites, we refer the reader to the monograph C. C. Chang–H. J. Keisler, *Model Theory* [C, K].

Definition 14.4. *Let L be a first-order language and let $\mathfrak{A} = \langle A, \mathcal{J} \rangle$ be a structure of the corresponding signature. Let $H \subset A$ and let $\prec$ be an ordering of the set H. We call H a set of* indiscernibles *in $\mathfrak{A}$ (with respect to the ordering $\prec$) if for every formula $\phi \in \mathrm{Wff}(L)$ with*

$$V(\phi) = \{u_0, \ldots, u_{n-1}\}, \qquad n < \omega$$

and for arbitrary sequences

$$\mathbf{x} = \langle x_0, \ldots, x_{n-1} \rangle \in {}^n H, \qquad \mathbf{y} = \langle y_0, \ldots, y_{n-1} \rangle \in {}^n H$$

with

$$x_0 \prec \cdots \prec x_{n-1}, \qquad y_0 \prec \cdots \prec y_{n-1}$$

we have

$$\mathfrak{A} \models \phi(\mathbf{x}) \iff \mathfrak{A} \models \phi(\mathbf{y}).$$

Definition 14.5. *Let $\mathfrak{A} = \langle A, \mathcal{J}_{\mathfrak{A}} \rangle$ and $\mathfrak{B} = \langle B, \mathcal{J}_{\mathfrak{B}} \rangle$ be structures of signatures corresponding to L. We say that the structure $\mathfrak{A}$ is an* elementary submodel *of $\mathfrak{B}$, or $\mathfrak{B}$ is an* elementary extension *of $\mathfrak{A}$, if $A \subset B$ and for an arbitrary formula $\phi \in \mathrm{Wff}(L)$ with $V(\phi) = \{u_0, \ldots, u_{n-1}\}$ $(n < \omega)$ and for each n-tuple $\mathbf{x} = \langle x_0, \ldots, x_{n-1} \rangle \in {}^n A$ we have*

$$\mathfrak{A} \models \phi(\mathbf{x}) \iff \mathfrak{B} \models \phi(\mathbf{x}).$$

The fact that $\mathfrak{A}$ is an elementary submodel of $\mathfrak{B}$ is denoted as $\mathfrak{A} \prec \mathfrak{B}$. (The symbol $\prec$ also has uses other than to indicate the elementary submodel property; e.g., we frequently use it as a generic symbol for orderings. We hope that the context will make it clear which use is meant.)

After these preliminaries we can formulate the theorem of model theory already hinted at above:

Ehrenfeucht–Mostowski Theorem 14.2. *Let L be an arbitrary first-order language, and let $\mathfrak{A} = \langle A, \mathcal{J}_{\mathfrak{A}} \rangle$ be a structure of the corresponding signature such that its underlying set A is infinite. Then $\mathfrak{A}$ has an elementary extension $\mathfrak{B} = \langle B, \mathcal{J}_{\mathfrak{B}} \rangle$ and there is an ordered set $\langle H, \prec \rangle$ such that*

$$H \subset B \setminus A, \qquad |H| = \omega,$$

and H is a set of indiscernibles in the structure $\mathfrak{B}$ with respect to the ordering $\prec$.

Proof. We may assume that $A = \kappa$ for some cardinal $\kappa \geq \omega$. Let L' be an extension of the language L by the addition of constant symbols

$$\{c_\alpha : \alpha < \kappa\} \qquad \text{and} \qquad \{d_n : n < \omega\}$$

not occurring in L. Consider the following sets of formulas:

$$\Gamma_0 = \{\phi(c_{\alpha_0}, \ldots, c_{\alpha_{n-1}}) : \phi \in \mathrm{Wff}(L)$$
$$\wedge |V(\phi)| = n \wedge \mathfrak{A} \models \phi(\alpha_0, \ldots, \alpha_{n-1}) \wedge \forall i < n\, (\alpha_i < \kappa)\};$$
$$\Gamma_1 = \{\phi(d_{i_0}, \ldots, d_{i_{n-1}}) \iff \phi(d_{j_0}, \ldots, d_{j_{n-1}}) : \phi \in \mathrm{Wff}(L)$$
$$\wedge |V(\phi)| = n \wedge i_0 < \cdots < i_{n-1} < \omega \wedge j_0 < \cdots < j_{n-1} < \omega\};$$
$$\Gamma_2 = \{d_i \neq d_j : i \neq j \wedge i,j < \omega\};$$
$$\Gamma_3 = \{c_\alpha \neq d_i : \alpha < \kappa \wedge i < \omega\}.$$

Let

$$\Gamma = \Gamma_0 \cup \Gamma_1 \cup \Gamma_2 \cup \Gamma_3.$$

We claim that it is sufficient to prove that the set Γ of formulas is consistent. To see this, let $\mathfrak{B}' = \langle B, \mathcal{J}' \rangle$ be a structure of signature corresponding to the language L' such that $\mathfrak{B}' \models \Gamma$. Let $\mathfrak{B} = \langle B, \mathcal{J}_\mathfrak{B} \rangle$ be the restriction of this structure with signature corresponding to the language L; that is, $\mathcal{J}_\mathfrak{B} = \mathcal{J}'|L$.

The embedding $\alpha \mapsto \mathcal{J}'(c_\alpha)$ is one-to-one, as $\mathfrak{B}' \models \Gamma_0$. Hence we may assume that $\kappa \subset B$. Again using the condition $\mathfrak{B}' \models \Gamma_0$, we can see that for arbitrary elements $\alpha_0, \ldots, \alpha_{n-1} < \kappa$ and for an arbitrary formula ϕ of L with n free variables, we have the relation

$$\mathfrak{A} \models \phi(\alpha_0, \ldots, \alpha_{n-1}) \equiv \mathfrak{B}' \models \phi(c_{\alpha_0}, \ldots, c_{\alpha_{n-1}})$$
$$\equiv \mathfrak{B} \models \phi(\mathcal{J}'(c_{\alpha_0}), \ldots, \mathcal{J}'(c_{\alpha_{n-1}})) \equiv \mathfrak{B} \models \phi(\alpha_0, \ldots, \alpha_{n-1}).$$

Hence $\mathfrak{A} \prec \mathfrak{B}$.

Put

$$H = \{\mathcal{J}'(d_n) : n < \omega\}.$$

Define the ordering $\prec$ of H by the stipulation

$$\mathcal{J}'(d_n) \prec \mathcal{J}'(d_m) \qquad \text{whenever} \qquad n < m < \omega;$$

$\prec$ is indeed an ordering, as $\mathfrak{B}' \models \Gamma_2$. As $\mathfrak{B}' \models \Gamma_1$, the set H is a set of indiscernibles in the structure $\mathfrak{B}$ with respect to the ordering $\prec$. Finally, the assumption $\mathfrak{B}' \models \Gamma_3$ ensures the requirement $H \subset B \setminus A$.

In view of what has been said, it is sufficient to prove that the set Γ of formulas is consistent. By the Compactness Theorem of First-Order Logic, it is sufficient to show that any finite subset of Γ is consistent.

Let Γ' be a finite subset of the set Γ of formulas. Let C be a finite subset of κ such that we have $\alpha \in C$ whenever the constant c_α occurs in any formula in Γ'. Let n be an integer such that we have $i < n$ whenever the constant d_i

occurs in any formula in Γ'. Let L'' be the extension of the language L with the constant symbols c_α $(\alpha \in C)$ and d_i $(i < n)$. Finally, put

$$\Gamma^* = \Big\{ \psi \in \mathrm{Wff}(L) : \exists k \big(k = |V(\psi)| \wedge \exists i_0, \ldots, i_{k-1}, j_0, \ldots, j_{k-1} < n$$
$$i_0 < \cdots < i_{k-1} \wedge j_0 < \cdots < j_{k-1} < n$$
$$\big(\psi(d_{i_0}, \ldots, d_{i_{k-1}}) \iff \psi(d_{j_0}, \ldots, d_{j_{k-1}}) \big) \in \Gamma' \big) \Big\}.$$

Γ^* is clearly a finite set.

Given an arbitrary $\phi \in \Gamma^*$, denote by m the cardinality of $V(\phi)$, and define an m-partition of κ with 2 colors by putting

$$f_\phi(\{\alpha_0, \ldots, \alpha_{m-1}\}) = 0 \iff \mathfrak{A} \models \phi(\alpha_0, \ldots, \alpha_{m-1})$$

for each increasing sequence $\alpha_0 < \cdots < \alpha_{m-1} < \kappa$ of length m or ordinals.

By repeated application of Ramsey's Theorem, we find an infinite set $D \subset \kappa$ that is homogeneous for the mapping f_ϕ for each formula $\phi \in \Gamma^*$.

Let $\mathfrak{A}'' = \langle A, \mathcal{J}'' \rangle$ be a structure with signature corresponding to the language L'' such that

$$\mathcal{J}''|L = \mathcal{J}|L, \qquad \mathcal{J}''(c_\alpha) = \alpha \quad \text{for} \quad \alpha \in C,$$

$$\mathcal{J}''(d_i) \in D \setminus C, \quad \text{and} \quad \mathcal{J}''(d_i) < \mathcal{J}''(d_j) \quad \text{whenever} \quad i < j < n.$$

It is easy to show that by the homogeneity of D we have

$$\mathfrak{A}'' \models \Gamma'.$$

$$* \qquad *$$
$$*$$

Next we turn to the discussion of the transfinite generalizations of Ramsey's Theorem. The question arises immediately whether Ramsey's Theorem remains true if we replace ω with an arbitrary infinite cardinal κ, that is, whether the assertion

$$\kappa \to (\kappa)^r_k \qquad (r, k < \omega)$$

or at least the assertion

$$\kappa \to (\kappa)^2_2$$

holds for an arbitrary cardinal $\kappa \geq \omega$.

Even before the above partition notation was invented, the Polish mathematician W. Sierpiński proved that this is not true in general, and we will next discuss his ingenious counterexample showing this. Later, in Section 18, we will return to the question, for which infinite cardinals the assertion of Ramsey's Theorem remains valid, and in Section 18 we will discuss possible generalizations of Sierpiński's example.

Sierpiński's Theorem 14.3. $2^{\aleph_0} \nrightarrow (\aleph_1)_2^2$.

Proof. Let $\langle \mathbb{R}, < \rangle$ be the set of real numbers with the ordering $<$ according to size, and let $\langle \mathbb{R}, \prec \rangle$ denote a wellordering of $\mathbb{R}$. Define the mapping $f : [\mathbb{R}]^2 \to 2$, i.e., a 2-partition of $\mathbb{R}$ with 2 colors, as follows: Given

$$x = \{x_0, x_1\} \in [\mathbb{R}]^2 \qquad \text{with} \qquad x_0 < x_1,$$

we put

$$f(x) = \begin{cases} 0 & \text{for} \quad x_0 \prec x_1, \\ 1 & \text{for} \quad x_1 \prec x_0. \end{cases}$$

If $H \subset R$ is homogeneous in color 0 with respect to f, then the ordering $<$ of the real numbers according to size wellorders H. If $H \subset R$ is homogeneous in color 1 with respect to f, then the ordering $>$, the reverse of the ordering of the real numbers according to size, wellorders H. The assertion therefore follows in view of Problem 8 in Section 7.

A natural generalization of Sierpiński's Theorem is the following:

Theorem 14.4. $2^\kappa \nrightarrow (\kappa^+)_2^2$ *for every* $\kappa \geq \omega$.

The simple proof of this theorem is outlined in Problems 14.1, 14.2, and 14.3.

In contrast to the somewhat misleading situation exemplified by Ramsey's Theorem, on a larger underlying set the validity of partition relations changes as we increase the superscript r.

First we discuss the case $r = 2$. For this, we will need a brief excursion to introduce some more notation.

Definition 14.6. *Let* $\langle \kappa_\nu : \nu < \gamma \rangle$ *be a sequence of cardinals. If* $\kappa_\nu = \tau$ *for* $1 \leq \nu < \tau$, *then we will write*

$$\kappa \to \left(\kappa_0, (\tau)_{\gamma-1} \right)^r \qquad \text{instead of} \qquad \kappa \to (\kappa_\nu)_{\nu<\gamma}^r.$$

Here $\gamma - 1$ *stands for the ordinal whose order type is* $\text{type}\langle \gamma \setminus \{0\}, < \rangle$. *If* γ *is finite, then we will write*

$$\kappa \to (\kappa_0, \ldots, \kappa_{\gamma-1})^r \qquad \text{instead of} \qquad \kappa \to (\kappa_\nu)_{\nu<\gamma}^r.$$

We will not announce the following theorem in its most general form. The ideas in its proof can be applied to prove all known partition relations of form $\kappa \to (\kappa_\nu)_{\nu<\gamma}^2$ such that κ is a regular cardinal greater than ω, $\kappa_0 = \kappa$, and $\kappa_\nu < \kappa$ for $1 \leq \nu < \gamma$.

Erdős–Rado Theorem 14.5. *Assume that* ρ *is a regular cardinal. Let* $\lambda < \rho$ *be a regular cardinal such that*

$$\forall \rho' < \rho \, \forall \lambda' < \lambda \, (\rho'^{\lambda'} < \rho)$$

holds, and let $\tau < \lambda$. *Then*

$$\rho \to \left(\rho, (\lambda)_{\tau-1} \right)^2.$$

Corollary 14.1. *If $\kappa \geq \omega$, then*

$$(2^\kappa)^+ \to \left((2^\kappa)^+, (\kappa^+)_\kappa\right)^2.$$

Proof. If $\rho = (2^\kappa)^+$, $\lambda = \kappa^+$, $\tau = \kappa$, then ρ is regular,

$$\rho' \leq 2^\kappa \qquad \text{for} \qquad \rho' < \rho,$$

$$\lambda' \leq \kappa \qquad \text{for} \qquad \lambda' < \lambda,$$

and so

$$\rho'^{\lambda'} \leq 2^\kappa < \rho.$$

Finally, $\lambda < \rho$, as $\kappa^+ \leq 2^\kappa < \rho$.

$$* \qquad *$$
$$*$$

Corollary 14.1 is the form most quoted in the literature of the Erdős–Rado Theorem; Theorem 14.5 is, however, more general and can be proved in the same way. For example, with the assumption of GCH, Theorem 14.5 shows that

$$\aleph_{\omega_1+1} \to (\aleph_{\omega_1+1}, \aleph_1)^2,$$

while the corollary has nothing to say about this relation, as $\aleph_{\omega_1}$ is not of the form 2^κ if GCH holds.

Proof of Theorem 14.5. Let $f : [\rho]^2 \to \tau$ be a 2-partition with τ colors of the cardinal ρ. We have to show that, with respect to f, either there is a homogeneous set of cardinality ρ in color 0 or there is a homogeneous set of cardinality λ of some other color. Let

$$A = S_{\rho,\lambda} = \{\alpha < \rho : \mathrm{cf}(\alpha) = \lambda\}.$$

As λ is regular and $\lambda < \rho$ according to our assumptions, the set A is a stationary subset of the cardinal ρ (cf. Problem 1 in Section 12).

Given $\alpha \in A$, we define an increasing sequence

$$\{\beta_\nu^\alpha : \nu < \xi_\alpha\} \subset \alpha$$

of some order type ξ_α by recursion on ν; the ordinal ξ_α will be determined in the course of this recursion.

Assume that for some α and ν, the sequence $\{\beta_\mu^\alpha : \mu < \nu\}$ has already been defined. Let

$$\delta_\nu^\alpha = \sup\{\beta_\mu^\alpha + 1 : \mu < \nu\}$$

and

$$A_\nu^\alpha = \{\beta < \alpha : \delta_\nu^\alpha \leq \beta \wedge f(\{\beta, \alpha\}) > 0$$
$$\wedge \, \forall \mu < \nu \, \left(f(\{\beta_\mu^\alpha, \beta\}) = f(\{\beta_\mu^\alpha, \alpha\})\right)\}.$$

If the set A_ν^α is empty, then we put $\xi_\alpha = \nu$ and we will not continue the sequence $\{\beta_\mu^\alpha : \mu < \nu\}$. If $A_\nu^\alpha \neq \emptyset$, then put

$$\beta_\nu^\alpha = \min_< A_\nu^\alpha.$$

This completes the definition of the above sequences for all $\alpha \in A$.

Assume first that there is an $\alpha \in A$ for which $\xi_\alpha \geq \lambda$. Then, for arbitrary ordinals $\mu < \nu < \lambda$, we have

$$f(\{\beta_\mu^\alpha, \beta_\nu^\alpha\}) = f(\{\beta_\mu^\alpha, \alpha\}) > 0.$$

Define the function $g : \lambda \to \tau \setminus \{0\}$ by stipulating that

$$g(\mu) = f(\{\beta_\mu^\alpha, \alpha\}).$$

As $\lambda \to (\lambda)_\tau^1$, there is a set

$$H \subset \lambda, \qquad |H| = \lambda,$$

and an ordinal ν with

$$1 \leq \nu < \tau,$$

such that

$$g(\mu) = \nu \qquad \text{for} \qquad \mu \in H.$$

In this case, however, $\{\beta_\mu^\alpha : \mu \in H\}$ is a set of cardinality λ that is homogeneous in color ν with respect to f.

We may therefore assume for the rest of the proof that $\xi_\alpha < \lambda$ for each $\alpha \in A$. Let

$$\delta_\alpha = \sup\{\beta_\nu^\alpha : \nu < \xi_\alpha\}.$$

As $\mathrm{cf}(\alpha) = \lambda$ for each $\alpha \in A$, we have

$$\delta_\alpha < \alpha \qquad \text{for} \qquad \alpha \in A.$$

According to Neumer's Theorem, there is a set $B \subset A$ and an ordinal $\delta < \rho$ such that

$$|B| = \rho \qquad \text{and} \qquad \delta_\alpha = \delta \quad \text{for} \quad \alpha \in B.$$

Observing that

$$|\delta|^{|\xi_\alpha|} < \rho \qquad \text{for} \qquad \alpha \in A,$$

an easy calculation shows that there is a set $C \subset B$ such that $|C| = \rho$ and there is an ordinal $\xi < \lambda$, a sequence $\{\beta_\nu : \nu < \xi\}$ of ordinals, and a function $g : \xi \to \tau$ such that for an arbitrary $\alpha \in C$ and $\nu < \xi$, we have

$$\xi_\alpha = \xi \qquad \text{and} \qquad \beta_\nu^\alpha = \beta_\nu.$$

Indeed, the number of possible sequences

$$\langle \beta_\nu^\alpha : \nu < \xi_\alpha \rangle$$

with $\xi_\alpha < \lambda$ and $\beta_\nu^\alpha < \delta$ is less than ρ in view of the above inequality. In this case, we also have

$$f(\{\beta_\nu, \alpha\}) = g(\nu)$$

in view of the definition of g above.

We claim that the set C is homogeneous in color 0 with respect to f. Let $\alpha, \alpha' \in C$ with $\alpha < \alpha'$. Assume, on the contrary, that $f(\{\alpha, \alpha'\}) > 0$. Then

$$f(\{\beta_\nu^\alpha, \alpha\}) = f(\{\beta_\nu^{\alpha'}, \alpha'\}) = f(\{\beta_\nu, \alpha\})$$

holds for an arbitrary ordinal $\nu < \xi_\alpha = \xi_{\alpha'} = \xi$. Hence we have $\alpha \in A_\xi^{\alpha'}$, but this contradicts the equality $\xi_{\alpha'} = \xi$.

$$* \qquad *$$
$$*$$

The following is probably the simplest partition relation involving singular cardinals as the cardinality of the underlying set. The paper [D, M] contains Erdős's proof of the case $\kappa > \mathrm{cf}(\kappa)$ for the theorem.

Theorem 14.6 (*Erdős; Dushnik–Miller*). *For every cardinal $\kappa \geq \omega$, we have*

$$\kappa \to (\kappa, \omega)^2.$$

Proof. Let

$$f : [\kappa]^2 \to 2$$

be a 2-partition with 2 colors. Let

$$F(x, i) = \{y \in \kappa \setminus \{x\} : f(\{x, y\}) = i\} \qquad \text{for} \qquad i < 2.$$

We clearly have

$$\kappa = \{x\} \cup F(x, 0) \cup F(x, 1).$$

First we prove the following assertion:

(1) If for each set $A \subset \kappa$ with $|A| = \kappa$ there is an $x \in A$ such that $|F(x, 1) \cap A| = \kappa$, then there is an infinite set $X \subset \kappa$ that is homogeneous in color 1 with respect to f.

To this end, suppose that the assumption of Assertion (1) is satisfied. By recursion on n, we define the elements $x_n \in \kappa$ and the sets A_n. Let $A_0 = \kappa$. Assume that the set $A_n \subset \kappa$ has been defined in such a way that

$$|A_n| = \kappa \qquad \text{and} \qquad x_i \notin A_n \quad \text{for} \quad i < n.$$

Let x_n be an element A_n for which

$$|F(x_n, 1) \cap A_n| = \kappa,$$

and let

$$A_{n+1} = |F(x_n, 1) \cap A_n|.$$

Then

$$|A_{n+1}| = \kappa \qquad \text{and} \qquad x_i \notin A_{n+1} \quad \text{for} \quad i \leq n.$$

This completes the recursive definition.

Let $X = \{x_n : n \in \omega\}$. It is clear that X is an infinite set homogeneous in color 1 with respect to f.

Next we prove the theorem in the case when κ is a regular cardinal. For this, it is enough to show that if κ does not contain a subset of cardinality κ that is homogeneous in color 0 with respect to f, then the assumption in Assertion (1) holds.

Let $A \subset \kappa$, $|A| = \kappa$. Let H be a maximal subset of A homogeneous in color 0. There is such an H according to the Teichmüller–Tukey Lemma; we may assume $|H| < \kappa$, since otherwise there is nothing to prove. In view of H being maximal, for an arbitrary element $y \in A \setminus H$, there is an $x \in H$ for which $y \in F(x, 1)$. Hence

$$A \setminus H = \bigcup \{F(x, 1) \cap A : x \in H\}.$$

As κ is regular, there is an $x \in H$ such that

$$|F(x, 1) \cap A| = \kappa.$$

Assume now that κ is singular. Let

$$\langle \kappa_\xi : \xi < \mathrm{cf}(\kappa) \rangle$$

be a strictly increasing sequence of cardinals less than κ for which $\kappa_0 = \mathrm{cf}(\kappa)$ and κ_ξ is a regular cardinal for each $\xi < \mathrm{cf}(\kappa)$, and, further,

$$\kappa = \sum_{\xi < \mathrm{cf}(\kappa)} \kappa_\xi.$$

Now assume that there is no infinite set $X \subset \kappa$ that is homogeneous in color 1 with respect to f. Then, according to (1) we may assume that there is a set $A \subset \kappa$ with $|A| = \kappa$ such that

$$|F(x,1) \cap A| < \kappa$$

for each $x \in A$.

Without loss of generality, we may assume that $A = \kappa$. We are going to show that then the following assertion is also true:

(2) For each regular cardinal λ with $\mathrm{cf}(\kappa) < \lambda < \kappa$ and every set $B \subset \kappa$ of cardinality λ, there is a set $C \subset B$ of cardinality λ such that

$$\left| \bigcup \{F(x,1) : x \in C\} \right| < \kappa.$$

Indeed, for each $\xi < \mathrm{cf}(\kappa)$, let

$$B_\xi = \{x \in B : |F(x,1)| \le \kappa_\xi\}.$$

Clearly,

$$B = \bigcup_{\xi < \mathrm{cf}(\kappa)} B_\xi.$$

Thus, by virtue of the assumptions about λ, there is a $\xi < \mathrm{cf}(\kappa)$ with $|B_\xi| = \lambda$.

An arbitrary $C \subset B_\xi$ with $|C| = \lambda$ satisfies the requirement of Assertion (2), since

$$\left| \bigcup \{F(x,1) : x \in C\} \right| \le \lambda \cdot \kappa_\xi < \kappa.$$

Now, by transfinite recursion on ξ, we define the sequence $\langle A_\xi : \xi < \mathrm{cf}(\kappa) \rangle$ of sets. Let $\xi < \mathrm{cf}(\kappa)$, and assume that the sets $A_\eta \subset \kappa$, $\eta < \xi$, have already been defined such that

$$\left| A_\eta \cup \bigcup \{F(x,1) : x \in A_\eta\} \right| < \kappa \qquad \text{for} \qquad \eta < \xi.$$

Then the set

$$\kappa \setminus \left(\bigcup \{F(x,1) : x \in A_\eta \wedge \eta < \xi\} \cup \bigcup_{\eta < \xi} A_\eta \right)$$

still has cardinality κ. Let B be a subset of cardinality κ_ξ of this set. By virtue of the already proven assertion for regular cardinals of the theorem, B has a subset C of cardinality κ_ξ that is homogeneous in color 0 with respect to f. According to Assertion (2) above, there is a subset A_ξ of cardinality κ_ξ of C such that

$$\left| A_\xi \cup \bigcup \{F(x,1) : x \in A_\xi\} \right| < \kappa$$

holds. Thus we have completed the definition of the sequence of sets A_ξ. It is clear from the construction that the sets A_ξ are pairwise disjoint and $|A_\xi| = \kappa_\xi$. Let $A = \bigcup_{\xi < cf(\kappa)} A_\xi$. Then

$$|A| = \sum_{\xi < cf(\kappa)} \kappa_\xi = \kappa.$$

Let $x, y \in A$ with $x \neq y$, $x \in A_\xi$, $y \in A_\eta$, $\eta \leq \xi$. If $\eta = \xi$, then $f(\{x, y\}) = 0$, since A_ξ is homogeneous in color 0 with respect to f. If $\eta < \xi$, then $f(\{x, y\}) = 0$ holds, because $x \notin F(y, 1)$ according to the construction. Hence A is homogeneous in color 0 with respect to f.

$$* \quad *$$
$$*$$

We would like to point out that there is no known necessary and sufficient condition to ensure $\kappa \to (\kappa, \omega_1)^2$ for any given singular cardinal κ. The newest results in this area are found in [Sh, S].

The last result of this section concerns r-partitions with $r > 2$.

Definition 14.7. *For each cardinal λ define the operation $\exp_i(\lambda)$ by recursion on $i < \omega$:*

$$\exp_0(\lambda) = \lambda, \qquad \exp_{i+1}(\lambda) = 2^{\exp_i(\lambda)} \quad for \quad i < \omega.$$

Recall the notation

$$2^{<\kappa} = \sum \{2^\tau : \tau < \kappa \wedge \tau \text{ is a cardinal}\}$$

introduced in Problem 7 of Section 11.

Erdős–Rado Theorem 14.7. *For each cardinal $\kappa \geq \omega$, for each ordinal $\gamma < cf(\kappa)$, and for every nonnegative integer $r < \omega$, we have*

$$\left[\exp_r(2^{<\kappa})\right]^+ \to (\kappa)_\gamma^{r+2}.$$

Before we set out to prove this result, we will describe some consequences and particular cases of this result.

Taking into account that $2^{<\lambda^+} = 2^\lambda$ provided $\lambda \geq \omega$, we have

$$\left[\exp_r(\kappa)\right]^+ \to (\kappa^+)_\kappa^{r+1}$$

for every $\kappa \geq \omega$. This assertion for $r = 1$, that is,

$$(2^\kappa)^+ \to (\kappa^+)_\kappa^2,$$

follows from Theorem 14.5, while Theorem 14.7 in the corresponding case $r = 0$ gives a result that is incomparable to this. To illustrate this, assume that GCH holds. Theorem 14.5 gives, among others, the following relations

$$\aleph_{\omega+1} \to \left(\aleph_{\omega+1}, (\aleph_0)_k\right)^2 \qquad (k < \omega)$$

and

$$\aleph_{\omega_1+1} \to \left(\aleph_{\omega_1+1}, (\aleph_1)_\gamma\right)^2 \qquad (\gamma < \omega_1),$$

while Theorem 14.7 gives the relations

$$\aleph_{\omega+1} \to (\aleph_\omega)_k^2 \qquad (k < \omega)$$

and

$$\aleph_{\omega_1+1} \to (\aleph_{\omega_1})_\gamma^2 \qquad (\gamma < \omega_1),$$

since GCH implies

$$\aleph_{\omega_1}^{\aleph_0} = \aleph_{\omega_1}, \qquad 2^{<\aleph_\omega} = \aleph_\omega, \qquad \text{and} \qquad 2^{<\aleph_{\omega_1}} = \aleph_{\omega_1}.$$

We point out that the above result for superscript greater than 2 cannot be improved. This is shown by the following.

Erdős–Hajnal–Rado Theorem 14.8. *For each $\kappa \geq \omega$ and each $r < \omega$ we have*

$$\exp_r(\kappa) \not\to (\kappa^+)_2^{r+1}.$$

The case of $r = 0$ of this theorem is obvious, and the case $r = 1$ is identical to Theorem 14.4. Below, in the problems, there are hints indicating how to prove this theorem.

The proof of Theorem 14.7 needs further preparations.

Definition 14.8. *Let ρ be a cardinal, γ an ordinal, $r < \omega$, and let $f : [\rho]^{r+1} \to \gamma$ be an $(r + 1)$-partition of ρ with γ colors. The set $H \subset \rho$ is said to be end-homogeneous with respect to f if for each $V \in [\rho]^r$ and for every $\alpha, \beta < \rho$ with $V < \alpha$ and $V < \beta$, the equation*

$$f(V \cup \{\alpha\}) = f(V \cup \{\beta\})$$

holds. Here $V < \alpha$ indicates that each element of V precedes α in the ordering $<$ of ordinals.

The main ingredient of the proof of Theorem 14.7 is the so-called Stepping-up Lemma, stated next, which reduces the study of $(r + 1)$-partitions to that of r-partitions.

Stepping-up Lemma 14.1. *Let $\kappa > \omega$ be a cardinal, and let $\gamma < \kappa$, $\rho = (2^{<\kappa})^+$, and $r < \omega$.*

Let, further, $f : [\rho]^{r+1} \to \gamma$ be an $(r+1)$-partition with γ colors of the cardinal ρ. Then there is a set $H \subset \rho$ that is end-homogeneous with respect to f with

$$\text{type } H(<) = \kappa \dotplus 1.$$

Proof. The construction used in the proof is similar to the one described in the proof of Theorem 14.5. For each ordinal $\alpha < \rho$, we define the sequence

$$\{\beta_\nu^\alpha : \nu < \xi_\alpha\} \subset \alpha$$

by transfinite recursion on ν, where the ordinals ξ_α will also be determined in the course of the construction. Let $\alpha < \rho$ be arbitrary, and assume that the sequence $\{\beta_\mu^\alpha : \mu < \nu\}$ has already been defined for some ordinal ν. Let

$$\delta_\nu^\alpha = \sup\{\beta_\mu^\alpha \dotplus 1 : \mu < \nu\}$$

and

$$A_\nu^\alpha = \{\beta < \alpha : \delta_\nu^\alpha \leq \beta \wedge$$
$$\forall V \in [\{\beta_\mu^\alpha : \mu < \nu\}]^r \left(f(V \cup \{\beta\}) = (f(V \cup \{\alpha\})) \right)\}.$$

If the set A_ν^α is empty, then let $\xi_\alpha = \nu$, and the sequence $\{\beta_\mu^\alpha : \mu < \alpha\}$ will not be continued. If $A_\nu^\alpha \neq \emptyset$, then let $\beta_\nu^\alpha = \min_< A_\nu^\alpha$. This completes the definition of the above sequence for each α.

If there is an ordinal $\alpha < \rho$ for which $\xi_\alpha \geq \kappa$, then

$$H = \{\beta_\nu^\alpha : \nu < \kappa\} \cup \{\alpha\}$$

is clearly a set of order type $\kappa \dotplus 1$ that is end-homogeneous with respect to f. Thus, proceeding by reductio ad absurdum, we assume that we have $\xi_\alpha < \kappa$ for each $\alpha < \rho$, and we will derive a contradiction.

This derivation is based on the following two assertions:

(1) For each α, α' with $\alpha' < \alpha < \rho$ and for each $\nu < \xi_\alpha$, the assumption $\alpha' = \beta_\nu^\alpha$ implies

$$\xi_{\alpha'} = \nu \qquad \text{and} \qquad \forall \mu < \nu \, (\beta_\mu^{\alpha'} = \beta_\mu^\alpha).$$

The assertion $\forall \mu < \nu \, (\beta_\mu^{\alpha'} = \beta_\mu^\alpha)$ easily follows by transfinite induction on μ, and then the equation $\xi_{\alpha'} = \nu$ follows directly from the definition given above.

(2) For each sequence $\{\beta_\nu : \nu < \xi\} \subset \rho$, the cardinality of the set

$$A = \{\alpha < \rho : \langle \beta_\nu^\alpha : \nu < \xi_\alpha \rangle = \langle \beta_\nu : \nu < \xi \rangle\}$$

is at most

$$|\gamma|^{|[\xi]^r|}.$$

Indeed, assume, on the contrary, that for a given $\{\beta_\nu : \nu < \xi\} \subset \rho$, the cardinality of A is larger than claimed. Then there are $\alpha, \alpha' \in A$ such that $\alpha' < \alpha$ and

$$f(V \cup \{\alpha'\}) = f(V \cup \{\alpha\})$$

holds for an arbitrary set $V \in [\{\beta_\nu : \nu < \xi\}]^r$. This, however, implies $\alpha' \in A_\xi^\alpha$, and so $A_\xi^\alpha \neq \emptyset$, but this contradicts the assumption $\xi_\alpha = \xi$.

Put

$$R_\xi = \{\alpha < \rho : \xi_\alpha = \xi\}$$

and

$$S_\xi = \{\langle \beta_\nu^\alpha : \nu < \xi \rangle : \alpha \in R_\xi\}$$

for an arbitrary $\xi \in \rho$. According to the first relation in Assertion (1), for each $\alpha \in R_\xi$ and each $\nu < \alpha$ we have $\beta_\nu^\alpha \in R_\nu$, and so for each ordinal ξ, we have

$$|S_\xi| \leq \prod_{\nu < \xi} |R_\nu|.$$

According to Assertion (2),

$$|R_\xi| \leq |S_\xi| \cdot |\gamma|^{|[\xi]^r|}$$

holds for an arbitrary ξ. From these two formulas one can immediately prove by transfinite induction on ξ that

$$|R_\xi| \leq 2^{(|\gamma| \cdot |\xi+1| + \omega)}$$

holds for every ξ. By our reductio ad absurdum assumption, we have $\xi_\alpha < \kappa$ for each $\alpha < \rho$, and so

$$\rho = \bigcup_{\xi < \kappa} R_\xi;$$

hence

$$\rho \leq \sum_{\xi < \kappa} |R_\xi| \leq 2^{<\kappa}.$$

$$* \qquad *$$
$$*$$

Next we turn to the

Proof of the Erdős–Rado Theorem 14.7. Let

$$\kappa_0 = \kappa, \qquad \text{and} \qquad \kappa_{r+1} = (2^{<\kappa_r})^+ \quad \text{for} \quad r < \omega.$$

We need to prove that

$$\kappa_r \to (\kappa)_\gamma^{r+1}$$

holds for $r < \omega$ and $\gamma < \mathrm{cf}(\kappa)$. For $r = 0$, this is the obvious partition relation $\kappa \to (\kappa)_\gamma^1$.

The assertion is proved by induction on r. Let $r < \omega$, and assume that the assertion is true for r. Let $\rho = \kappa_{r+1}$, and let $f : [\rho]^{r+1} \to \gamma$ be an $(r+1)$-partition of ρ with γ colors. According to the Stepping-up Lemma, there is a set $H \subset \rho$ end-homogeneous with respect to f such that

$$\mathrm{type}\, H = \kappa_r \dot{+} 1.$$

Write H in the form $K \cup \{\alpha\}$, where $K < \alpha$.

Define the r-partition f' of the set K with γ colors by stipulating that $f'(V) = f(V \cup \{\alpha\})$ for every $V \in [K]^r$. According to the induction hypothesis, there is a set $L \subset K$ and an ordinal $\nu < \gamma$ such that $|L| = \kappa$ and L is homogeneous in color ν with respect to f'. We claim that L is also homogeneous with respect to f. Indeed, let $V \cup \{\mu\} \in [L]^{r+1}$, $V < \mu$. As H is end-homogeneous with respect to f, we have

$$f(V \cup \{\mu\}) = f(V \cup \{\alpha\}) = f'(V) = \nu.$$

$$* \qquad *$$
$$*$$

The proof of the theorem gives that $L \cup \{\alpha\}$ is also homogeneous in color ν with respect to f, and so, in fact, we obtained a somewhat stronger statement ensuring the existence of a homogeneous set of order type $\kappa + 1$. We will not introduce a symbol to denote this stronger statement here.

Before closing this section, we introduce yet another general notion.

Definition 14.9. *The partially ordered set $\langle T, \prec \rangle$ is called a tree if for each $x \in T$ the set*

$$T| \prec x = \{y \in T : y \prec x\}$$

is wellordered.

In the proof of Theorem 14.7, we tried to avoid all extraneous ingredients. We feel it is still fair to inform the reader of the implicit role trees played in the proof:

Using the notation introduced in the proof, define the relation $\prec$ by the stipulation

$$\alpha' \prec \alpha \iff \alpha' < \alpha \wedge \exists \nu < \xi_\alpha \, (\alpha' = \beta_\nu^\alpha);$$

then the partially ordered set $\langle \rho, \prec \rangle$ is a tree.

Assertion (1) stated in the proof shows exactly that this is the case. The tree $\langle \rho, \prec \rangle$ is called the *canonical partition tree* for the $(r+1)$-partition f. We will return to the role of this tree in Section 18.

Problems

1. Let $\{\lambda_\nu : \nu < \kappa\}$ be an arbitrary sequence of cardinals, and let $P = \bigtimes_{\nu < \kappa} \lambda_\nu$. If $f, g \in P$ with $f \neq g$ are arbitrary functions, then let

$$\delta(f, g) = \min\{\nu < \kappa : f(\nu) \neq g(\nu)\}$$

and

$$f \prec g \iff f(\delta(f, g)) < g(\delta(f, g)).$$

Prove that $\langle P, \prec \rangle$ is an ordered set. The ordering $\prec$ is called the *lexicographic ordering* of P.

2. Let $\lambda \geq \omega$ be a regular cardinal, and let $\{f_\alpha : \alpha < \lambda\} \subset P$ be a sequence such that $f_\alpha \prec f_\beta$ for $\alpha < \beta < \lambda$. (Here P denotes the ordered set defined in the preceding problem.) Prove that there is a set $L \in [\lambda]^\lambda$ and a sequence $\{\nu_\alpha : \alpha \in L\} \subset \kappa$ of ordinals such that $\nu_\alpha \leq \nu_\beta$ and $\nu_\alpha = \delta(f_\alpha, f_\beta)$ whenever $\alpha < \beta$ and $\alpha, \beta \in L$.

3. Prove that $2^\kappa \not\rightarrow (\kappa^+)^2_2$ for $\kappa \geq \omega$.

4. Prove that $2^\kappa \not\rightarrow (3)^3_\kappa$ for $\kappa \geq \omega$.

5. Prove that $\kappa^{\aleph_0} \not\rightarrow (\kappa^+, \aleph_1)$ for $\kappa \geq \omega$.

6. Prove that if $\kappa \geq \omega$ is a regular cardinal and we have $\lambda^{\aleph_0} < \kappa$ for all $\lambda < \kappa$, then $\kappa \rightarrow (\kappa, \aleph_1)^2$.

7. Prove that

$$\kappa^{\operatorname{cf}(\kappa)} \not\rightarrow \left(\kappa^+, (\aleph_0)_{\operatorname{cf}(\kappa)}\right)^2$$

holds for $\kappa \geq \omega$.

8.* Assume GCH holds. Show that if κ is a singular cardinal, then

$$\kappa^+ \not\rightarrow \left(\kappa^+, (3)_{\operatorname{cf}(\kappa)}\right)^2.$$

9. Prove that GCH implies

$$\aleph_{\omega+1} \rightarrow (\aleph_\omega)^2_n$$

for all $n < \omega$.

10. Prove that if κ is a singular cardinal and

$$\operatorname{cf}(\kappa) \not\rightarrow \left(\operatorname{cf}(\kappa), \lambda\right)^2$$

holds, then

$$\kappa \not\rightarrow (\kappa, \lambda)^2$$

also holds.

11.* Prove that if κ is a strong limit singular cardinal (see Section 10, after the proof of Corollary 10.5), then the assertion

$$\operatorname{cf}(\kappa) \rightarrow \left(\operatorname{cf}(\kappa), \lambda\right)^2$$

implies that

$$\kappa \to (\kappa, \lambda)^2.$$

12. Let $P = {}^\kappa 2$ for some $\kappa > \omega$, let $\prec$ be the lexicographic ordering of P, and let $\prec_0$ be a wellordering of P. Let $\{f, g, h\} \in [P]^3$, $f \prec_0 g \prec_0 h$. Define the set $K \subset [P]^3$ by the stipulation

$$f, g, h \in K \iff f \prec g \quad \text{and} \quad g \succ h.$$

Prove that if $X \in [P]^\kappa$ and $[X]^3 \cap K = \emptyset$, then there is a set $Y \in [X]^\kappa$ such that either

$$f \prec_0 g \iff f \prec g \qquad \text{for} \qquad f, g \in Y$$

or

$$f \prec_0 g \iff f \succ g \qquad \text{for} \qquad f, g \in Y.$$

13.* Prove that

$$2^{2^{\aleph_0}} \nrightarrow (\aleph_1)^3_2.$$

14.* Prove that for every $\kappa \geq \aleph_0$

$$\kappa \nrightarrow (\aleph_0)^{\aleph_0}_2.$$

15. INACCESSIBLE CARDINALS. MAHLO CARDINALS

We defined the concepts of inaccessible and strongly inaccessible cardinals in Section 10. In Section A9 of the Appendix, we discussed some consistency results and problems connected with inaccessible cardinals greater than ω. We mentioned there that the assumption of the existence of such cardinals appears to be a natural extension of the axiom system of set theory. In what follows, we will study assumptions that lead to the existence of larger and larger cardinals. The consistency questions for such extensions are similar to those discussed in the Appendix. Roughly speaking, the nonexistence of new, larger cardinals is always relatively consistent, while the relative consistency of their existence is unprovable.

Before we start searching for cardinals larger than merely inaccessible, we will sketch a model-theoretic characterization of inaccessible cardinals. We will not use this characterization afterwards, so readers not conversant in mathematical logic may well want to skip this discussion.

In what follows, we will mainly study the set-theoretical properties of large cardinals. In order to express "how large" these cardinals are, it is, however, best to use methods from mathematical logic. Theorem 15.1 is perhaps the simplest example for the application of such methods.

Definition 15.1. *Let L' be the language of set theory supplemented with a one-place relation symbol* **A**.

We call the ordinal $\alpha > 0$ first-order strongly indescribable if the following condition holds: For each sentence (i.e., formula with no free variables) ϕ of L' and for every set $A \subset R_\alpha$, if

$$\langle R_\alpha, \in, A \rangle \models \phi,$$

then there is a β with $0 < \beta < \alpha$ such that

$$\langle R_\alpha, \in, A \cap R_\beta \rangle \models \phi$$

(cf. Definition A7.1 for the meaning of R_α). (This is understood in the sense that A is represented by **A** *in the structure; that is,*

$$\left(\langle R_\alpha, \in, A \rangle \models \mathbf{A}(x) \right) \iff x \in A$$

for each $x \in R_\alpha$.)

*We call the ordinal $\alpha > 0$ first-order indescribable if the above require-
ment holds for sets $A \subset \alpha$ and structures $\langle \alpha, \in, A \rangle$ instead of $A \subset R_\alpha$ and
$\langle R_\alpha, \in, A \rangle$.*

The reason for the adjective "first-order" in the above definition is that ϕ
is a first-order formula. By using higher order formulas, one can formulate
stronger indescribability properties; we will not discuss them here.

To say that α is indescribable means, heuristically, that "whatever one
says about α with a first-order formula, the same thing can also be said
about a smaller ordinal."

Theorem 15.1. *The ordinal $\alpha > 0$ is a strongly inaccessible cardinal
greater than ω if and only if it is first-order strongly indescribable.*

The theorem remains valid if we omit both occurrences of the modifier
"strongly."

Proof. 1. Assume that $\kappa > \omega$ is a strongly inaccessible cardinal. Let
$\phi \in \text{Wff}(L')$ be a sentence such that

$$\langle R_\kappa, \in, A \rangle \models \phi.$$

Then the following assertion holds in view of a well-known result in model
theory:
  (1) There is an integer $k < \omega$ and a function $f : {}^k R_\kappa \to R_\kappa$ such that the
following holds. Let $X \subset R_\kappa$. If X is closed with respect to f, that is, if
$f``({}^k X) \subset X$, then

$$\langle X, \in, A \cap X \rangle \models \phi.$$

f is called a *Skolem function* of ϕ in the structure $\langle R_\kappa, \in, A \rangle$.
  In view of Assertion (1), it is enough to prove that there is an ordinal β
with $0 < \beta < \kappa$ such that R_β is closed with respect to f, since then

$$\langle R_\beta, \in, A \cap R_\beta \rangle \models \phi.$$

Define a sequence

$$\{\beta_n : n < \omega\} \subset \kappa$$

of ordinals by recursion on n as follows:
  Let β_0 with $0 < \beta_0 < \kappa$ be an arbitrary ordinal. If β_n has been defined
for some $n < \omega$, let

$$\beta_{n+1} = \min\{\beta : f``({}^k R_{\beta_n}) \subset R_\beta\}.$$

We will prove $\beta_n < \kappa$ by induction on n. If $\beta_n < \kappa$ holds for some $n < \omega$,
then, as we have proved in Section A9, $|R_{\beta_n}| < \kappa$ also holds in view of κ
being strongly inaccessible. In this case we have

$$|f``R_{\beta_n}| < \kappa, \qquad f``R_{\beta_n} \subset R_\kappa,$$

and so, by the regularity of κ and by Assertion 5 in Theorem A7.1, we can see that

$$f"R_{\beta_n} \in R_\gamma \qquad \text{for some} \qquad \gamma < \kappa.$$

Thus $\beta_{n+1} < \kappa$ also holds. Taking into account that $\kappa > \omega$ and κ is regular, we obtain that

$$\beta \overset{def}{=} \sup\{\beta_n : n < \omega\} < \kappa.$$

It is easy to check that R_β is closed with respect to the function f.

2. Assume now that $\alpha > 0$ and α is not an inaccessible cardinal greater than ω. We then have to prove that α is not first-order strongly indescribable. If α is not a limit ordinal, then $\alpha = \beta + 1$ for some $\beta < \alpha$. Let $A = \{\beta\} \subset R_\alpha$, and let $\phi = \exists x_0 \, \mathbf{A}(x_0)$. It is obvious that

$$\langle R_\alpha, \in, A \rangle \models \phi$$

and

$$\langle R_\beta, \in, A \cap R_\beta \rangle \models \neg\phi$$

for any $\beta < \alpha$. We may therefore assume that α is a limit ordinal in what follows. It is easy to see that in this case an aggregate of finitely many relations with finitely many places each can be encoded with a single one-place relation. Hence it is sufficient to describe a language L'' containing finitely many finite-place relation symbols and a sentence ϕ in this language such that

$$\langle R_\alpha, \in, \mathcal{J} \rangle \models \phi$$

and

$$\langle R_\beta, \in, \mathcal{J}|R_\beta \rangle \models \neg\phi \qquad \text{for} \qquad \beta < \alpha.$$

(See Definition 14.3 for the function $\mathcal{J}$; here $\mathcal{J}|R_\beta$, by a slight abuse of notation, denotes the function on the relation and function symbols of the language whose values are the restrictions to R_β of the values of $\mathcal{J}$.)

According to our assumption, one of the following possibilities must hold:

(I) $\alpha = \omega$,

(II) α is singular,

(III) $\exists \beta < \alpha \quad |\alpha| \leq 2^{|\beta|}$.

In case (I), the formula $\phi = \forall x \exists y (x \in y)$ shows that α is not strongly indescribable.

In case (II), let $\beta = \mathrm{cf}(\alpha) < \alpha$. Let $f : \beta \to \alpha$ be a mapping of β onto a cofinal subset of α. Let r be the two-place relation on R_α for which

$$r(x, y) \iff x \in \mathrm{D}(f) \wedge y = f(x).$$

Let s_0 be a one-place relation symbol, and let s_1 be a two-place relation symbol in L'', and let $\mathcal{J}(s_0) = \beta$ and $\mathcal{J}(s_1) = r$. Let

$$\phi = \exists x \left(\mathrm{Ordinal}(x) \wedge \neg s_0(x) \right) \wedge \forall x \left(s_0(x) \implies \exists y \, s_1(x, y) \right).$$

Here Ordinal(x) denotes the formula defined in Section A5, describing the concept of ordinal. We will use the following fact about this formula: For each ordinal $\gamma > 0$ and for every $x \in R_\gamma$

$$\big(\langle R_\gamma, \in \rangle \models \text{Ordinal}(x) \big) \iff \text{Ordinal}(x).$$

This result can easily be verified by using the methods discussed in the Appendix. It is clear that

$$\langle R_\alpha, \in, \mathcal{J} \rangle \models \phi,$$

as $\beta < \alpha$. Now let $\gamma < \alpha$. If

$$\langle R_\gamma, \in, \mathcal{J}|R_\gamma \rangle \models \phi$$

then, according to our remarks above, we have

$$\beta < \gamma < \alpha,$$

since ϕ says that there is an ordinal not less than β in R_γ. As $\gamma < \alpha$, there is an ordinal $\delta < \beta$ such that

$$f(\delta) > \gamma.$$

In this case, however,

$$\langle R_\gamma, \in, \mathcal{J}|R_\gamma \rangle \not\models \exists y \, \big(s_1(\delta, y) \big),$$

and so

$$\langle R_\gamma, \in, \mathcal{J}|R_\gamma \rangle \not\models \phi.$$

(III) Let $\beta < \alpha$, and let f be a mapping of $R_{\beta+1}$ onto α. As α is a limit ordinal, we have $\beta \dot{+} 1 < \alpha$, and so $R_{\beta+1} \in R_\alpha$. Let

$$r(x, y) \iff x \in R_{\beta+1} \wedge y = f(x).$$

Let L'' be the language described in the discussion of Case (II), let ϕ be the formula described there, and

$$\mathcal{J}(s_0) = R_{\beta+1}, \qquad \mathcal{J}(s_1) = r.$$

Arguments similar to those given in the discussion of Case (II) show that α is not strongly indescribable in this case, either.

$$* \qquad *$$
$$*$$

It is easy to formulate a condition stronger than that there is a strongly inaccessible cardinal greater than ω:

$(*)$ For every cardinal λ, there is an inaccessible cardinal greater than λ. Assuming $(*)$, denote by

$$\kappa_0 < \cdots < \kappa_\xi < \ldots$$

the increasing sequence of strongly inaccessible cardinals. Then $\kappa_0 = \omega$, and κ_1 is the first inaccessible cardinal greater than ω.

If $(*)$ holds, then for each ordinal ξ we can define κ_ξ, but it is possible that $\kappa_\xi > \xi$ for every ordinal ξ.

It is, however, possible to imagine that there is a strongly inaccessible cardinal κ_ξ such that $\kappa_\xi = \xi$.

An even stronger condition is the following:

$(**)$ For every cardinal λ, there is a strongly inaccessible cardinal $\kappa_\xi > \lambda$ such that $\kappa_\xi = \xi$.

Assuming $(**)$, denote by κ^1_ξ the increasing sequence of strongly inaccessible cardinals such that $\kappa_\xi = \xi$. Condition $(**)$ ensures that κ^1_ξ is defined for every ordinal ξ. An even stronger condition would be to require that there is a cardinal κ^1_ξ such that $\kappa^1_\xi = \xi$.

It is clear that this procedure can be continued even in a transfinite way.

If ξ is the smallest ordinal such that $\kappa_\xi = \xi$, then the function f defined on the set $A = \{\kappa_\eta : \eta < \xi\}$ of strongly inaccessible cardinals less than κ_ξ by the equation $f(\kappa_\eta) = \eta$ is a one-to-one regressive function; thus A is not κ_ξ-stationary. This observation motivates the following definition.

Definition 15.2. *A cardinal $\kappa > \omega$ is called a (strongly) Mahlo cardinal if the set*

$$\{\lambda : \lambda < \kappa \wedge \lambda \text{ is (strongly) inaccessible}\}$$

is κ-stationary.

The assumption of the existence of Mahlo cardinals is much stronger than Conditions $(*)$, $(**)$, $\ldots$ above.

To describe strengthenings of this assumption, it will be convenient to work with classes, which were discussed above, in Section A8.

Definition 15.3. 1. *The class of all sets is denoted by* V, *and the class of all ordinals is denoted by* On.

2. *The class of inaccessible cardinals greater than ω is denoted by $\overline{M}_0$, and the class of strongly inaccessible cardinals is denoted by M_0. Clearly, $M_0 \subset \overline{M}_0$.*

Definition 15.4 (*Mahlo operation*). *For an arbitrary class $X \subset \overline{M}_0$, we put*

$$M(X) = \{\kappa \in X : \kappa \cap X \text{ is } \kappa\text{-stationary}\}.$$

According to the definition, $M(\overline{M}_0)$ is the class of Mahlo cardinals, and $M(M_0)$, of the strongly Mahlo cardinals.

Definition 15.5. *We define the classes* $\overline{M}_\alpha$ *and* M_α *by transfinite recursion. We defined* $\overline{M}_0$ *and* M_0 *in Definition 15.3. If* $\alpha = \beta \dotplus 1$ *then*

$$\overline{M}_\alpha = M(\overline{M}_\beta), \qquad M_\alpha = M(M_\beta).$$

If α *is a limit ordinal then*

$$\overline{M}_\alpha = \bigcap_{\beta < \alpha} \overline{M}_\beta \qquad \text{and} \qquad M_\alpha = \bigcap_{\beta < \alpha} M_\beta.$$

$\overline{M}_\alpha$ *and* M_α *are called the class of* α-*Mahlo and of strongly* α-*Mahlo cardinals, respectively.*

The proofs of the following assertions are left to the reader.

Lemma 15.1. 1. $\overline{M}_0 \supset \cdots \supset \overline{M}_\alpha \supset \ldots;$ $\quad M_0 \supset \cdots \supset M_\alpha \supset \ldots.$
2. *If* $\overline{M}_\alpha \neq \emptyset$ $(M_\alpha \neq \emptyset)$, *then*

$$\overline{M}_{\alpha+1} \subsetneqq \overline{M}_\alpha \qquad (M_{\alpha+1} \subsetneqq M_\alpha).$$

Thus $\overline{M}_\alpha \neq \emptyset$ $(M_\alpha \neq \emptyset)$ are increasingly stronger "large cardinal" assumptions.

We formulate one more possible strengthening.

Definition 15.6.

$$\overline{M}^\infty = \mathop{\Delta}_{\alpha \in \mathrm{On}} \overline{M}_\alpha = \{\kappa : \forall \alpha < \kappa \quad \kappa \in \overline{M}_\alpha\},$$

$$M^\infty = \mathop{\Delta}_{\alpha \in \mathrm{On}} M_\alpha = \{\kappa : \forall \alpha < \kappa \quad \kappa \in M_\alpha\}.$$

It is interesting to point out that P. Mahlo introduced and studied the notion of Mahlo cardinals by using a direct definition in 1911, well before the appearance of the concept of stationary set. To this date, this is a useful concept that describes the largeness of a cardinal without the use of mathematical logic. In the next few sections, we are going to study set-theoretical assumptions that ensure the existence of much "larger" cardinals.

16. MEASURABLE CARDINALS

In the beginning of the 1930s, the Polish mathematician S. Banach raised the following problem:

(∗) For which infinite cardinals κ is there a nontrivial measure $\mu : \mathrm{P}(\kappa) \to [0,1]$ defined on all subsets of κ? The word "nontrivial" here means that $\mu(\kappa) = 1$ and $\mu(\{\alpha\}) = 0$ for $\alpha < \kappa$.

As usual, measures are required to be σ-additive, i.e., $\aleph_1$-additive (that is, the measure of the union of countably many pairwise disjoint measurable sets must be the sum of the measures of these sets; here, of course, all subsets of κ are required to be measurable). We will return to a detailed discussion of this problem in Section 17. Soon after the problem had been raised, it turned out that to clear up the question it was of primary importance to answer the following.

(∗′) For which cardinals κ does there exist a nontrivial measure $\mu : \mathrm{P}(x) \to \{0,1\}$.

This problem is equivalent to the following:

(∗″) For which cardinals κ is there an ω_1-complete ultrafilter $\mathcal{U} \subset \mathrm{P}(\kappa)$ that is not a principal filter.

The equivalence of problems (∗′) and (∗″) is shown by the canonical correspondence

$$X \in \mathcal{U} \iff \mu(X) = 1$$

between μ and $\mathcal{U}$. The actual proof of the equivalence is left to the reader. After giving some preliminary results, we will return to the discussion of the history of the problem.

Lemma 16.1. *If κ is the smallest cardinal for which there is a σ-complete ultrafilter that is not a principal filter, then every such filter is also κ-complete.*

Proof. Let $\mathcal{U} \subset \mathrm{P}(\kappa)$ be a σ-complete ultrafilter that is not a principal filter. We need to prove that $\mathcal{U}$ is also κ-complete; that is, that the ideal $\mathcal{I} = \mathrm{co}(\mathcal{U})$, which is a prime ideal but not a principal ideal on κ, is κ-complete.

Assume, on the contrary, that $\lambda < \kappa$ and

$$A = \bigcup \{A'_\alpha : \alpha < \lambda\}$$

is such that $A \notin \mathcal{I}$ and

$$\{A'_\alpha : \alpha < \lambda\} \subset \mathcal{I}.$$

Let

$$A_\alpha = A'_\alpha \setminus \bigcup_{\beta < \alpha} A'_\beta \qquad \text{for} \qquad \alpha < \lambda.$$

Then

$$A_\alpha \in \mathcal{I}, \qquad \bigcup\{A_\alpha : \alpha < \lambda\} = A,$$

and the set A_α are pairwise disjoint.

For each $Y \subset \lambda$ put

$$K(Y) = \bigcup\{A_\alpha : \alpha \in Y\}$$

and let

$$\tilde{\mathcal{I}} = \{Y \subset \lambda : K(Y) \in \mathcal{I}\}.$$

As the sets A_α are pairwise disjoint, the mapping K preserves unions and intersections, and, further, $K(\lambda) = A$. Hence it follows that $\tilde{\mathcal{I}}$ is an ω_1-complete prime ideal in $P(\lambda)$. If $\tilde{\mathcal{I}}$ were a principal ideal, then we would have $\bigcup \tilde{\mathcal{I}} \in \tilde{\mathcal{I}}$. In this case, we would have $\lambda \setminus \bigcup \tilde{\mathcal{I}} = \{\alpha\}$ for some $\{\alpha\}$, as $\tilde{\mathcal{I}}$ is a prime ideal. This is, however, impossible, since $A_\alpha \in \mathcal{I}$, and so $\{\alpha\} \in \tilde{\mathcal{I}}$. This contradicts the minimality of κ.

$$* \qquad *$$
$$*$$

The following definition is motivated by Lemma 16.1.

Definition 16.1. *An infinite cardinal κ is called a* measurable cardinal *if there is a κ-complete ultrafilter in $P(\kappa)$ that is not a principal filter.*

In the terminology of Definition 16.1, ω is a measurable cardinal (even though, obviously, it does not carry a σ-additive measure); thus, using this terminology, the question is whether there is a measurable cardinal greater than ω. The answer to this question is not known to this date. It will, however, turn out below that the assumption that there is a measurable cardinal greater than ω is a typical "large cardinal" assumption, and, as for the consistency of this assumption, the remarks of Sections A9 and 15 apply the same way as they applied to the question of existence of inaccessible cardinals.

The following theorem was already known in the beginning of the 1930s.

Theorem 16.1 *(A. Tarski). If κ is a measurable cardinal, then κ is strongly inaccessible.*

Proof. Let $\mathcal{I} \subset P(\kappa)$ be a κ-complete prime ideal on κ that is not principal. Then $\{\alpha\} \in \mathcal{I}$ for each $\alpha < \kappa$, and so, by κ-completeness,

$$[\kappa]^{<\kappa} \subset \mathcal{I}.$$

We will establish the following two assertions:

(1) κ is not singular.

(2) If $\lambda < \kappa$, then $2^\lambda < \kappa$.

Let

$$\kappa = \bigcup\{A_\alpha : \alpha < \mathrm{cf}(\kappa)\},$$

where

$$A_\alpha \in [\kappa]^{<\kappa}$$

holds for $\alpha < \mathrm{cf}(\kappa)$. Assuming $\mathrm{cf}(\kappa) < \kappa$, the κ-completeness of $\mathcal{I}$ implies $\kappa \in \mathcal{I}$, which contradicts the assumption that $\mathcal{I}$ is an ideal on κ. This shows that (1) is true.

Now assume that, for some $\lambda < \kappa$, (2) does not hold. As $|{}^\lambda 2| = 2^\lambda$, we may assume that there is a set $A \subset {}^\lambda 2$ with $|A| = \kappa$, and there is a κ-complete ultrafilter $\mathcal{U}$ in $\mathrm{P}(A)$ that is not principal.

Let

$$A_{\alpha,i} = \{f \in A : f(\alpha) = i\} \qquad \text{for} \qquad \alpha < \lambda \quad \text{and} \quad i < 2.$$

Then

$$A = A_{\alpha,0} \cup A_{\alpha,1} \qquad \text{and} \qquad A_{\alpha,0} \cap A_{\alpha,1} = \emptyset$$

for every $\alpha < \lambda$. As $\mathcal{U}$ is an ultrafilter, for each $\alpha < \lambda$ there is exactly one $i < 2$ such that $A_{\alpha,i} \in \mathcal{U}$. Let $g \subset {}^\lambda 2$ be the function defined by $A_{\alpha,g(\alpha)} \in \mathcal{U}$ for each $\alpha < \lambda$. We have

$$\bigcap\{A_{\alpha,g(\alpha)} : \alpha < \lambda\} = \{g\} \in \mathcal{U}$$

by the κ-completeness of $\mathcal{U}$; this, however, contradicts the assumption that the ultrafilter $\mathcal{U}$ is not principal.

* *
*

For the further study of ideals and filters we need several results of a technical nature. These are formulated in Lemmas 16.2, 16.3, and 16.4.

Definition 16.2. *Let X be a class and* R, *a property with two variables;* R *is said to be well-founded on X if for an arbitrary set $x \subset X$ with $x \neq 0$ there is a $u \in x$ such that $v \notin x$ holds whenever vRu.*

In words: Each nonempty subset of X has an R-minimal element.

Examples: For an arbitrary wellordered set $\langle A, \prec \rangle$, the relation $\prec$ is well-founded on A. Axiom $\mathbf{A}_7$, which was discussed in Section A7 in detail, says precisely that the property $\in$ is well-founded on the class of all sets V.

In what follows, we will use the following fact, already observed implicitly: R is well-founded on the class X if and only if there is no sequence $\{x_n : n < \omega\} \subset X$ such that x_{n+1}Rx_n holds for every $n < \omega$.

Definition 16.3. *Let X be an arbitrary set, and $\mathcal{I} \subset P(X)$, an ideal on X. Denote by $\mathcal{F}$ the dual filter $\mathrm{CO}(\mathcal{I})$. Consider the class $^X\mathrm{On}$ of ordinal-valued functions on X, and define three fundamentally important properties on this class: For arbitrary functions $f, g \in {}^X\mathrm{On}$ put*

1. $f \equiv_{\mathcal{F}} g \iff \{u \in X : f(u) = g(u)\} \in \mathcal{F}$,
2. $f \prec_{\mathcal{F}} g \iff \{u \in X : f(u) < g(u)\} \in \mathcal{F}$,
3. $f \preceq_{\mathcal{F}} g \iff \{u \in X : f(u) \leq g(u)\} \in \mathcal{F}$.

If there is no danger of misunderstanding, these properties will also be denoted as

$$f \equiv_{\mathcal{I}} g, \qquad f \prec_{\mathcal{I}} g, \qquad f \preceq_{\mathcal{I}} g,$$

respectively.

The following lemma describes the basic facts about these properties.

Lemma 16.2. *Let $\mathcal{F}$ be a filter in $P(X)$. Then*

1. *$f \equiv_{\mathcal{F}} g$ is an equivalence property on the class $^X\mathrm{On}$;*
2. *$f \prec_{\mathcal{F}} g$ is irreflexive and transitive on the class $^X\mathrm{On}$;*
3. *$f \preceq_{\mathcal{F}} g$ is reflexive and transitive on the class $^X\mathrm{On}$;*
4. *If the filter $\mathcal{F}$ is ω_1-complete, then $\prec_{\mathcal{F}}$ is well-founded on the class $^X\mathrm{On}$;*
5. *If $\mathcal{F}$ is an ultrafilter, then for each $f, g \in {}^X\mathrm{On}$ we have*

$$\text{either} \quad f \prec_{\mathcal{F}} g \quad \text{or} \quad g \prec_{\mathcal{F}} f \quad \text{or} \quad f \equiv_{\mathcal{F}} g,$$

and so

$$f \preceq_{\mathcal{F}} g \iff f \prec_{\mathcal{F}} g \vee f \equiv_{\mathcal{F}} g.$$

Proof. 1.

a) $$\{u \in X : f(u) = f(u)\} = X \in \mathcal{F},$$

and so $\equiv_{\mathcal{F}}$ is reflexive on the class $^X\mathrm{On}$.

b) If $f \equiv_{\mathcal{F}} g$, then $g \equiv_{\mathcal{F}} f$, since

$$\{u \in X : f(u) = g(u)\} = \{u \in X : g(u) = f(u)\},$$

and so $\equiv_{\mathcal{F}}$ is symmetric on the class $^X\mathrm{On}$.

c) If $f \equiv_{\mathcal{F}} g$ and $g \equiv_{\mathcal{F}} h$, then

$$\{u \in X : f(u) = h(u)\} \supset$$
$$\{u \in X : f(u) = g(u)\} \cap \{u \in X : g(u) = h(u)\} \in \mathcal{U}.$$

Thus $f \equiv_{\mathcal{F}} g$, and so the property $\equiv_{\mathcal{F}}$ is transitive on the class $^X\mathrm{On}$.

In cases 2 and 3, the proofs of irreflexivity and reflexivity, respectively, are straightforward; the proof of transitivity is quite similar to that in case 1.$c)$, and it will be left to the reader.

4. Assume, on the contrary, that there is a sequence $\{f_n : n < \omega\} \subset {}^X \mathrm{On}$ such that

$$f_{n+1} \prec_{\mathcal{F}} f_n$$

holds for each $n \in \omega$.

Let

$$A_n = \{u \in X : f_{n+1}(u) < f_n(u)\}.$$

We have $A_n \in \mathcal{F}$ for each n according to our assumption. As $\mathcal{F}$ is ω_1-complete, we have

$$\bigcap_{n<\omega} A_n \in \mathcal{F},$$

and so

$$\bigcap_{n<\omega} A_n \neq \emptyset.$$

If $u \in \bigcap_{n<\omega} A_n$, then

$$f_0(u) > \cdots > f_n(u)\ldots$$

holds. This is a contradiction, since the ordering of the ordinals is a wellordering.

5. Let $f, g \in {}^X \mathrm{On}$. Then

$$X = \{u \in X : f(u) < g(u)\}$$
$$\cup \{u \in X : g(u) < f(u)\} \cup \{u \in X : f(u) = g(u)\} \in \mathcal{U}.$$

As $\mathcal{F}$ is an ultrafilter, exactly one of these three sets is an element of $\mathcal{F}$.

*   *
*

It is noteworthy that the innocent-looking Assertion 4 of Lemma 16.2 is the key ingredient of the proofs below. It is also to be observed that if a property $\leq$ is transitive and reflexive on a set A, then $x \leq y \wedge y \leq x$ is an equivalence property. A transitive and reflexive property $\leq$ is called a *pre-ordering*. On the set of all equivalence classes, a pre-ordering defines an ordering. We did not call attention to this in Lemma 16.2, since ${}^X \mathrm{On}$ is a proper class (i.e., not a set), so it is not even clear if one is allowed to consider equivalence classes. However, as we pointed out in Theorem A7.3, with the assumption of the Axiom of Regularity we can define a good substitute for the equivalence classes. In fact, put

$$[f]_{\mathcal{F}} = \{g \in {}^X \mathrm{On} : g \equiv_{\mathcal{F}} f \wedge \forall h \in {}^X \mathrm{On}\, (h \equiv_{\mathcal{F}} g \implies \mathrm{rk}(g) \leq \mathrm{rk}(h))\};$$

that is, $[f]_{\mathcal{F}}$ is the *set* of elements g of minimal rank that are equivalent to f with respect to $\mathcal{F}$. It is easy to verify that

$$[f]_{\mathcal{F}} = [g]_{\mathcal{F}} \iff f \equiv_{\mathcal{F}} g$$

holds for all elements $f, g \in {}^X\text{On}$.

Lemma 16.2 shows that if $\mathcal{F}$ is an ω_1-complete ultrafilter in $P(X)$ then the property

$$[f]_{\mathcal{F}} \prec [g]_{\mathcal{F}} \; (\iff f \prec_{\mathcal{F}} g)$$

is a wellordering on the class

$$\{[f]_{\mathcal{F}} : f \in {}^X\text{On}\}.$$

Definition 16.4. Let $f : X \to Y$ and $\mathcal{I} \subset P(X)$. The operation $*$ will be defined as follows:

$$f * \mathcal{I} = \{B \subset Y : f^{-1}(B) \in \mathcal{I}\}.$$

Lemma 16.3. If $f : X \to Y$, $\mathcal{I} \subset P(X)$, and $\tilde{\mathcal{I}} = f * \mathcal{I}$ then
1. $\tilde{\mathcal{I}}$ is an ideal on Y, provided that $\mathcal{I}$ is an ideal on X.
2. $\tilde{\mathcal{I}}$ is a κ-complete ideal, provided that $\mathcal{I}$ is a κ-complete ideal on X.
3. $\tilde{\mathcal{I}}$ is a prime ideal, provided that $\mathcal{I}$ is a prime ideal on X.

Proof. 1. *a)* $f^{-1}(Y) = X \notin \mathcal{I}$, and so $Y \notin \tilde{\mathcal{I}}$.
b) If $B \subset B'$ and $B' \in \tilde{\mathcal{I}}$ then

$$f^{-1}(B) \subset f^{-1}(B') \in \mathcal{I},$$

and so

$$f^{-1}(B) \in \mathcal{I}.$$

c) If $A, B \in \tilde{\mathcal{I}}$ then

$$f^{-1}(A \cup B) = f^{-1}(A) \cup f^{-1}(B) \in \mathcal{I},$$

and so

$$A \cup B \in \tilde{\mathcal{I}}.$$

2. The proof is similar to the proof of 1.*c)*.
3. If $B, B' \subset Y$ and $B \cap B' = \emptyset$, then

$$f^{-1}(B) \cap f^{-1}(B') = \emptyset,$$

and so either $f^{-1}(B) \in \mathcal{I}$ or $f^{-1}(B') \in \mathcal{I}$. Thus $B \in \tilde{\mathcal{I}}$ or $B' \in \tilde{\mathcal{I}}$; this property implies that $\tilde{\mathcal{I}}$ is a prime ideal. (Prime ideals were defined in Definition 12.3 as maximal ideals. As pointed out there, they are duals of ultrafilters, and an alternative definition can be given, patterned on Definition 9.1 of ultrafilters; the property corresponding to Convention 2 in the latter definition is confirmed by the above argument.)

$$* \qquad *$$
$$*$$

Lemma 16.4. *If $\kappa > \omega$ is a measurable cardinal, then there is a nonprincipal κ-complete prime ideal $\tilde{\mathcal{I}} \subset \mathrm{P}(\kappa)$ that is a normal ideal.*

Proof. Let $\mathcal{I} \subset \mathrm{P}(\kappa)$ be a κ-complete prime ideal that is not principal, and put $\mathcal{F} = \mathrm{co}(\mathcal{I})$; further, let

$$Y = \{f \in {}^{\kappa}\kappa : \forall \alpha < \kappa \, f^{-1}(\{\alpha\}) \in \mathcal{I}\}.$$

Then Y is a *subset* of the class ${}^{\kappa}\mathrm{On}$. As $\mathcal{I}$ is not a principal ideal, we have

$$\mathrm{Id}_{\kappa}^{-1}(\{\alpha\}) = \{\alpha\} \in \mathcal{I}$$

for every α. Hence $\mathrm{Id}_{\kappa} \in Y$, and so, certainly, $Y \neq \emptyset$. According to Assertion 4 of Lemma 16.2, there is an element of Y that is minimal with respect to $\prec_{\mathcal{F}}$. Denote such an element by f_0. By the definition of Y, we have

$$f_0 : \kappa \to \kappa.$$

Let $\tilde{\mathcal{I}} = f_0 * \mathcal{I}$. We claim that $\tilde{\mathcal{I}}$ satisfies the requirements of the theorem. According to Lemma 16.3, $\tilde{\mathcal{I}}$ is a κ-complete prime ideal. Since $f_0 \in Y$, for an arbitrary $\alpha < \kappa$

$$f_0^{-1}(\{\alpha\}) \in \mathcal{I}$$

holds, and so we have

$$\{\alpha\} \in \tilde{\mathcal{I}};$$

thus $\tilde{\mathcal{I}}$ is not a principal ideal.

We now establish that $\tilde{\mathcal{I}}$ is normal. Put

$$\tilde{\mathcal{F}} = \{B \subset \kappa : f_0^{-1}(B) \in \mathcal{F}\}.$$

As $\mathcal{I}$ and $\tilde{\mathcal{I}}$ are prime ideals, we have

$$B \notin \tilde{\mathcal{I}} \equiv B \in \tilde{\mathcal{F}} \equiv f_0^{-1}(B) \in \mathcal{F} \equiv f_0^{-1}(B) \notin \mathcal{I}.$$

Assume now that $B \notin \tilde{\mathcal{I}}$, and let g be a function on κ that is regressive on B. We may assume that $0 \notin B$, as $\{0\} \in \mathcal{I}$, and so

$$B \setminus \{0\} \notin \tilde{\mathcal{I}}.$$

Define the function $h \in {}^\kappa\kappa$ by putting $h(\alpha) = g(f_0(\alpha))$ for all $\alpha < \kappa$. As g is regressive on B and $0 \notin B$, we have

$$\{\alpha < \kappa : h(\alpha) < f_0(\alpha)\} \supset \{\alpha < \kappa : f_0(\alpha) \in B\} = f_0^{-1}(B) \in \mathcal{F};$$

thus $h \prec_{\mathcal{F}} f_0$. Then, by the minimality of f_0, there is a $\rho < \kappa$ such that

$$h^{-1}(\{\rho\}) \notin \mathcal{I}.$$

But

$$h^{-1}(\{\rho\}) = \{\alpha < \kappa : g(f_0(\alpha)) = \rho\} = f_0^{-1}(g^{-1}(\{\rho\})),$$

and so

$$g^{-1}(\{\rho\}) \notin \tilde{\mathcal{I}}.$$

*   *
*

The attentive reader may realize that, for the function f_0 defined in the proof of Lemma 16.4, the equivalence class $[f_0]$ is the κth element in the wellordering $\prec_{\mathcal{I}}$ of $\{[g]_{\mathcal{I}} : g \in {}^\kappa\text{On}\}$. For a normal ideal $\mathcal{I}$, $[\text{Id}_\kappa]_{\mathcal{I}}$ is the κth equivalence class; furthermore, for an arbitrary $f \prec_{\mathcal{I}} \text{Id}_\kappa$, we have $[f]_{\mathcal{I}} = [\alpha_{\mathcal{I}}]$ for some $\alpha < \kappa$, where α denotes the constant function with value α.

After these preparations, we are in a position to establish the first result saying that a measurable cardinal must be a large inaccessible cardinal:

Theorem 16.2. *If κ is a measurable cardinal greater than ω, then κ is a strong α-Mahlo cardinal for each $\alpha < \kappa$.*

Proof. According to Lemma 16.4, there is a nonprincipal κ-complete prime ideal $\mathcal{I} \subset P(\kappa)$ that is normal. By Corollary 12.2, we then have $\text{NS}(\kappa) \subset \mathcal{I}$. We will use the notation $\mathcal{F} = \text{CO}(\mathcal{I})$. As $\mathcal{I}$ is a prime ideal, we have

$$A \notin \mathcal{I} \iff A \in \mathcal{F}.$$

As $\mathcal{I}$ is a normal ideal, for an arbitrary function that is regressive on a set $A \notin \mathcal{I}$, there is a uniquely determined ordinal γ such that

$$\{\xi \in A : f(\xi) = \gamma\} \in \mathcal{F}.$$

We will denote this ordinal by $\alpha(f)$.

We are going to prove the following assertion:

(1) $$\kappa \setminus M_\alpha \in \mathcal{I} \qquad \text{for each} \qquad \alpha < \kappa;$$

the assertion of the theorem follows from this, as we will show next.

Indeed, assuming that there is a $\beta < \kappa$ such that $\kappa \notin M_\beta$, denote the least such β by α. According to Theorem 16.1 we have $\kappa \in M_0$, and so $\alpha > 0$. Hence, by Definition 15.5 of the classes M_α, α has form $\beta \dot{+} 1$; then we have $\kappa \in M_\beta$. As $\kappa \notin M_\alpha = M(M_\beta)$, it follows from Definition 15.4 of the Mahlo operation that the set $\kappa \cap M_\beta$ is not κ-stationary; hence $\kappa \cap M_\beta \in \mathcal{I}$. Then we have

$$\kappa = (\kappa \setminus M_\beta) \cup (\kappa \cap M_\beta) \in \mathcal{I}$$

by virtue of (1). This is a contradiction.

We prove Assertion (1) by transfinite induction on α. Let $\alpha < \kappa$ be an ordinal such that $\kappa \setminus M_\beta \in \mathcal{I}$ for every $\beta < \alpha$.

We distinguish three cases:

(I) $\alpha = 0$,
(II) α is a limit ordinal,
(III) $\alpha = \beta \dot{+} 1$.

(I) By the definition of M_0, we have

$$\kappa \setminus M_0 = A_0 \cup A_1 \cup A_2,$$

where

$$A_0 = [0, \omega],$$
$$A_1 = \{\xi < \kappa : \xi \text{ is singular}\},$$
$$A_2 = \{\xi < \kappa : \exists \eta < \xi \, (|\xi| \le 2^{|\eta|})\}.$$

It is enough to show that $A_i \in \mathcal{I}$ for $i < 3$. For $i = 0$ this is obvious.

Assume, by contradiction, that $A_1 \notin \mathcal{I}$. The function $\mathrm{cf}(\xi)$ is regressive on A_1, as $\mathrm{cf}(\xi) < \xi$ if ξ is singular. Thus there exists a cardinal $\lambda < \kappa$ and a set $B \subset A_1$ with $B \notin \mathcal{I}$ such that $\mathrm{cf}(\xi) = \lambda$ whenever $\xi \in B$.

For each $\xi \in B$, let

$$\{f_\nu(\xi) : \nu < \lambda\}$$

be a sequence of ordinals strictly increasing in ν such that

$$\sup\{f_\nu(\xi) : \nu < \lambda\} = \xi.$$

Then f_ν is a regressive function on B for every $\nu < \lambda$. For each $\nu < \lambda$, let

$$\alpha_\nu = \alpha(f_\nu)$$

and

$$B_\nu = \{\xi \in B : f_\nu(\xi) = \alpha_\nu\}.$$

Clearly, $B_\nu \in \mathcal{F}$. Put

$$\alpha = \sup\{\alpha_\nu : \nu < \lambda\}.$$

As κ is a regular cardinal, we have $\alpha < \kappa$. The filter $\mathcal{F}$ is κ-complete, and so

$$\bigcap_{\nu < \lambda} B_\nu \in \mathcal{F}.$$

Hence there is a ξ such that

$$\xi \in \bigcap_{\nu < \lambda} B_\nu \setminus \{\alpha\}.$$

In this case, however,

$$\xi = \sup\{f_\nu(\xi) : \nu < \lambda\} = \sup\{\alpha_\nu : \nu < \lambda\} = \alpha,$$

which is a contradiction, and so

$$A_1 \in \mathcal{I}.$$

Next assume, by contradiction, that $A_2 \notin \mathcal{I}$. Define the function g on A_2 by

$$g(\xi) = \min\{|\eta| : |\xi| \le 2^{|\eta|}\};$$

clearly, g is a regressive function of A_2. There is therefore a set $C \subset A_2$ with $C \notin \mathcal{I}$ and a cardinal λ such that

$$f(\xi) = \lambda \qquad \text{for} \qquad \xi \in C.$$

Thus

$$|\xi| \le 2^\lambda$$

holds for each $\xi \in C$. As C is cofinal in κ, we have $|\xi| \le 2^\lambda$ for every $\xi < \kappa$. Hence

$$\kappa \le (2^\lambda)^+$$

follows, which contradicts κ's being strongly inaccessible. Therefore $A_2 \in \mathcal{I}$.

(II)

$$\kappa \setminus M_\alpha = \bigcup_{\beta < \alpha} (\kappa \setminus M_\beta)$$

and so

$$\kappa \setminus M_\alpha \in \mathcal{I}.$$

(III)

$$\kappa \setminus M_\alpha = (\kappa \setminus M_\beta) \cup \left(\kappa \cap (M_\beta \setminus M(M_\beta))\right).$$

By the induction hypothesis, it is sufficient to show that

$$A = \left(\kappa \cap \left(M_\beta \setminus M(M_\beta) \right) \right) \in \mathcal{I}.$$

Assume, on the contrary, that $A \notin \mathcal{I}$.

For each $\lambda \in A$, we have $\lambda \in M_\beta$, and, according to Definition 15.4 of the Mahlo operation, the set $\lambda \cap M_\beta$ is not stationary in λ. Let the set $C_\lambda \subset \lambda$ be a witness to this, that is, assume that

$$(2) \qquad C_\lambda \subset \lambda \quad \text{is a } \lambda\text{-club, and} \quad C_\lambda \cap M_\beta = \emptyset.$$

Define the set $C \subset \kappa$ as follows:

$$C = \big\{ \xi < \kappa : \{ \lambda \in A : \xi \in C_\lambda \} \in \mathcal{F} \big\}.$$

We are going to show that

$$(3) \qquad\qquad\qquad C \text{ is a } \kappa\text{-club.}$$

As we clearly have $C \cap M_\beta = \emptyset$, (3) implies that

$$A \subset M_\beta \in \mathrm{NS}(\kappa),$$

which contradicts our initial assumption $A \notin \mathcal{I}$.

For the proof of (3), we first show that C is cofinal in κ. Let $\eta < \kappa$. Then

$$A \setminus (\eta \dot{+} 1) \notin \mathcal{I}.$$

Define the regressive function g_η on $A \setminus (\eta \dot{+} 1)$ by putting

$$g_\eta(\lambda) = \min(C_\lambda \setminus \eta).$$

This definition is sound, since we have $\eta < \lambda$ for $\lambda \in A \setminus (\eta \dot{+} 1)$, and, further, C_λ is cofinal in λ. g_η is a regressive function on $A \setminus (\eta \dot{+} 1)$, and, clearly, $\alpha(g_\eta) \geq \eta$ and $\alpha(g_\eta) \in C$.

We have yet to see that C is closed in κ. To this end, pick the set D with

$$\emptyset \neq D \subset C, \qquad \sup D = \xi < \kappa.$$

For each $\eta \in D$, put

$$A_\eta = \{ \lambda \in A : \eta \in C_\lambda \}.$$

By the definition of C, we have $A_\eta \in \mathcal{F}$ whenever $\eta \in D$. $\mathcal{F}$ is κ-complete, and so

$$E = \bigcap_{\eta \in D} A_\eta \in \mathcal{F}.$$

Given any $\lambda \in E$ with $\lambda > \xi$, we have

$$D \subset C_\lambda,$$

and thus

$$\xi \in C_\lambda$$

since C_λ is λ-closed according to (2). Hence $\xi \in C$.

$$* \qquad *$$
$$*$$

Corollary 16.1. *If $\kappa > \omega$ is a measurable cardinal, then for every non-principal normal κ-complete prime ideal we have*

$$\kappa \setminus M^\infty \in \mathcal{I}.$$

Proof.

$$\kappa \setminus M^\infty = \kappa \setminus \underset{\alpha < \kappa}{\Delta} (\kappa \cap M_\alpha).$$

For each $\alpha < \kappa$

$$\kappa \cap M_\alpha \in \mathcal{F} = \mathrm{co}(\mathcal{I})$$

holds by virtue of (1) in the proof of the above theorem, and so

$$\underset{\alpha < \kappa}{\Delta} (\kappa \cap M_\alpha) \in \mathcal{F}.$$

$$* \qquad *$$
$$*$$

After the discovery of Theorem 16.1, the problem of whether the first strongly inaccessible cardinal could be measurable remained open for about thirty years. In 1961, W. P. Hanf, a student of A. Tarski, established a model-theoretic conjecture formulated by Tarski that implied the answer immediately. The proof given above is due to H. J. Keisler and A. Tarski. It is clear that Corollary 16.1 can be sharpened, for example, by iterating the operation M^∞. These strengthenings, however, still do not give a real indication of the size of measurable cardinals. A more accurate picture can only be gained by tools from model theory, and this is beyond the scope of the present book. Theorem 16.2 can be established under much weaker assumptions on κ (for example, for the much "smaller" weakly compact cardinals, defined in Section 18).

Problems

Definition. *Given a property* R *on the class* X, *denote by* $[u]_R$ *the class* $\{v \in X : vRu\}$. R *is called set-like if* $[u]_R$ *is always a set;* R *is called extensional if* $u = v$ *whenever* $u, v \in X$ *and* $[u]_R = [v]_R$.

1. *(Theorem on Well-Founded Induction)* Prove that if R is a well-founded set-like property on X, and Φ is a property such that

$$\forall u \left(\forall v \in [u]_R \, \Phi(v) \implies \Phi(u)\right)$$

holds, then we have

$$\forall u \, \Phi(u).$$

2. *(Theorem on Well-Founded Recursion)* Prove that if R is a well-founded set-like property on X, and $\mathcal{G}$ is an arbitrary operation, then there is a uniquely determined operation $\mathcal{F}$ on X such that, for every $u \in X$, we have

$$\mathcal{F}(u) = \mathcal{G}\big(\mathcal{F}\big|[u]_R\big).$$

3. *(Mostowski's Collapsing Lemma)* Prove that if R is an extensional well-founded set-like property on the class X, then there is a transitive class M and an operation $\mathcal{F}$ such that $X \sim_{\mathcal{F}} M$ and

$$uRv \iff \mathcal{F}(u) \in \mathcal{F}(v)$$

holds for every $u, v \in X$.

4.* Prove that if κ is a measurable cardinal $> \omega$ and

$$(\forall \lambda : \omega \leq \lambda < \kappa)(2^\lambda = \lambda^+),$$

then

$$2^\kappa = \kappa^+.$$

5.* Let $\kappa \geq \omega$. Show that a discrete topological space of cardinality κ can be embedded as a closed subspace into a power of the real line if and only if κ is smaller than the first measurable cardinal greater than ω (here the word *embedding* means a homeomorphic map into the space).

17. REAL-VALUED MEASURABLE
CARDINALS, SATURATED IDEALS

In this section, we consider Problem $(*)$ of S. Banach formulated in Section 16.

Definition 17.1. *Let S be a set system, $S \subset P(X)$, and $\lambda \geq \omega$. The function $\mu : S \to [0,1]$ is said to be λ-additive on S if for every system*

$$\{A_\alpha : \alpha < \tau\} \subset S$$

of pairwise disjoint sets such that

$$A = \bigcup_{\alpha < \tau} A_\alpha \in S \quad \text{and} \quad \tau < \lambda$$

we have

$$\mu(A) = \sum_{\alpha < \tau} \mu(A_\alpha).$$

It is to be noted that the sum $\sum_{\alpha < \tau} \mu(A_\alpha)$ can be finite only if $\mu(A_\alpha) = 0$ except for countably many values of the subscript α. Indeed, the sum of uncountably many positive numbers is infinite, since one can find a positive rational number that is smaller than infinitely many of these numbers. It is easy to verify the following analogue of Lemma 16.1:

Lemma 17.1. *If κ is the least cardinal such that there is an $\aleph_1$-additive function*

$$\mu : P(\kappa) \to [0,1]$$

that is not trivial, i.e., is such that

$$\mu(\kappa) = 1 \quad \text{and} \quad \mu(\{\alpha\}) = 0 \quad \text{for} \quad \alpha < \kappa,$$

then μ is κ-additive on $P(\kappa)$.

The proof is completely analogous to that of Lemma 16.1, and we leave it to the reader. This lemma justifies the following definition.

Definition 17.2. *The cardinal κ is called* a real-valued measurable cardinal *if there is a function*

$$\mu : P(\kappa) \to [0, 1]$$

that is κ-additive on $P(\kappa)$ and is nontrivial, *i.e,*

$$\mu(\kappa) = 1 \qquad \text{and} \qquad \mu(\{\alpha\}) = 0 \quad \text{for} \quad \alpha < \kappa.$$

We remark that, according to our definition, every measurable cardinal (including ω) is also real-valued measurable. The problem now is to determine which cardinals $\kappa > \omega$ can be real-valued measurable. The main goal of this section is to prove the analogue of Theorem 16.2: If $\kappa > \omega$ is real-valued measurable, then κ is an α-Mahlo cardinal for every $\alpha < \kappa$. This κ is therefore just as large in relation to the classes $\overline{M}_\alpha$ as uncountable measurable cardinals are in relation to the classes M_α.

As further pieces of information, we also mention the following results established by R. Solovay in 1966.

a) $\text{Con}(\text{ZFC} \cup \{\exists \kappa > \omega \quad \kappa \text{ is real-valued measurable}\}) \iff \text{Con}(\text{ZFC} \cup \{\exists \kappa > \omega \quad \kappa \text{ is measurable}\}).$

b) $\text{Con}(\text{ZFC} \cup \{\exists \kappa > \omega \quad \kappa \text{ is measurable}\}) \iff \text{Con}(\text{ZFC} \cup \{2^{\aleph_0} \text{ is real-valued measurable}\}).$

Below we will prove results stronger than the just mentioned analogue of Theorem 16.2.

Lemma 17.2. *If $\kappa > \omega$ is a real-valued measurable cardinal and $\mu : P(\kappa) \to [0, 1]$ is a measure witnessing this, then*

$$\mathcal{I} = \{A \subset \kappa : \mu(A) = 0\}$$

is a κ-complete $\aleph_1$-saturated ideal such that

$$[\kappa]^{<\kappa} \subset \mathcal{I}.$$

Proof. The concept of λ-saturated ideal was described in Definition 12.15. If we had a system

$$\mathcal{G} \subset P(\kappa) \setminus \mathcal{I} \qquad |\mathcal{G}| \geq \aleph_1$$

of pairwise disjoint sets, then the sum

$$\sum_{A \in \mathcal{G}} \mu(A)$$

would be infinite.

The other assertions about $\mathcal{I}$ are immediate.

Definition 17.3. *The cardinal $\kappa \geq \omega$ is called a* saturated cardinal *if there is a κ-complete, κ-saturated ideal $\mathcal{I}$ on κ such that $[\kappa]^{<\kappa} \subset \mathcal{I}$.*

According to this definition and Lemma 17.2, real-valued measurable cardinals are also saturated cardinals. We will show that if κ is a saturated cardinal, then κ is very large. We would like to point out that the term "saturated cardinal" is not generally accepted in the literature; we introduce it here for the sake of brevity.

Before we turn to the proofs of the main results, we establish a classical result from the 1930s, which serves as a supplement to the quoted results of Solovay, and shows that the "true" real-valued measurable cardinals are to be sought among cardinals not greater than $2^{\aleph_0}$.

Theorem 17.1 *(A. Tarski). If $\kappa > \omega$, $\mathcal{I} \subset \mathrm{P}(\kappa)$ is a κ-complete, λ-saturated ideal on κ with $[x]^1 \subset \mathcal{I}$, and, further, $2^{<\lambda} < \kappa$ holds, then κ is measurable.*

Corollary 17.1. *If $\kappa > 2^{\aleph_0}$ and κ is real-valued measurable, then κ is also measurable.*

The corollary is immediate from Lemma 17.2 and Theorem 17.1.

Proof of Theorem 17.1. In view of the κ-completeness, we also have $[\kappa]^{<\kappa} \subset \mathcal{I}$. For this reason, we have $|A| = \kappa$ whenever case $A \notin \mathcal{I}$. Thus, if the ideal $\mathcal{I} + (\kappa \setminus A)$ is 2-saturated then, by Lemma 12.3, the cardinal κ is measurable. Therefore, by contradiction, we assume that

(1) for each set $A \subset \kappa$, there are sets B_A^0 and B_A^1 such that

$$B_A^0 \cup B_A^1 = A, \quad B_A^0 \cap B_A^1 = \emptyset, \quad \text{and} \quad B_A^i \notin \mathcal{I} \quad \text{for } A \notin \mathcal{I} \text{ and } i < 2.$$

Define the sets

$$A_f \subset \kappa \qquad (f \in {}^{\nu}2, \quad \nu \leq \lambda)$$

by transfinite recursion on ν. Put $A_\emptyset = \kappa$. Let $\nu > 0$, and assume that the sets A_f have already been defined for $f \in {}^{\mu}2$ with $\mu < \nu$. If ν is a limit ordinal, then put

$$A_f = \cap\{A_{f|\mu} : \mu < \nu\}.$$

If $\nu = \mu \dotplus 1$ for some μ and $f \in {}^{\nu}2$, then put

$$A_f = B_{A_{f|\mu}}^i \qquad \text{if} \qquad f(\mu) = i \quad (i < 2).$$

Next we prove the following assertion.

(2) For each $\nu \leq \lambda$, we have

$$\kappa = \bigcup\{A_f : f \in {}^{\nu}2\},$$
$$A_{f_0} \cap A_{f_1} = \emptyset \quad \text{whenever} \quad f_0 \neq f_1 \quad \text{with} \quad f_0, f_1 \in {}^{\nu}2,$$

and, for each $\mu < \nu$ and $f \in {}^{\nu}2$,

$$\alpha \in A_f \qquad \text{implies} \qquad \alpha \in A_{f|\mu}.$$

We prove (2) by transfinite induction on ν. For $\nu = 0$ the assertion is obvious. Assume that $\nu > 0$ and the assertion is true for each ordinal $\mu < \nu$. If $\nu = \mu + 1$ and $\alpha \in \kappa$, then by the induction hypothesis there is a uniquely determined function $g \in {}^{\mu}2$ such that $\alpha \in A_g$ and $\alpha \in A_{g|\mu'}$ for $\mu' < \mu$. Thus, by (1) there is a uniquely determined $i < 2$ such that

$$f = g \cup \{\langle \mu, i \rangle\} \in {}^{\nu}2 \qquad \text{and} \qquad \alpha \in A_f.$$

Then

$$\alpha \in A_{g|\mu'} = A_{f|\mu'} \qquad \text{for} \qquad \mu' \le \mu.$$

If ν is a limit ordinal, then for each $\mu < \nu$ there is a uniquely determined $g_\mu \in {}^{\mu}2$ for which $\alpha \in A_{g_\mu}$. By the induction hypothesis, we have $g_{\mu'} = g_\mu|\mu'$ for $\mu' < \mu$. Thus

$$f = \bigcup_{\mu < \nu} g_\mu \in {}^{\nu}2, \qquad \alpha \in A_f,$$

and

$$\alpha \in A_{f|\mu}$$

for $\mu < \nu$.

Write

$$ {}^{<\lambda}2 = \bigcup \{{}^{\nu}2 : \nu < \lambda\} \qquad \text{and} \qquad B = \bigcup\{A_f : f \in {}^{<\lambda}2 \wedge A_f \in \mathcal{I}\}.$$

By the assumption on the cardinal κ and by the κ-completeness of $\mathcal{I}$, we have $B \in \mathcal{I}$. Let $\alpha \in \kappa \setminus B$. By (2), we have $\alpha \in A_f$ for some $f \in {}^{\lambda}2$. Using (2) again, by the definition of B we obtain

$$A_{f|\nu+1} \notin \mathcal{I} \qquad \text{for} \qquad \nu < \lambda.$$

For $\nu < \lambda$, write

$$C_\nu = A_{f|\nu} \setminus A_{f|\nu+1}.$$

By (1), $C_\nu \notin \mathcal{I}$, and, by (2), the sets C_ν are pairwise disjoint. This contradicts the assumption that $\mathcal{I}$ is λ-saturated.

$$*\qquad *$$
$$*$$

Next we prove the classical analogue of Theorem 16.1, dating back to the 1930s.

Theorem 17.2 *(S. Ulam). If $\kappa > \omega$ is a saturated cardinal, then κ is inaccessible.*

Proof. If κ is a singular cardinal then, as we saw in the proof of Theorem 16.1, there is no κ-complete ideal $\mathcal{I}$ on κ such that $[\kappa]^1 \subset \mathcal{I}$.

What we have to prove, therefore, is that if $\kappa = \lambda^+$ for some $\lambda \geq \omega$, then κ is not a saturated cardinal. To this end, we will construct a set system

$$\{A_{\xi,\eta} : \xi < \lambda \wedge \eta < \lambda^+\} \subset P(\lambda^+)$$

satisfying the following conditions.
  1. For each $\xi < \lambda$,
$$A_{\xi,\eta} \cap A_{\xi,\eta'} = \emptyset$$

holds whenever $\eta \neq \eta'$ and $\eta, \eta' < \lambda^+$.
  2. For each $\eta < \lambda^+$ we have

$$\left| \lambda^+ \setminus \bigcup \{A_{\xi,\eta} : \xi < \lambda\} \right| < \lambda^+.$$

A set system with these conditions is called a λ^+ Ulam matrix. Informally, in the Ulam matrix, each row consists of pairwise disjoint sets, and the union of each column almost covers the whole set. The above set system is constructed as follows.

Let $f_\beta : \beta \to \lambda$ be a one-to-one function for each $\beta < \lambda^+$. There is such a function, since $|\beta| \leq \lambda$ for $\beta < \lambda^+$. For every $\xi < \lambda$, $\eta < \lambda^+$, and $\beta < \lambda^+$ put

$$\beta \in A_{\xi,\eta} \iff \eta < \beta \wedge f_\beta(\eta) = \xi.$$

We claim that this set system satisfies Conditions 1 and 2.
  1. Let $\xi < \lambda$ and $\eta, \eta' < \lambda^+$, $\eta \neq \eta'$. If there were a β with

$$\beta \in A_{\xi,\eta} \cap A_{\xi,\eta'},$$

then we would have

$$\eta, \eta' < \beta \quad \text{and} \quad f_\beta(\eta) = f_\beta(\eta') = \xi,$$

but this contradicts that f_β is one-to-one.
  2. If $\eta < \lambda^+$, then
$$\eta \in D(f_\beta)$$

for each $\beta > \eta$; hence there is an ordinal $\xi < \lambda$ such that

$$\xi = f_\beta(\eta).$$

Thus

$$\lambda^+ \setminus (\eta + 1) \subset \bigcup \{A_{\xi,\eta} : \xi < \lambda\}.$$

For the proof of Theorem 17.2, now consider a λ^+ Ulam matrix satisfying Conditions 1 and 2, and let $\mathcal{I} \subset P(\lambda^+)$ be a λ^+-complete ideal for which $[\lambda^+]^{<\lambda^+} \subset \mathcal{I}$. Then, by Condition 2, we have

$$\lambda^+ \setminus \bigcup\{A_{\xi,\eta} : \xi < \lambda\} \in \mathcal{I},$$

and thus

$$\bigcup\{A_{\xi,\eta} : \xi < \lambda\} \notin \mathcal{I},$$

for each $\eta < \lambda^+$.

As the ideal $\mathcal{I}$ is λ^+-complete, there is a $\xi < \lambda$ such that $A_{\xi,\eta} \notin \mathcal{I}$. Define the function $f : \lambda^+ \to \lambda$ by putting

$$f(\eta) = \min_{<}\{\xi < \lambda : A_{\xi,\eta} \notin \mathcal{I}\}$$

for each $\eta < \lambda^+$. Then there clearly exists a set $L \subset \lambda^+$ with $|L| = \lambda^+$ and an ordinal $\xi < \lambda$ such that

$$f(\eta) = \xi \qquad \text{for} \qquad \eta \in L.$$

According to the definition of f, the set system $\{A_{\xi,\eta} : \eta \in L\}$ consists of sets not belonging to $\mathcal{I}$, and these sets are pairwise disjoint by virtue of Condition 1. As $|L| = \lambda^1$, the ideal $\mathcal{I}$ is not λ^+-saturated.

$$* \qquad *$$
$$*$$

Our next task is to verify the analogue of Lemma 16.4.

Lemma 17.3. *If $\kappa > \omega$ a saturated cardinal, then there is a normal κ-saturated ideal $\tilde{\mathcal{I}}$ on κ such that*

$$[\kappa]^{<\kappa} \subset \tilde{\mathcal{I}}.$$

Proof. Let $\mathcal{I} \subset P(\kappa)$ be a κ-complete, κ-saturated ideal such that

$$[\kappa]^{<\kappa} \subset \mathcal{I};$$

write $\mathcal{F} = \mathrm{co}(\mathcal{I})$. Let

$$Y = \left\{f \in {}^{\kappa}\mathrm{On} : \forall \rho < \kappa \left(f^{-1}(\{\rho\}) \in \mathcal{I} \wedge f(\rho) < \kappa\right)\right\}.$$

Then Y is a subset of the class ${}^{\kappa}\mathrm{On}$ for which $\mathrm{Id}_{\kappa} \in Y$, since

$$\mathrm{Id}_{\kappa}^{-1}(\{\rho\}) = \{\rho\} \in \mathcal{I}$$

holds for each $\rho < \kappa$. Let f_0 be a minimal element of Y in the partial ordering $\prec_\mathcal{F}$; such a minimal element exists according to Assertion 4 of Lemma 16.2. Let $\tilde{\mathcal{I}} = f_0 * \mathcal{I}$. By Lemma 16.3, $\tilde{\mathcal{I}}$ is a κ-complete ideal. If $\rho < \kappa$, then, by the definition of f_0, we have $f_0^{-1}(\{\rho\}) \in \mathcal{I}$, and so $\{\rho\} \in \tilde{\mathcal{I}}$. Hence

$$[\kappa]^{<\kappa} \in \tilde{\mathcal{I}}.$$

We claim that $\tilde{\mathcal{I}}$ is κ-saturated. Indeed, if $\mathcal{G} \subset \mathrm{P}(\kappa)$ consists of pairwise disjoint sets, then the same is true about the set system

$$\{f_0^{-1}(A) : A \in \mathcal{G}\}.$$

As $\mathcal{I}$ is κ-saturated, if $|\mathcal{G}| \geq \kappa$ we have $f^{-1}(A) \in \mathcal{I}$ for some $A \in \mathcal{G}$, and then $A \in \tilde{\mathcal{I}}$. Write

$$\tilde{\mathcal{F}} = \mathrm{co}(\tilde{\mathcal{I}}) = f_0 * \mathcal{F}.$$

We claim that $\tilde{\mathcal{I}}$ satisfies the following:

(1) If $A \in \tilde{\mathcal{F}}$ and g is a regressive function on A, then there is an ordinal ρ such that

$$g^{-1}(\{\rho\}) \notin \tilde{\mathcal{I}}.$$

Claim (1) can be verified as follows, similarly as the analogous statement was established in the proof of Lemma 16.4:

We may assume that $0 \notin A$, since $A \setminus \{0\} \in \tilde{\mathcal{F}}$ provided $A \in \tilde{\mathcal{F}}$. Let g be a regressive function on the set A, and let $h = g \circ f_0$. Then

$$\{\alpha < \kappa : h(\alpha) < f_0(\alpha)\} \supset \{\alpha < \kappa : f_0(\alpha) \in A\} \in \tilde{\mathcal{F}},$$

and so $h \prec_\mathcal{F} f_0$. In view of the minimality of f_0, there is a $\rho < \kappa$ such that

$$h^{-1}(\{\rho\}) \notin \mathcal{I};$$

here

$$h^{-1}(\{\rho\}) = \{\alpha < \kappa : g(f_0(\alpha)) = \rho\} = f_0^{-1}(g^{-1}(\{\rho\})).$$

Thus

$$g^{-1}(\{\rho\}) \notin \tilde{\mathcal{I}}.$$

In the remaining part of the proof, we will show that an arbitrary κ-complete ideal $\tilde{\mathcal{I}}$ with $[\kappa]^{<\kappa} \subset \tilde{\mathcal{I}}$ satisfying (1) can be extended to a normal ideal $\hat{\mathcal{I}}$. As $\tilde{\mathcal{I}} \subset \hat{\mathcal{I}}$, the ideal $\hat{\mathcal{I}}$ is clearly also κ-saturated, and we have $[\kappa]^{<\kappa} \subset \hat{\mathcal{I}}$ as well; by Lemma 12.2, $\hat{\mathcal{I}}$ is also κ-complete.

The method to be described can also be used to construct the ideal $\hat{\mathcal{I}} = \mathrm{NS}(\mathcal{I})$ by taking $\tilde{\mathcal{I}}$ to be $[\kappa]^{<\kappa}$ (instead of the $\tilde{\mathcal{I}}$ defined above), if beforehand we prove that, for each $\xi < \kappa$ and for every regressive function f on $\kappa \setminus \xi$, there is a $\rho < \kappa$ such that $|f^{-1}(\{\rho\})| = \kappa$. In this way, we can obtain a proof of Fodor's Theorem from Neumer's Theorem. In Section 12 we derived Fodor's Theorem directly, and then we used it to deduce Neumer's Theorem; however, a direct proof of the latter is simpler than one of the former.

Let

$$\hat{\mathcal{I}} = \{A \subset \kappa : \exists f \, (f \text{ is regressive on } A \quad \text{and} \quad \forall \rho < \kappa \, f^{-1}(\{\rho\}) \in \tilde{\mathcal{I}})\}.$$

We claim that $\hat{\mathcal{I}}$ satisfies the requirements stated above. Indeed, as $\kappa \in \tilde{\mathcal{F}}$, we also have $\kappa \notin \hat{\mathcal{I}}$ according to Assertion (1). If $A_0, A_1 \in \hat{\mathcal{I}}$ ($A_0 \cap A_1 = \emptyset$), and f_0, f_1 are regressive functions witnessing this, then the regressive function $f = f_0 \cup f_1$ is a witness that $A \notin \hat{\mathcal{I}}$ for the set $A = A_0 \cup A_1$.

Thus we only have to verify the normality of $\hat{\mathcal{I}}$. To this end, assume $A \notin \hat{\mathcal{I}}$, and let f be a regressive function on A ($D(f) = A$). For each $\rho < \alpha$, write

$$f^{-1}(\{\rho\}) = A_\rho.$$

Assume, by contradiction, that $A_\rho \in \hat{\mathcal{I}}$ for each $\rho < \alpha$, and let f_ρ be a regressive function on the set A_ρ such that

$$f_\rho^{-1}(\{\alpha\}) \in \tilde{\mathcal{I}} \qquad \text{for} \qquad \alpha < \kappa.$$

Clearly,

$$A = \bigcup \{A_\rho : \rho < \kappa\},$$

and the sets A_ρ are pairwise disjoint. Define the function g on the set A by putting

$$g(\alpha) = \max\{\rho, f_\rho(\alpha)\} \qquad \text{for} \qquad \alpha \in A_\rho.$$

g is a regressive function on A, since, given $\alpha \in A_\rho$ with $\alpha > 0$, we have $\rho = f(\alpha) < \alpha$ and $f_\rho(\alpha) < \alpha$.

Now fix $\sigma < \kappa$, and consider the α's in A with

$$\sigma = g(\alpha) = \max\{f(\alpha), f_{f(\alpha)}(\alpha)\}$$

(the second equation holds by the definition of g). Then either $\sigma = f(\alpha)$ and $f_\sigma(\alpha) \leq \sigma$ or $f_{f(\alpha)}(\alpha) = \sigma$ and $f(\alpha) \leq \sigma$. Hence

$$g^{-1}(\{\sigma\}) \subset \bigcup \{f_\rho^{-1}(\sigma \dot{+} 1) : \rho \leq \sigma\}.$$

Then

$$g^{-1}(\{\rho\}) \in \tilde{\mathcal{I}},$$

as $\tilde{\mathcal{I}}$ is κ-complete. Thus g witnesses $A \in \hat{\mathcal{I}}$. This is a contradiction.

$$* \qquad *$$
$$*$$

Next we prove the analogue of Theorem 16.2.

Theorem 17.3 *(R. Solovay)*. *If $\kappa > \omega$ is a saturated cardinal, then κ is an α-Mahlo cardinal for each $\alpha < \kappa$.*

Proof. According to Lemma 17.3, there is a normal ideal $\mathcal{I} \subset P(\kappa)$ such that $[\kappa]^{<\omega} \subset \mathcal{I}$. This $\mathcal{I}$ is also κ-complete. By Corollary 12.2 we then have $\text{NS}(\kappa) \subset \mathcal{I}$.

We prove the following assertion.

(1) $\kappa \setminus \overline{M}_\alpha \in \mathcal{I}$ holds each $\alpha < \kappa$.

As we have $\kappa \in \overline{M}_0$ by Theorem 17.2, the assertion of the theorem follows from this in exactly the same way as that of Theorem 16.2 did from Assertion (1) there.

For the proof, we will use some ideas formulated in the proof of Theorem 12.5.A. For an arbitrary $A \subset \kappa$ and an arbitrary function f defined on A put

$$A(f \geq \rho) = \{\xi \in A : f(\xi) \geq \rho\}.$$

We say that the function $f \in {}^{\kappa}\kappa$ is essentially bounded on A if there is an ordinal $\alpha < \kappa$ such that $A(f \geq \alpha) \in \mathcal{I}$.

In the last paragraph of the proof of Theorem 12.5.A, we in effect showed that if $\mathcal{I}$ is a normal, κ-complete, and κ-saturated ideal in $P(\kappa)$, then

(2) For each $A \subset \kappa$, every regressive function f on A is essentially bounded on A.

The adaptation of that proof to the present situation is left to the reader. For an arbitrary function f regressive on A, we will denote by $\alpha_A(f)$ its least essential bound. That is,

$$\alpha_A(f) = \min\{\alpha : A(f \geq \alpha) \in \mathcal{I}\}.$$

We prove Assertion (1) by transfinite induction on α. We distinguish three cases.

(I) $\alpha = 0$.
(II) α is a limit ordinal.
(III) $\alpha = \beta \dotplus 1$.

(I) According to the definition of $\overline{M}_0$ we have $\kappa \setminus \overline{M}_0 = A_0 \cup A_1 \cup A_2$, where

$$A_0 = \{\xi < \kappa : \xi \text{ is a successor ordinal}\} \cup \{0, \omega\},$$
$$A_1 = \{\xi < \kappa : \xi \text{ is a singular limit ordinal}\},$$
$$A_2 = \{\xi < \kappa : \xi \text{ is a successor cardinal}\}.$$

It is sufficient to show that $A_i \in \mathcal{I}$ for $i < 3$. It is obvious that

$$A_0 \in \text{NS}(\kappa) \subset \mathcal{I}.$$

The assertion for A_1 can be proved by using (2) in exactly the same way as Theorem 12.5.A was proved. Thus this proof will be left to the reader.

Assume now that $A_2 \notin \mathcal{I}$. Define the function f by putting

$$f(\xi) = \min\{\lambda < \xi : \xi = \lambda^+\} \qquad \text{for} \qquad \xi \in A_2.$$

Then f is regressive and one-to-one on A_2, and so

$$A_2 \in \text{NS}(\kappa) \subset \mathcal{I}.$$

(II) $\kappa \setminus \overline{M}_\alpha = \bigcup_{\beta < \alpha}(\kappa \setminus \overline{M}_\beta) \in \mathcal{I}.$

(III) Analogously as in the proof of Theorem 16.2, we have to show that

$$A = \kappa \cap \left(\overline{M}_\beta \setminus M(\overline{M}_\beta)\right) \in \mathcal{I}.$$

Taking Definition 15.4 of the Mahlo operation into account, we have

$$\lambda \in \overline{M}_\beta \quad \text{and} \quad \lambda \cap \overline{M}_\beta \subset \text{NS}(\lambda) \qquad \text{for} \qquad \lambda \in A.$$

As $A \subset \overline{M}_\beta$, from this it follows also that

$$\lambda \cap A \in \text{NS}(\lambda) \quad \text{for} \quad \lambda \in A.$$

Next we are going to establish an auxiliary result, which will be used also in the proof of another theorem below.

(3) Under the assumptions on the ideal $\mathcal{I}$, for an arbitrary set $B \subset \kappa$ consisting of regular cardinals greater than ω, the set

$$C = \{\lambda \in B : B \cap \lambda \in \text{NS}(\lambda)\}$$

belongs to $\mathcal{I}$.

From (3), it immediately follows that $A \in \mathcal{I}$, and so from now on we will confine ourselves to the proof of (3). Proceeding by reductio ad absurdum, we assume that $C \notin \mathcal{I}$, and so $B \notin \mathcal{I}$ as well. By Neumer's Theorem, for each $\lambda \in C$ there is a function f_λ such that

(4) f_λ is regressive on $B \cap \lambda$, and for each ρ, we have

$$|f_\lambda^{-1}(\{\rho\})| < \lambda \qquad \text{for} \qquad \lambda \in C.$$

We right away formulate also a consequence of (4):

For each $\rho < \kappa$, define the function $g_\rho : C \to \kappa$ by

$$g_\rho(\lambda) = \sup f_\lambda^{-1}(\{\rho\}).$$

Then, taking into account that λ is regular for $\lambda \in C$, we have $g_\rho(\lambda) < \lambda$, and so g_ρ is regressive on C. Put $\alpha_\rho = \alpha_C(g_\rho)$ for $\rho < \kappa$.

Turning back to the proof of (3), let $\xi \in B$ be arbitrary. Define the function h_ξ on the set $C \setminus (\xi \dotplus 1)$ by putting $h_\xi(\lambda) = f_\lambda(\xi)$. Then

$$h_\xi(\lambda) = f_\lambda(\xi) < \xi < \lambda,$$

and so h_ξ is regressive on $C \setminus (\xi \dotplus 1)$. There is therefore a smallest ordinal $\eta < \xi$ such that

$$C_\xi \overset{\text{def}}{=} \{\lambda \in C \setminus (\xi \dotplus 1) : f_\lambda(\xi) = \eta\} \notin \mathcal{I}.$$

Put $f(\xi) = \eta$; this defines f as a regressive function on B. We are going to show that $|f^{-1}(\{\rho\})| < \kappa$ for every ρ; from this the conclusion will follow, since this implies $B \in \text{NS}(\kappa)$ by Neumer's Theorem, and so $B \in \mathcal{I}$ also holds.

To this end, pick an arbitrary $\rho < \kappa$. If $f(\xi) = \rho$ for some $\xi < \kappa$ then

$$f_\lambda(\xi) = \rho \qquad \text{for} \qquad \lambda \in C_\xi \notin \mathcal{I},$$

and so

$$\xi \in f_\lambda^{-1}(\{\rho\}) \qquad \text{for} \qquad \lambda \in C_\xi \notin \mathcal{I}.$$

Thus, if $\lambda \in C_\xi \notin \mathcal{I}$, we have

$$g_\rho(\lambda) \geq \xi;$$

this implies $\xi < \alpha_\rho$. Thus

$$f^{-1}(\{\rho\}) \subset \alpha_\rho < \kappa.$$

$$*\qquad*$$
$$*$$

Corollary 17.2. If $\kappa > \omega$ is a saturated cardinal and $\mathcal{I} \subset \text{P}(\kappa)$ is a normal, κ-saturated ideal with $[\kappa]^{<\kappa} \subset \mathcal{I}$ then $\kappa \setminus \overline{M}^\infty \in \mathcal{I}$.

Proof. We have

$$\kappa \setminus \overline{M}^\infty = \kappa \setminus \underset{\alpha < \kappa}{\triangle} \overline{M}_\alpha \in \mathcal{I}$$

according to Theorem 17.3.

$$*\qquad*$$
$$*$$

Further iteration of the Mahlo operation is possible here as well, but that will not give any significantly new results. We remark that the model-theoretic characterization of saturated cardinals is almost as nice as that of measurable cardinals – except that it uses Boolean-valued models, whereas no model-theoretic characterization at all seems to exist for real-valued measurable cardinals; we will not discuss the model-theoretic characterization of saturated cardinals here.

In concluding this section, we prove R. Solovay's Theorem 12.5 announced previously, according to which, *for each regular cardinal $\kappa > \omega$, every κ-stationary set $A \subset \kappa$ can be represented as a union of κ many pairwise disjoint κ-stationary sets.*

Proof of Theorem 12.5. In view of Theorem 12.5.A, we may assume that $\kappa \in \overline{M}_0$, i.e., that κ is an inaccessible cardinal. Let $A \subset \kappa$ be κ-stationary. Assume, on the contrary, that the assertion is not true. As we saw in the proof of Theorem 12.5.A, this means exactly that the ideal $\mathcal{I} = \mathrm{NS} + (\kappa \setminus A)$ is κ-saturated. We know that $\mathcal{I}$ is normal and $[\kappa]^{<\kappa} \subset \mathcal{I}$. Thus, by Theorem 17.3 we have $\kappa \setminus \overline{M}_0 \in \mathcal{I}$, and so may assume that $A \subset \overline{M}_0$. According to Assertion (3) in the proof of Theorem 17.3 we obtain that

(1) $$\{\lambda \in A : A \cap \lambda \in \mathrm{NS}(\lambda)\} \in \mathcal{I},$$

that is, the set

$$\{\lambda \in A : A \cap \lambda \in \mathrm{NS}(\lambda)\}$$

is not stationary in κ.

In the remaining part of the proof, we will show that (1) cannot hold for any set $A \in \mathrm{Stat}(\kappa)$ consisting of regular cardinals greater than ω.

To this end, first consider a set $X \subset \lambda > \omega$ (λ is regular), and let X' be the set of limit points of X, i.e.,

$$X' = \{\xi < \kappa : 0 < \xi = \sup X \cap \xi\}.$$

We claim that if X is cofinal in λ, then $X' \cap \lambda$ is a λ-club.

The closedness of $X' \cap \lambda$ is obvious; its being cofinal also follows, since for every sequence

$$x_0 < x_1 < \cdots < x_n < \ldots, \qquad \{x_n : n < \omega\} \subset X$$

we have $\sup\{x_n : n < \omega\} < \lambda$. Assume now, on the contrary, that (1) holds for some κ-stationary set $A \subset \kappa$. Let B be a κ-club such that

$$B \cap \{\lambda \in A : A \cap \lambda \in \mathrm{NS}(\lambda)\} = \emptyset.$$

Let B' be the set of limit points of the set B just defined. Then B' is also a κ-club and $B' \subset B$. As the set A is κ-stationary, we have $A \cap B' \neq \emptyset$.
Let

$$\lambda = \min(A \cap B').$$

Then λ is a regular cardinal greater than ω according to our assumptions. As $\lambda \in B'$, the set $B \cap \lambda$ is cofinal in λ. Thus $(B \cap \lambda)'$ is a λ-club, according to our remark above. However, $(B \cap \lambda)' \cap \lambda = B' \cap \lambda$. As $\lambda \in B' \subset B$ and $\lambda \in A$, we have

$$A \cap \lambda \notin \mathrm{NS}(\lambda),$$

that is, $A \cap \lambda$ is λ-stationary. Therefore

$$\emptyset \neq A \cap \lambda \cap B' \cap \lambda = A \cap B' \cap \lambda,$$

which contradicts the minimality of λ.

*   *
*

Problems

Definition. *Let $f, g \in {}^\omega\omega$, and write $f \prec g$ if there is an $n_0 \in \omega$ such that $\forall n > n_0 \, \big(f(n) < g(n)\big)$.*

A set $\{f_\alpha : \alpha < \kappa\}$ is said to be a κ-scale in ${}^\omega\omega$ if we have $f_\alpha \prec f_\beta$ whenever $\alpha < \beta < \kappa$, and for every $g \in {}^\omega\omega$ there is an $\alpha < \kappa$ such that $g \prec f_\alpha$.

1. Prove that if for a cardinal $\kappa > \omega$ there is a κ-scale in ${}^\omega\omega$, then κ is not real-valued measurable.

2. Prove that if $2^{\aleph_0} = \aleph_1$, then there is an $\aleph_1$-scale in ${}^\omega\omega$.

3. Change the definition of λ-saturated ideal as follows: *$\mathcal{I} \subset \mathrm{P}(\kappa)$ is said to be λ-saturated if for each system*

$$\mathcal{F} \subset \mathrm{P}(\kappa) \setminus \mathcal{I}$$

almost disjoint for $\mathcal{I}$ we have $|\mathcal{F}| < \lambda$, where to say that $\mathcal{F}$ is almost disjoint for $\mathcal{I}$ means that we have $F_1 \cap F_2 \in \mathcal{I}$ whenever $F_1, F_2 \in \mathcal{F}$ and $F_1 \neq F_2$.

Prove that, given $\kappa \geq \omega$, $2 \leq \lambda \leq \kappa$, and a κ-complete ideal $\mathcal{I}$ in $\mathrm{P}(\kappa)$, the ideal $\mathcal{I}$ is λ-saturated if and only if it is λ-saturated in the new sense.

18. WEAKLY COMPACT AND RAMSEY CARDINALS

In this section, we will study the question, for which cardinals Ramsey's Theorem will remain valid, and we will establish a number of important properties of these cardinals.

Definition 18.1. *The cardinal κ is said to be* weakly compact *if*

$$\kappa \to (\kappa)_2^2$$

holds.

This name, generally accepted in the literature, is justified by an equivalent model-theoretic formulation of the property for which the name is natural.

Theorem 18.1. *If κ is weakly compact, then κ is strongly inaccessible.*

Proof. If κ is not strongly inaccessible, then one of the following two assertions holds:

(1) There is a $\lambda < \kappa$ such that $2^\lambda \geq \kappa$;
(2) κ is singular.

If (1) holds, then by Theorem 14.4 we have

$$2^\lambda \nrightarrow (\lambda^+)_2^2,$$

hence, *a fortiori*,

$$\kappa \nrightarrow (\kappa)_2^2.$$

Therefore, it is enough to prove that

$$\lambda \nrightarrow (\lambda)_2^2$$

holds for each singular λ.

Let

$$\{A_\alpha : \alpha < \mathrm{cf}(\lambda)\}$$

be a set system consisting of pairwise disjoint sets such that

$$\bigcup \{A_\alpha : \alpha < \mathrm{cf}(\lambda)\} = \lambda$$

and

$$|A_\alpha| < \lambda \quad \text{for} \quad \alpha < \text{cf}(\lambda).$$

Define a 2-partition f of λ with 2 colors as follows. For each $\{x, y\} \in [\lambda]^2$ put

$$f(\{x, y\}) = 0 \iff \exists \alpha < \text{cf}(\lambda)(x, y \in A_\alpha).$$

It is easy to show that the partition f establishes the relation

$$\lambda \not\to (\lambda, \text{cf}(\lambda)^+)^2.$$

* *
*

In order to characterize weakly compact cardinals, we will need some further definitions. According to Definition 14.9, a partially ordered set $\langle T, \prec \rangle$ is called a tree if the set

$$T| \prec x = \{y \in T : y \prec x\}$$

is wellordered for each $x \in T$.

Definition 18.2. Let $\langle T, \prec \rangle$ be a tree.
1. For each $x \in T$, put

$$\alpha_T(x) = \alpha(x) = \text{type}(T| \prec x(\prec)).$$

2. For each α, write

$$T_\alpha = \{x \in T : \alpha_T(x) = \alpha\}.$$

3. $\langle T, \prec \rangle$ is said to have height α if

$$\alpha = \min\{\beta : T_\beta = \emptyset\}.$$

It is clear that if $x, y \in T_\alpha$ and $x \neq y$, then $x \not\prec y$, since otherwise we would have $\alpha(x) < \alpha(y)$.

4. Given $\kappa \geq \omega$, we call $\langle T, \prec \rangle$ a κ-tree if it has height κ and for each $\alpha < \kappa$ we have $|T_\alpha| < \kappa$.

5. A maximal ordered subset of $\langle T, \prec \rangle$ is called a branch of this tree.

6. The cardinal $\kappa \geq \omega$ has the tree property if each κ-tree $\langle T, \prec \rangle$ has a branch of cardinality κ.

7. Given $x \in T$ and $\beta \leq \alpha_T(x)$, we denote by $x|_T\beta$ the unique element of T_β for which $y \preceq x$.

8. For each $x, y \in T$ with $x \neq y$, we write

$$\delta_T(x, y) = \min\{\beta : x|_T\beta \neq y|_T\beta\}.$$

The next lemma says that for each tree $\langle T, \prec \rangle$ the partial ordering $\prec$ can be extended to an ordering of T in a natural way.

Definition 18.3. Let $\langle T, \prec \rangle$ be a tree of height α and for each $\beta < \alpha$ let $\prec_\beta$ be an ordering of T_β. Define the relation $\prec^*$ on T by putting

$$x \prec^* y \iff (x \prec y) \vee \left(y \nprec x \wedge y \nprec x \wedge x|_T \delta_T(x,y) \prec_{\delta_T(x,y)} y|_T \delta_T(x,y) \right)$$

for each $x, y \in T$.

$\prec^*$ is called the squashing of T with respect to $\langle \prec_\beta : \beta < \alpha \rangle$.

Lemma 18.1. Using the notation of Definition 18.3, $\prec^*$ is an ordering of the set T.

Proof. Enlarge the set T_β by adding a new element, i.e., put

$$T'_\beta = T_\beta \cup \{\infty_\beta\},$$

and define the ordering $\prec'_\beta$ of T'_β as an extension of $\prec_\beta$ such that $x \prec'_\beta \infty_\beta$ for each $x \in T_\beta$. Given an arbitrary $x \in T$ with $\alpha_T(x) = \beta$, let

$$x^* \in \underset{\gamma < \alpha}{\bigtimes} T'_\gamma$$

be a sequence such that

$$x^*(\gamma) = x|_T \gamma \qquad \text{for} \qquad \gamma \le \beta$$

and

$$x^*(\gamma) = \infty_\gamma \qquad \text{for} \qquad \beta < \gamma < \alpha.$$

Let $\prec^{**}$ be the lexicographic ordering of $\bigtimes_{\beta < \alpha} T'_\beta$, that is, the ordering such that

$$g \prec^{**} f \iff g(\beta) \prec'_\beta f(\beta)$$

with β being the least ordinal γ for which

$$g(\gamma) \ne f(\gamma).$$

We will leave it to the reader to show that $\prec^{**}$ is an ordering. As we have

$$x \prec^* y \iff x^* \prec^{**} y^*$$

for each $x, y \in T$, it follows that $\prec^*$ is also an ordering. $\blacksquare$

$$* \qquad *$$
$$*$$

Theorem 18.2. The following assertions are equivalent.

    a) κ is weakly compact.

    b) κ is strongly inaccessible and has the tree property.

    c) $\forall \gamma < \kappa \, \forall r < \omega \quad \kappa \to (\kappa)^r_\gamma$.

Proof. As the implication *c)* $\Longrightarrow$ *a)* is obvious, it is sufficient to verify the implications *a)* $\Longrightarrow$ *b)* and *b)* $\Longrightarrow$ *c)*.

a) $\Longrightarrow$ *b)*. Assume that $\kappa \to (\kappa)_2^2$. Then κ is strongly inaccessible according to Theorem 18.1. Let $\langle T, \prec \rangle$ be a κ-tree, let $\prec_\alpha$ be an arbitrary ordering of T_α, and let $\prec^*$ be the corresponding squashing of T, as described in Definition 18.3. It is clear that $|T| = \kappa$; we may therefore assume that $T = \kappa$. Define the 2-partition f of κ with 2 colors similarly to the way it was done in the proof of Sierpiński's Theorem 14.3. That is, put

$$f(\{x, y\}) = 0 \iff x \prec^* y \qquad \text{whenever} \qquad x < y < \kappa.$$

The assumption $\kappa \to (\kappa)_2^2$ implies the existence of a set $H \subset \kappa$ with $|H| = \kappa$ that is homogeneous with respect to f. For reasons of symmetry, we may assume that H is homogeneous in color 0; that is, we have $x \prec^* y$ whenever $x, y \in H$ and $x < y$.

For each $\alpha < \kappa$, write

$$\hat{T}_\alpha = \bigcup \{T_\beta : \beta < \alpha\}.$$

As κ is regular, we have

$$|\hat{T}_\alpha| < \kappa$$

for every $\alpha < \kappa$. Let

$$x, y \in \kappa \setminus \hat{T}_\alpha \qquad \text{with} \qquad x|_T\alpha \neq y|_T\alpha.$$

Then, as can immediately be seen from the definition of $\prec^*$, we have

$$x \prec^* y \iff x|_T\alpha \prec^* y|_T\alpha.$$

Now fix $\alpha < \kappa$. For each $z \in T_\alpha$, put

$$H_z = \{x \in H \setminus \hat{T}_\alpha : x|_T\alpha = z\}.$$

As κ is regular and $|T_\alpha| < \kappa$, there is a $z \in T_\alpha$ for which

$$|H_z| = \kappa.$$

We show that there is only one such z. Indeed, if we have $|H_{z_0}| = |H_{z_1}| = \kappa$ and, for example, $z_0 \prec^* z_1$, then first pick an $x_1 \in H_{z_1}$ and then pick an $x_0 \in H_{z_0}$ with $x_1 < x_0$. Then we have $x_1 \prec^* x_0$ by the homogeneity of H, and this contradicts the assumption $z_0 \prec^* z_1$. Note that from the uniqueness of z, it follows that for each $\alpha < \kappa$ there is a $z_\alpha \in T_\alpha$ such that $|H \setminus H_{z_\alpha}| < \kappa$.

We claim that

$$\{z_\alpha : \alpha < \kappa\}$$

is an ordered subset of the tree $\langle T, \prec \rangle$. Indeed, let $\alpha, \beta < \kappa$ with $\alpha \neq \beta$. In view of the assertion just proved, there is an x such that

$$x \in H_{z_\alpha} \cap H_{z_\beta}.$$

Then $z_\alpha \prec x$ and $z_\beta \prec x$ and so z_α and z_β are comparable in the tree $\langle T, \prec \rangle$. It is also clear that $\{z_\alpha : \alpha < \kappa\}$ is a branch of the tree T.

$b) \implies c)$. Assume that κ is strongly inaccessible and has the tree property. By induction on r we are going to prove the relation $\kappa \to (\kappa)_\gamma^r$. For $r = 1$ this is obvious in view of the regularity of κ (see the remark made after Theorem 14.4). Assume that this relation is true for some r, and let

$$f : [\kappa]^{r+1} \to \gamma$$

be an $(r+1)$-partition of κ in γ colors. Here we follow the proof of Theorem 14.7, substituting κ for the ρ used there. For each $\alpha < \kappa$, we define the sequence

$$\{\beta_\nu^\alpha : \nu < \xi_\alpha\},$$

and, as we did in Definition 14.9, for $\alpha' \prec \alpha$ we define the partition tree $T = \langle \kappa, \prec \rangle$ by putting

$$\alpha' \prec \alpha \iff \alpha' < \alpha \wedge \exists \nu < \xi_\alpha \, (\alpha' = \beta_\nu^\alpha).$$

The sets R_ξ defined in the proof of the Stepping-up Lemma 14.1 correspond to the sets T_ξ in the tree T for $\xi < \kappa$. As we proved there, we have

$$|T_\xi| = |R_\xi| \leq 2^{|\gamma| \cdot |\xi+1| + \omega} < \kappa$$

for every $\xi < \kappa$, as κ is strongly inaccessible. Thus the tree $\langle \kappa, \prec \rangle$ is a κ-tree, and so it has a branch H of cardinality κ.

We claim that H is end-homogeneous with respect to f. To see this, assume $V \in [H]^r$, $V < \alpha, \alpha' \in H$, and, for example, $\alpha' < \alpha$. Then $\alpha' = \beta_\nu^\alpha$ for some ordinal $\nu < \xi_\alpha$, and

$$f(V \cup \{\alpha'\}) = f(V \cup \{\beta_\nu^\alpha\}) = f(V \cup \{\alpha\}).$$

In view of the existence of the end-homogeneous set H of cardinality κ, the assertion follows from the induction hypothesis, similarly as in the proof of Theorem 14.7.

As a remark, we add that the implication $b) \implies c)$ was proved by P. Erdős and A. Tarski in 1942, while the implication $a) \implies b)$ is a consequence of a theorem of W. P. Hanf proved in 1961. The result that ω has the tree property is a theorem of D. König.

$$* \qquad *$$
$$*$$

Erdős and Tarski already proved in 1942 that every measurable cardinal is weakly compact. We will prove a stronger result.

Theorem 18.3 *(F. Rowbottom). Let $\kappa > \omega$ be a measurable cardinal and let $\mathcal{F} \subset P(\kappa)$ be a nonprincipal normal ultrafilter, $r < \omega$, $\gamma < \omega$, and let $f : [\kappa]^r \to \gamma$ be an r-partition of the cardinal κ in γ colors. Then there is a set $H \in \mathcal{F}$ that is homogeneous with respect to f.*

Proof. We will prove the assertion by induction on r. For $r = 1$, the assertion is straightforward. Assume that the assertion is true for some r. Consider

$$f : [\kappa]^{r+1} \to \gamma.$$

For each $\alpha < \kappa$, we define the mapping

$$f_\alpha : [\kappa \setminus (\alpha + 1)]^r \to \gamma$$

by the stipulation

$$\alpha < V \in [\kappa]^r \implies f_\alpha(V) = f(\{\alpha\} \cup V).$$

By the induction hypothesis for each α, there exist

$$H_\alpha \subset \kappa \setminus (\alpha + 1) \quad (H_\alpha \in \mathcal{F}) \qquad \text{and} \qquad \nu_\alpha \quad (\nu_\alpha < \gamma)$$

such that H_α is homogeneous in color ν_α with respect to f_α. Put

$$H' = \mathop{\Delta}_{\alpha < \kappa} H_\alpha.$$

Then $H' \in \mathcal{F}$. Therefore, there is an $H \subset H'$ $(H \in \mathcal{F})$ and an ordinal $\nu < \gamma$ such that $\nu_\alpha = \nu$ for $\alpha \in H$. We claim that H is homogeneous with respect to f in color ν. Let

$$\{\alpha\} \cup V \in [H]^{r+1}, \qquad \alpha < V.$$

Then, by the definition of diagonal intersection, we have $V \subset H_\alpha$, and so $\alpha \in H$ implies

$$f_\alpha(V) = \nu_\alpha = \nu;$$

on the other hand, we have

$$f_\alpha(V) = f(\{\alpha\} \cup V).$$

$$* \quad *$$
$$*$$

Definition 18.4. *Let κ, λ be cardinals and let γ be an ordinal. The symbol*

$$\kappa \to (\lambda)^{<\omega}_\gamma$$

indicates that the following assertion holds: If, for each $r < \omega$, $f_r : [\kappa]^r \to \gamma$ is an r-partition of κ in γ colors, then there exists a set $H \subset \kappa$ with $|H| = \kappa$ that is homogeneous with respect to each f_r. To indicate that this assertion is not true, we write

$$\kappa \nrightarrow (\lambda)^{<\omega}_\gamma.$$

If $\kappa \to (\kappa)^{<\omega}_\gamma$ holds for each $\gamma < \kappa$, then we say that κ is a Ramsey cardinal.

Corollary 18.1. *If κ is a measurable cardinal greater than ω then κ is a Ramsey cardinal.*

Proof. Let $\mathcal{I}$ be a normal prime ideal that is not principal, and let $\mathcal{F} = \operatorname{co}(\mathcal{I})$. Let

$$f_r : [\kappa]^r \to \gamma, \qquad r < \omega, \quad \gamma < \kappa$$

be given. According to Theorem 18.3, for each $r < \omega$ there is a set $H_r < \omega$ with $H_r \in \mathcal{F}$ that is homogeneous with respect to f_r. As $\kappa > \omega$, we have

$$H \overset{def}{=} \bigcap_{r < \omega} H_r \in \mathcal{F},$$

and so

$$|H| = \kappa.$$

$$*\qquad*$$
$$*$$

Corollary 18.1 had been established by P. Erdős and A. Hajnal in [E, H; 1] even before the concept of normal ideal was introduced.

Theorem 18.4. *If $\kappa > \omega$ is a measurable cardinal, then there is a $\lambda < \kappa$ that is a Ramsey cardinal.*

It is to be noted that this theorem can be obtained as a corollary to a much more general result. In particular, it is a consequence of a result for measurable cardinals involving a higher order "indescribability" property of the type outlined after Definition 15.1.

Proof of Theorem 18.4. Let κ be a measurable cardinal, and let $\mathcal{I} \subset \mathrm{P}(\kappa)$ be a nonprincipal normal prime ideal, and put $\mathcal{F} = \operatorname{co}(\mathcal{I})$. We will prove the following much stronger assertion than the one announced in the theorem:

$$\{\lambda < \kappa : \lambda \text{ is a Ramsey cardinal}\} \in \mathcal{F}.$$

Assume that this assertion is false. Then

$$A'' = \{\lambda < \kappa : \lambda \text{ is not a Ramsey cardinal}\} \in \mathcal{F}.$$

Put

$$A' = \{\lambda \in A'' : \lambda \text{ is regular}\}.$$

Then $A' \in \mathcal{F}$ according to the statement $A_1 \in \mathcal{I}$ in the proof of Theorem 16.2.

For each $\lambda \in A'$ there is a $\gamma < \lambda$ such that

$$\lambda \not\to (\lambda)^{<\omega}_{\gamma}$$

holds, and so, by the normality of $\mathcal{I}$, there is a $\gamma < \kappa$ such that

$$A \stackrel{def}{=} \{\lambda \in A' : \lambda \nrightarrow (\lambda)_\gamma^{<\omega}\} \in \mathcal{F}.$$

For each $\lambda \in A$, let

$$f_{\lambda,r} : [\lambda]^r \to \gamma \qquad r < \omega$$

be a sequence of partitions witnessing the relation

$$\lambda \nrightarrow (\lambda)_\gamma^{<\omega}.$$

Fix an arbitrary $r < \omega$. For every $V \in [\kappa]^r$ and $\nu < \gamma$, let

$$A_{V,\nu} = \{\lambda \in A : V \subset \lambda \wedge f_{\lambda,r}(V) = \nu\}.$$

As

$$A \setminus \sup V = \bigcup\{A_{V,\nu} : \nu < \gamma\}$$

holds for each $V \in [\kappa]^r$, we can define a partition

$$f_r : [\kappa]^r \to \gamma$$

by stipulating that

$$f_r(V) = \nu \iff A_{V,\nu} \in \mathcal{F}$$

for every $V \in [\kappa]^r$. Note that $r < \omega$ was arbitrary; that is, such an f_r was defined for each $r < \omega$.

As κ is measurable, by Corollary 18.1 it is a Ramsey cardinal. Hence there is a set $H \in \kappa$ with $|H| = \kappa$ that is homogeneous with respect to each f_r.
First we claim that

$$B = \{\lambda \in A : |\lambda \cap H| = \lambda\} \in \mathcal{F}.$$

Assuming this is not true, we have

$$C = \{\lambda \in A : \lambda \cap H \text{ is not cofinal in } \lambda\} \in \mathcal{F},$$

since the elements of A are regular cardinals. Let g be a function such that

$$\lambda \cap A \subset g(\lambda) < \lambda$$

for each $\lambda \in C$. As g is regressive on the set $C \in \mathcal{F}$, we have

$$\alpha(g) = \alpha < \kappa$$

for some $\alpha < \kappa$ with the notation introduced in the proof of Theorem 16.2. This is, however, impossible, since then

$$\{\lambda \in A : H \cap \lambda \subset \alpha\} \in \mathcal{F}$$

would hold; thus we would have $H \subset \alpha < \kappa$. Hence we indeed have $B \in \mathcal{F}$.

For each $\lambda \in B$, the set $\lambda \cap H$ is not homogeneous with respect to at least one of the partitions $f_{\lambda,r}$, $r < \omega$. That is, for each $\lambda \in B$, there is an integer $r_\lambda < \omega$ and there are ordinals

$$\alpha_0(\lambda), \ldots, \alpha_{r_\lambda - 1}(\lambda) \in H \cap \lambda \quad \text{with} \quad \alpha_0(\lambda) < \cdots < \alpha_{r_\lambda - 1}(\lambda),$$

$$\beta_0(\lambda), \ldots, \beta_{r_\lambda - 1}(\lambda) \in H \cap \lambda \quad \text{with} \quad \beta_0(\lambda) < \cdots < \beta_{r_\lambda - 1}(\lambda)$$

such that

$$V_\lambda = \{\alpha_i(\lambda) : i < r_\lambda\},$$
$$V'_\lambda = \{\beta_i(\lambda) : i < r_\lambda\},$$
$$f_{\lambda,r_\lambda}(V_\lambda) \neq f_{\lambda,r_\lambda}(V'_\lambda).$$

By the normality of $\mathcal{F}$, there is an $r < \omega$ and there are ordinals

$$\alpha_0 < \cdots < \alpha_{r-1}, \qquad \beta_0 < \cdots < \beta_{r-1}$$

such that

$$D \overset{def}{=} \left\{ \lambda \in B : r_\lambda = r \wedge \forall i < r \left(\alpha_i(\lambda) = \alpha_i \wedge \beta_i(\lambda) = \beta_i \right) \right\} \in \mathcal{F}.$$

In this case, however, for

$$V = \{\alpha_0, \ldots, \alpha_{r-1}\}$$

and

$$V' = \{\beta_0, \ldots, \beta_{r-1}\}$$

we have

$$f_r(V) \neq f_r(V').$$

This is a contradiction.

$$* \qquad *$$
$$*$$

We will announce another theorem; this theorem is perhaps the best illustration of the extraordinary strengths of positive relations involving the symbol introduced in Definition 18.4.

Theorem 18.5 *(J. Silver).* If $\kappa \to (\omega)_2^{\leq \omega}$, then there is a $\lambda < \kappa$ that is *weakly compact.*

This is a special case of a more general result given by Silver in [Si; 1]. The proof, relying on methods of model theory, will be omitted. We do not know of a proof based on combinatorial methods. Even if such a proof could be found, it would not properly show the power of the model-theoretic method. For more recent developments in this area, we refer to the book [Ka] by A. Kanamori.

As we have mentioned several times, weakly compact cardinals are also "very large." In fact, the following strengthening of Theorem 16.2 holds.

Theorem 18.6. *If $\kappa > \omega$ is weakly compact, then κ is a strongly α-Mahlo cardinal for each $\alpha < \kappa$.*

The main idea of the proof is similar to that of Theorem 16.2, but the proof is technically much more complicated; hence we omit the proof here. We will, however, present the key theorems and definitions needed for the proof. After a thorough understanding of the methods given in the preceding sections, the reader may be able to establish these results and in the end arrive at a proof of Theorem 18.6.

Theorem 18.7 *(J. Baumgartner).* *A cardinal $\kappa > \omega$ is weakly compact if and only if for each sequence*

$$\{f_\beta : \beta < \alpha\} \subset {}^\kappa\kappa$$

of regressive functions there is a sequence

$$\{\rho_\beta : \beta < \kappa\}$$

of ordinals such that

$$\{\xi < \kappa : \forall \beta \in D\left(f_\beta(\xi) = \rho_\beta\right)\} \in \mathrm{Stat}(\kappa)$$

holds for each $D \in [\kappa]^{<\kappa}$.

As we indicated above, we will omit the proof.

Definition 18.5. *Given a weakly compact cardinal $\kappa > \omega$, denote by $\mathcal{F}_\kappa^*$ the set of those subsets X of κ for which Theorem 18.6 remains valid when replacing ${}^\kappa\kappa$ with ${}^X\kappa$, that is, $X \in \mathcal{F}_\kappa^*$ holds if and only if for each sequence*

$$\{f_\beta : \beta < \alpha\} \subset {}^X\kappa$$

of regressive functions on X there is a sequence

$$\{\rho_\beta : \beta < \kappa\}$$

of ordinals such that

$$\forall D \in [\kappa]^{<\kappa}\{\xi \in X : \forall \beta \in D\left(f_\beta(\xi) = \rho_\beta\right)\} \in \mathrm{Stat}(\kappa).$$

Lemma 18.2. *If $\kappa > \omega$ is a weakly compact cardinal, then $\mathcal{F}_\kappa^*$ is a normal filter on κ such that*

$$[\kappa]^1 \cap \mathcal{F}_\kappa^* = \emptyset.$$

$\mathcal{F}_\kappa^*$ is called the *weakly compact filter* on κ. We have $\kappa \in \mathcal{F}_\kappa^*$ by Theorem 18.7. The other assertions of the lemma can be verified by an easy calculation.

Lemma 18.3. *If $\kappa > \omega$ is a weakly compact and $\alpha < \kappa$, then*

$$\kappa \setminus M_\alpha \in \mathcal{F}_\kappa^*.$$

The proof of this lemma is similar to that of Theorem 16.2. Theorem 18.6 now easily follows from Lemma 18.3.

Problems

1. Prove that, given $\lambda, \kappa \geq \omega$ and $\gamma \geq 2$, we have $\lambda \to (\kappa)_\gamma^{<\omega}$ if and only if, given arbitrary n-partitions $[\lambda]^n = \bigcup_{\xi<\gamma} I_\xi^n$ for $n < \omega$, there is $n_0 < \omega$ and an $X \subset \lambda$ with $|X| = \kappa$ that is homogeneous with respect to the partition $\bigcup_{\xi<\gamma} I_\xi^n$ for each $n \geq n_0$.

2. Prove the following assertions:

a) $2^{\aleph_0} \nrightarrow (\omega)_2^{<\omega}$.

b)* $\lambda \nrightarrow (\omega)_2^{<\omega}$ if λ is smaller than the least inaccessible cardinal greater than ω.

c) If for all cardinals $\lambda' < \lambda$ we have

$$\lambda' \nrightarrow (\omega)_2^{<\omega},$$

then

$$\lambda \nrightarrow (\omega \dotplus 1)_2^{<\omega}.$$

(For partition relations involving ordinals or order types, see the remark given after Definition 14.2.)

3.* Prove that if α is a limit ordinal and $\kappa \to (\alpha)_2^{<\omega}$, then we have $\kappa \to (\alpha)_{2^{\aleph_0}}^{<\omega}$.

4. Prove that $\kappa \geq \omega$ is weakly compact if and only if

$$\kappa \to (\kappa, 4)^3$$

holds.

Definition. $\mathcal{S} \subset \mathrm{P}(\kappa)$ is said to be a κ-complete field of sets if

1. If $A, B \in \mathcal{S}$, then $A \setminus B \in \mathcal{S}$;
2. For each set $\mathcal{S}' \in [\mathcal{S}]^{<\kappa}$, we have $\bigcup \mathcal{S}' \subset \mathcal{S}$.

Given a field of sets $\mathcal{S}$, a set $\mathcal{F} \subset \tilde{\mathcal{S}}$ is called a filter in $\mathcal{S}$ if it is not empty, $\emptyset \notin \mathcal{F}$, and the following two properties are satisfied:

(I) $A, B \in \mathcal{F} \implies A \cap B \in \mathcal{F}$;

(II) $A \in \mathcal{F}$ and $A \subset B \in \mathcal{S} \implies B \in \mathcal{F}$.

$\mathcal{F}$ is called an ultrafilter in $\mathcal{S}$ if it is a filter in $\mathcal{S}$ and for arbitrary sets $A, B \in \mathcal{S}$ with $A \cup B \in \mathcal{F}$ we have either $A \in \mathcal{F}$ or $B \in \mathcal{F}$.

5.* We say that $\mathcal{S}' \subset \mathcal{S}$ generates the κ-complete field $\mathcal{S}$ of sets if $\mathcal{S}$ is the κ-complete field of sets including $\mathcal{S}'$ that is the smallest with respect to inclusion. Prove that κ is weakly compact if and only if, in every κ-complete field $\mathcal{S}$ of at least κ sets generated by one of its subsets of cardinality $\leq \kappa$, every κ-complete filter can be extended to a κ-complete ultrafilter.

Definition. The topological space $\langle X, \tau \rangle$ is said to be κ-compact if every open cover of X has a subcover of cardinality less than κ (here τ denotes the *set of open subsets of X). The weight of a topological space $\langle X, \tau \rangle$ is the smallest cardinal λ such that $\langle X, \tau \rangle$ has a base of cardinality λ.*

6.[*][+] Assume that $\kappa^{<\kappa} = \kappa$. Prove that κ is weakly compact if and only if for each sequence $\{X_\nu : \nu < \kappa\}$ of Hausdorff topological spaces where, for every $\nu < \kappa$, X_ν is κ-compact and its weight is at most κ, the topological product

$$\underset{\nu<\kappa}{\mathsf{X}}\, X_\nu$$

is κ-compact.

7.[*] Prove that if κ is not weakly compact, then there is an ordered set $\langle R, \prec \rangle$ with $|R| = \kappa$ that has no wellordered subset of type κ in the ordering $\prec$ or $\succ$.

8.[*] Prove that if $\kappa^{<\kappa} = \kappa$ and κ is not weakly compact, then there is a κ-compact topological space X of cardinality κ and of weight κ such that X^2 is not κ-compact.

9.[*] Show that if $\kappa \geq \omega$ is weakly compact and $n < \omega$, then

$$\kappa^2 \to (\kappa^2, n)^2.$$

Here κ^2 denotes the order type of the anti-lexicographic ordering of $\kappa \times \kappa$.

19. SET MAPPINGS

Definition 19.1. 1. *The mapping* $F : X \to P(X)$ *on the set* X *is said to be a set mapping if* $x \notin F(x)$ *for each* $x \in X$.

2. *We say that the above set mapping has order* λ *if* $|F(x)| < \lambda$ *for each* $x \in X$.

3. *We call the set* $S \subset X$ *free with respect to* F *if* $x \notin F(y)$ *for each* $x, y \in S$.

We observe that, given a set mapping F of order λ on X, if $Y \subset X$ and $F_Y(X) = Y \cap F(X)$, then the set mapping is also of order λ, and every set $S \subset Y$ that is free with respect to F_Y is also free with respect to F. We will frequently use this observation without calling attention to it explicitly.

It was P. Turán who first pointed out that if the order of F is small and X is large, then often there is a large free subset. He proved the existence of a free subset of cardinality $2^{\aleph_0}$ in the case when $\lambda = \omega$ and $|X| = 2^{\aleph_0}$. The next lemma was established by D. Lázár in 1936.

Lemma 19.1. *Assume* $\kappa \geq \omega$ *is regular,* $\kappa > \lambda$, *and* F *is a set mapping of order* λ *on* κ. *Then there is a set* $S \subset \kappa$ *of cardinality* κ *that is free with respect to* F.

Before setting out to prove this result it is worth noting that the assumption $\kappa \geq \lambda$ would not be enough, since if we put $F(\alpha) = \alpha$ for $\alpha < \kappa$, then F is a set mapping of order κ on κ and there is no two-element set that is free with respect to α.

Proof. By transfinite recursion, we define the sequence

$$\{X_\alpha : \alpha < \lambda\} \subset P(\kappa)$$

of sets. Let $\alpha < \lambda$ and assume that the set X_β has already been defined for each $\beta < \alpha$. Let X_α be a maximal subset of

$$\kappa \setminus \bigcup \{X_\beta : \beta < \alpha\}$$

that is free with respect to F. Such an X_α exists according to the Teichmüller–Tukey Lemma. Thus we have defined the sets X_α for each $\alpha < \lambda$.

We claim that at least one of the sets X_α has cardinality κ. Assume, on the contrary, that

$$|X_\alpha| < \kappa \qquad \text{for each} \qquad \alpha < \lambda.$$

Write

$$Y_\alpha = \bigcup\{F(y) : y \in X_\alpha\} \cup X_\alpha.$$

Then

$$|Y_\alpha| \le |X_\alpha| \cdot \lambda + |X_\alpha| < \kappa.$$

Put

$$Y = \bigcup\{Y_\alpha : \alpha < \lambda\}.$$

Then we have $|Y| < \kappa$ by the regularity of κ. Assume that $x \in \kappa \setminus Y$ and $\alpha < \lambda$. Then

$$x \in \kappa \setminus \bigcup\{X_\beta : \beta < \alpha\} \qquad \text{and} \qquad x \notin F(y) \quad \text{for each} \quad y \in X_\alpha.$$

By the maximality of X_α, we therefore have

$$F(x) \cap X_\alpha \ne \emptyset.$$

As the sets X_α are pairwise disjoint, this implies that

$$|F(x)| \ge \lambda,$$

which is a contradiction.

$$* \qquad *$$
$$*$$

In what follows we prove two theorems concerning set mappings. The possibilities of generalizing the results and ideas discussed here will be indicated in the problems below.

Theorem 19.1 *(G. Fodor).* If $\lambda \ge \omega$ and $F : X \to \mathrm{P}(X)$ *is a set mapping of order λ on the set X, then X can be represented as the union of at most λ sets free with respect to F.*

For the proof, we will need the following:

Definition 19.2. *The set $Y \subset \kappa$ is said to be* closed with respect to the *set mapping F if for each $u \in Y$ we have $F(u) \subset Y$.*

Proof. Without loss of generality, we may assume that $X = \kappa$ for some cardinal κ. We may assume that $\kappa > \lambda$. For every $\gamma < \kappa$, there is a set $Y = Y_\gamma \subset \kappa$ with $|Y_\gamma| \le \lambda$ and $\gamma \in Y_\gamma$ such that Y_γ is closed with respect to the set mapping F. Indeed, define Y_γ^n by recursion on $n < \omega$ with

$Y_\gamma^0 = \{\gamma\}$, $Y_\gamma^{n+1} = Y_\gamma^n \cup \{F(z) : z \in Y_\gamma^n\}$. Then $Y_\gamma = \bigcup_{n<\omega} Y_\gamma^n$ satisfies the requirements.

We now define the sets X_α for $\alpha < \kappa$ by recursion on α. Assume X_β for $\beta < \alpha$ has already been defined. Put $Z_\alpha = \bigcup_{\beta<\alpha} X_\beta$, $\gamma_\alpha = \min(\kappa \setminus Z_\alpha)$, and $X_\alpha = Y_{\gamma_\alpha} \setminus Z_\alpha$. Clearly, the X_α's are pairwise disjoint and $|X_\alpha| \leq \lambda$ for each $\alpha < \kappa$. Moreover, if $\beta < \alpha$ and $y \in X_\beta$, then $F(y) \subset Y_{\gamma_\beta}$, and so $F(y) \cap X_\alpha = \emptyset$. Finally, it is immediate that $\kappa = \bigcup_{\alpha<\kappa} X_\alpha$.

Using the fact that $|X_\alpha| \leq \lambda$, we can see that there is a decomposition $\kappa = \bigcup_{\nu<\lambda} S_\nu$ such that $|S_\nu \cap X_\alpha| \leq 1$ for every $\alpha < \kappa$ and $\nu < \lambda$. It is sufficient to decompose each S_ν into the union of λ sets free for F.

For $\nu < \lambda$, define a mapping $\Phi_\nu : S_\nu \to \lambda$. Given $\alpha < \kappa$, if $x \in S_\nu \cap X_\alpha$ and Φ_ν has already been defined for $y \in S_\nu \cap \bigcup_{\beta<\alpha} X_\beta$, let

$$\Phi_\nu(x) = \min\left(\lambda \setminus \{\Phi_\nu(y) : y \in F(x) \cap S_\nu \cap \bigcup_{\beta<\alpha} X_\beta\}\right).$$

The sets $\Phi_\nu^{-1}(\mu)$, $\mu < \lambda$ are free for F and

$$S_\nu = \bigcup\{\Phi_\nu^{-1}(\mu) : \mu < \lambda\}.$$

*   *
✦

We now prove the generalization of Lemma 19.1 for singular cardinals. It needs to be observed that this does not follow from Fodor's Theorem 19.1, since, if $\lambda \geq \mathrm{cf}(\kappa)$, all the free sets there may have cardinalities less than κ.

Theorem 19.2 *(Hajnal's Set Mapping Theorem).* Assume $\kappa \geq \omega$, $\kappa > \lambda$, and let F be a set mapping of order λ on κ. Then there is a set $S \subset \kappa$ of cardinality κ that is free with respect to F.

Proof. In view of Lemma 19.1, we may assume that κ is singular. Let τ be a regular cardinal that is smaller than κ but greater than $\mathrm{cf}(\kappa)$ and λ, e.g.,

$$\tau = \left(\max\left(\lambda, \mathrm{cf}(\kappa)\right)\right)^+.$$

Let

$$\{\kappa_\alpha : \alpha < \mathrm{cf}(\kappa)\}$$

be an increasing sequence of cardinals such that $\kappa_0 > \tau$, κ_α is regular for $\alpha < \mathrm{cf}(\kappa)$, and

$$\kappa = \sum_{\alpha<\mathrm{cf}(\kappa)} \kappa_\alpha.$$

We define the sequence

$$\{S_\alpha : \alpha < \mathrm{cf}(\kappa)\} \subset P(\kappa)$$

of sets by transfinite recursion on α as follows. Assume that $\alpha < \mathrm{cf}(\kappa)$ and that, for each $\beta < \alpha$, we have already defined the sets S_β, and we have also defined a partition

$$S_\beta = \bigcup \{S_{\beta,\nu} : \nu < \tau\}$$

of these sets into pairwise disjoint sets such that

$$|S_\beta| = |S_{\beta,\nu}| = \kappa_\beta \qquad \text{for} \qquad \beta < \alpha \quad \text{and} \quad \nu < \tau.$$

Put

$$Z_\alpha = \bigcup \{S_\beta : \beta < \alpha\} \cup \bigcup \{F(y) : y \in S_\beta \wedge \beta < \alpha\};$$

then $|Z_\alpha| \leq \sum_{\beta<\alpha} \kappa_\beta \cdot \lambda < \kappa$. Pick

$$X_\alpha \subset \kappa \setminus Z_\alpha$$

with $|X_\alpha| = \kappa_\alpha$. According to Lemma 19.1, there is a set

$$Y_\alpha \subset X_\alpha \qquad \text{with} \qquad |Y_\alpha| = |X_\alpha|$$

that is free with respect to F. For each $y \in Y_\alpha$, put

$$\nu(y) = \sup\{\nu < \tau : \exists \beta < \alpha \, (S_{\beta,\nu} \cap F(y) \neq \emptyset)\}.$$

Noting that we have $|F(y)| < \lambda$, the sets $S_{\beta,\nu}$ are pairwise disjoint for $\beta < \alpha$ and $\nu < \tau$, and τ is regular, we can conclude that

$$\nu(y) < \tau.$$

As $\kappa_\alpha > \tau$ and κ_α is regular, there is a set $S_\alpha \subset Y_\alpha$ and an ordinal $\nu_\alpha < \tau$ such that

$$|S_\alpha| = \kappa_\alpha$$

and

$$\nu_\alpha = \nu(y) \qquad \text{for} \qquad y \in S_\alpha.$$

Let

$$S_\alpha = \{S_{\alpha,\nu} : \nu < \tau\}$$

be an arbitrary partition of S_α into pairwise disjoint sets such that

$$|S_{\alpha,\nu}| = \kappa_\alpha \qquad \text{for} \qquad \nu < \tau.$$

This completes the definition of the sequence $\{S_\alpha : \alpha < \kappa\}$.
    Let

$$\nu = \sup\{\nu_\alpha \dotplus 1 : \alpha < \mathrm{cf}(\kappa)\}.$$

As τ is regular and $\text{cf}(\kappa) < \tau$, we have $\nu < \tau$. Put

$$S = \bigcup \{S_{\alpha,\nu} : \alpha < \text{cf}(\kappa)\}.$$

Then

$$|S| = \sum_{\alpha < \text{cf}(\kappa)} \kappa_\alpha = \kappa.$$

We claim that S is a free set with respect to F. Indeed, let $x, y \in S$, $x \in S_{\alpha,\nu}$, $y \in S_{\beta,\nu}$, $\alpha \leq \beta$.

If $\alpha = \beta$, then $x \notin F(y)$ and $y \notin F(x)$, as $S_{\alpha,\nu} \subset Y_\alpha$ and Y_α is free with respect to F.

If $\alpha < \beta$, then $y \notin F(x)$, since $y \in X_\beta$, $F(x) \subset Z_\beta$, and $X_\beta \cap Z_\beta = \emptyset$. On the other hand, $x \notin F(y)$; indeed, we have $\nu_\beta < \nu$, and so

$$F(y) \cap S_{\alpha,\nu} = \emptyset$$

in view of the choice of ν_β.

*   *
*

Problems

Definition. *Let κ, λ, and τ be cardinals and let*

$$F : [\kappa]^\tau \to [\kappa]^{<\lambda}$$

or

$$F : [\kappa]^{<\tau} \to [\kappa]^{<\lambda}.$$

We call F a set mapping of type τ or of type $< \tau$, respectively, and of order λ on κ if

$$F(V) \cap V = \emptyset$$

for each $V \in D(F)$. The set $S \subset \kappa$ is said to be free with respect to F if for each $V \subset S$ with $V \in D(F)$ we have $F(V) \cap S = \emptyset$.

The meaning of the term set mapping of type 1 will be ambiguous, since in addition to the description given in this definition, a set mapping in the ordinary sense, that is, in the sense it was introduced in Definition 19.1, will also be called a set mapping of type 1. Furthermore, a mapping

$$F : [\kappa]^\tau \to \kappa \qquad \text{or} \qquad F : [\kappa]^{<\tau} \to \kappa$$

such that

$$F(V) \notin V$$

for each $V \in D(F)$ will also be called a set mapping of order 2. There is no reason that these ambiguities should lead to a confusion.

1. Prove that if $k, l < \omega$ and if F is a set mapping of type k and order l on ω, then there exists an infinite subset of ω that is free with respect to f.

2.* If $F : X \to P(X)$ is a set mapping, and there is a $k < \omega$ such that each finite subset of X is the union of k sets that are free with respect to F, then X itself is the union of k sets free with respect to F.

3.* Prove that if $k < \omega$ and $F : X \to P(X)$ is a set mapping of order k on the set X, then X is the union of at most $2k$ sets free with respect to F.

4.* Let F be a set mapping defined on the set $\mathbb{R}$ of real numbers. Prove that if $F(x)$ is a nowhere dense set for each $x \in R$, then for each $\alpha < \omega_1$ there is a set of order type α that is free with respect to F.

5.* Assume that $2^{\aleph_0} = \aleph_1$. Prove that then there exists a set mapping F defined on the set $\mathbb{R}$ of real numbers such that for each $x \in \mathbb{R}$ the elements of $F(x)$ form a convergent ω-sequence tending to x, and there is no subset of cardinality $\aleph_1$ of $\mathbb{R}$ that is free with respect to F. (Hence, no free subset of cardinality $\aleph_1$ needs to exist under the assumptions of Problem 4 if $2^{\aleph_0} = \aleph_1$.)

6.* Prove the following generalization of Theorem 19.2: Given the assumptions of Theorem 19.2, let $\mu < \kappa$ and let $X_\xi \subset \kappa$ for $\xi < \mu$ be sets of cardinality κ. Then there is a set S free with respect to F such that

$$|S \cap X_\xi| = \kappa \qquad \text{whenever} \qquad \xi < \mu.$$

7.* Prove that for each $k < \omega$ there is a set mapping F of type $k + 1$ and order ω_1 on ω_{k+1} such that there is no set of $k + 2$ elements that is free with respect to F. Show that there is no such set mapping on ω_{k+2}.

8.* Prove that for each $\kappa \geq \omega$ there is a set mapping of type ω and of order 2 with respect to which there is no infinite free set.

9.* Prove that if $\kappa \geq \omega$ and F is a set mapping of type 2 and of order κ on $(2^\kappa)^+$, then there is a set of cardinality κ^+ that is free with respect to F.

10.* Prove that if $2^{\aleph_0} = \aleph_1$, then there is a set mapping F of type 2 and of order 2 on ω_1 with respect to which there is no free set of cardinality ω_1.

11. Prove that if $\kappa > \omega$ is a real-valued measurable cardinal and μ is a measure witnessing this, and $\mathcal{F} \subset P(\kappa)$ is such that $|\mathcal{F}| = \lambda = \text{cf}(\lambda) > \omega$ for some $\lambda < \kappa$ and $\mu(F) > 0$ for each $F \in \mathcal{F}$, then there is a set $\mathcal{F}' \subset \mathcal{F}$ with $|\mathcal{F}'| = \lambda$ such that $\bigcap \mathcal{F}' \neq \emptyset$.

12.* Prove that if $\kappa > \omega$ is a real-valued measurable cardinal and F is a set mapping of type $< \omega$ and order λ on κ for some $\lambda < \kappa$, then there is a set of cardinality κ that is free with respect to F.

13.* Prove that if κ is real-valued measurable, then κ has the tree property.

20. THE SQUARE-BRACKET SYMBOL. STRENGTHENINGS OF THE RAMSEY COUNTEREXAMPLES

In Theorem 14.3, we showed that Ramsey's Theorem cannot be generalized by writing an arbitrary infinite cardinal κ in place of ω. According to the former theorem, we have

$$2^{\aleph_0} \not\to (\aleph_1)^2_2.$$

It occurred to P. Erdős that this counterexample may possibly be further strengthened by dividing the pairs of $2^{\aleph_0}$ into, say, three classes in such a way that every set of cardinality $\aleph_1$ should include a pair from each of the three classes. The questions raised by pursuing this idea have not been completely resolved even today. Before we can outline the present state of knowledge in this area, we would like to introduce some new concepts.

Definition 20.1. Let λ be a cardinal, $\gamma \geq 2$, and let $f : [A]^\lambda \to \gamma$ be a λ-partition of A with γ colors. We say that the set $B \subset A$ is completely inhomogeneous with respect to f if $f\,"[B]^\lambda = \gamma$.

Definition 20.2. Given cardinals κ, λ, and μ, and an ordinal γ, the symbol

$$\kappa \to [\mu]^\lambda_\gamma$$

indicates that the following assertion holds:

For an arbitrary set X of cardinality κ and for an arbitrary partition $f : [X]^\lambda \to \gamma$, there is a set Y with $|Y| = \mu$ that is not completely inhomogeneous with respect to f.

The negation of this assertion is denoted as

$$\kappa \not\to [\mu]^\lambda_\gamma,$$

in the spirit of Definition 14.2.

We remark that, similarly to the way it was done in Definition 14.2, the symbol $\kappa \to [\kappa_\nu]^\lambda_{\nu<\gamma}$ can also be introduced. We will, however, not discuss this latter symbol here.

Similarly as before, this symbol describes an assertion of some relevance only in the case when $\lambda \geq 2$, $\gamma > 2$, and $\mu \geq \lambda$. Furthermore, the following assertions are straightforward:

$$\kappa \to (\mu)_2^\lambda \iff \kappa \to [\mu]_2^\lambda,$$
$$\kappa \to [\mu]_\gamma^\lambda \implies \kappa \to [\mu]_{\gamma'}^\lambda, \quad \text{if} \quad \gamma \leq \gamma'.$$

For this reason, for every negative partition relation

$$\kappa \nrightarrow (\mu)_2^\lambda$$

we may study the question of which of its strengthenings

$$\kappa \nrightarrow [\mu]_\gamma^\lambda \quad (\gamma > 2)$$

hold. In this section, we will only study questions of the form $2^\kappa \nrightarrow [\kappa^+]_\gamma^2$. Some information about results of different forms will be contained in the problems below.

Theorem 20.1. *If $\kappa \geq \omega$ and $2^\kappa = \kappa^+$, then*

$$\kappa^+ \nrightarrow [\kappa^+]_{\kappa^+}^2.$$

Proof. Let $[\kappa^+]^\kappa = \{A_\alpha : \alpha < \kappa\}$ be a wellordering of type κ^+ of the set $[\kappa^+]^\kappa$; such a wellordering exists according to the assumptions on κ. Write

$$S_\alpha = \{A_\beta : \beta < \alpha \wedge A_\beta \subset \alpha\}.$$

For each $\alpha < \kappa^+$ and $\beta < \alpha$, we define $f(\{\alpha, \beta\})$ such that

(1) For each $\delta < \alpha$ and for each $\gamma < \alpha$ with $A_\gamma \in S_\alpha$, there is a $\beta \in A_\gamma$ such that $f(\{\alpha, \beta\}) = \delta$.

This is possible according to Problem 1 *b)* in Section 10, as $S_\alpha \subset [\kappa^+]^\kappa$ and $|S_\alpha| \leq \kappa$. Let $X \subset \kappa^+$ with $|X| = \kappa^+$ and $\delta < \kappa$ be arbitrary. Pick $\gamma < \kappa^+$ such that $A_\gamma \subset X$. Choose $\alpha \in X$ such that $\max(\gamma, \delta) < \alpha$ and $A_\gamma \subset \alpha$. Then we have $f(\{\alpha, \beta\}) = \delta$ for some $\beta \in A_\gamma$ according to (1), and so $\delta \in f``[X]^2$.

$$* \qquad *$$
$$*$$

We will establish a strengthening of this result. The frequently reusable method of the proof of this strengthening, namely, the choice of the auxiliary functions $g \in {}^\kappa\kappa$ is due to W. Sierpiński.

Theorem 20.2. *Assume* $\kappa \geq \omega$ *and* $2^\kappa = \kappa^+$. *Then there is a 2-partition* $f : [\kappa^+]^2 \to \kappa^+$ *with* κ^+ *colors such that for each* $X \subset \kappa^+$ *with* $|X| = \kappa^+$ *and for each* $Y \subset X$ *with* $|Y| = \kappa$, *there is a* $y \in Y$ *for which*

$$\{f(\{\alpha, y\}) : \alpha \in X\} = \kappa^+.$$

Proof. Let $\{g_\beta : \beta < \kappa^+\}$ be a wellordering of type κ^+ of the set of all functions of cardinality κ with domains and ranges that are subsets of κ^+ – such a wellordering exists according to the assumptions on κ – and put

$$S_\alpha = \{g_\beta : \beta < \kappa \wedge D(g_\beta) \subset \alpha\}.$$

For each $\alpha < \kappa^+$ and $\beta < \alpha$, define $f(\{\alpha, \beta\})$ such that

(2) $\forall g_\gamma \in S_\alpha \, \exists \beta \in D(g_\beta) \, f(\{\alpha, \beta\}) = g_\gamma(\beta).$

Again, this is possible according to Problem 1 *b)* in Section 10. Assume we are given $X \subset \kappa^+$ with $|X| = \kappa^+$ and $Y \subset X$ with $|Y| = \kappa$. Proceeding via reductio ad absurdum, assume that the assertion to be proved is false, and for each $\beta \in Y$, define $g : Y \to \kappa^+$ as

$$g(\beta) = \min\{\delta < \kappa^+ : \forall \alpha \in X \setminus (\beta \dot{+} 1) \, f(\{\alpha, \beta\}) \neq \delta\}.$$

Then $g = g_\gamma$ for some $\gamma < \kappa^+$. Let $\alpha \in X$ be such that $\alpha > \gamma$ and $D(g_\gamma) \subset \alpha$. Then we have $f(\{\alpha, \beta\}) = g(\beta)$ for some $\beta \in Y$ according to (2); this contradicts the definition of g.

$$* \qquad *$$
$$*$$

As a surprising turn of events, it was noticed only much later, in 1986, that the assertion of Theorem 20.1 can be established for many cardinals κ even without the assumption $2^\kappa = \kappa^+$ (see [To]):

Todorčević's Theorem 20.3. *If* $\kappa = \mathrm{cf}(\kappa) \geq \omega$, *then*

$$\kappa^+ \not\to [\kappa^+]^2_{\kappa^+}.$$

In what follows we are going to present the method S. Todorčević invented in order to establish this result, or, rather, we will present a generalization of this method developed by S. Shelah. Then we will establish Theorem 20.4 below, and Theorem 20.3 will be an immediate consequence of this. To begin with, we will need a lemma that is a consequence of a basic result about elementary chains – this latter being a general tool of model theory and mathematical logic. We will announce this lemma only in the special case needed for the proof of Theorem 20.3, and we will give a direct proof of this special case.

Lemma 20.1. *Let* $\lambda = \mathrm{cf}(\lambda) > \omega$, *and for each* $\underline{\epsilon} \in {}^{<\omega}\lambda \overset{def}{=} \bigcup_{n<\omega} {}^{n}\lambda$, *let* $R_{\underline{\epsilon}}(x,y)$ *be a two-place relation on* $\lambda \setminus \epsilon$, *where* $\epsilon = \max \mathrm{R}(\underline{\epsilon}) \dotplus 1$. *Then there is a* λ-*large set* $D \subset \lambda$ *such that the following assertion holds:*

If we have $\delta_0, \delta_1 \in D$, $\underline{\epsilon} \in {}^{<\omega}\delta_0$, $\epsilon < \delta_0 \le \alpha_0 < \delta_1 \le \alpha_1 < \lambda$, *and* $R_{\underline{\epsilon}}(\alpha_0, \alpha_1)$, *then*

$$\forall \xi_0 < \delta_0 \, (\exists \eta_0 : \xi_0 < \eta_0 < \delta_0) \, \forall \xi_1 < \delta_0 \, (\exists \eta_1 : \xi_1 < \eta_1 < \delta_0) \, R_{\underline{\epsilon}}(\eta_0, \eta_1).$$

Proof. It is sufficient to establish the result for a single relation $R(x,y)$. Indeed, if for each $\underline{\epsilon}$ there is a set $D_{\underline{\epsilon}}$ that satisfies the requirements for the single relation $R_{\underline{\epsilon}}(x,y)$, then

$$D = \{\delta \in \bigcap \{D_{\underline{\epsilon}} : \underline{\epsilon} \in {}^{<\omega}\delta\} : \delta < \lambda\}$$

is a λ-large set that satisfies the requirements of the lemma.

For each $X \subset \lambda$ let

$$\tilde{X} = \begin{cases} X' & \text{if } \sup X = \lambda, \\ \lambda \setminus (\sup X \dotplus 1) & \text{otherwise;} \end{cases}$$

recall that here X' denotes the set of limit points of X. Let

$$R_\alpha = \{\beta < \lambda : R(\alpha, \beta)\},$$
$$D_0 = \underset{\alpha<\lambda}{\triangle} \tilde{R}_\alpha, \quad R = \{\alpha < \lambda : \sup R_\alpha = \lambda\},$$
$$D_1 = \tilde{R}, \quad \text{and} \quad D = D_0 \cap D_1.$$

Let $\delta_0 \le \alpha_0 < \delta_1 \le \alpha_1 < \lambda$, $\delta_0, \delta_1 \in D$, and assume that $R(\alpha_0, \alpha_1)$ holds. Then $\delta_1 \in D_0$ implies that $\sup R_{\alpha_0} = \lambda$, and so $\alpha_0 \in R$; hence $\delta_0 \in D_1$ implies that $\sup(R \cap \delta_0) = \delta_0$. If $\alpha \in R \cap \delta_0$, then δ_0 is a limit point of the set $R_\alpha \cap \delta_0$. This proves the assertion.

$$* \qquad *$$
$$*$$

Returning to our preparations for the proof of Theorem 20.3, fix the cardinal $\lambda = \mathrm{cf}(\lambda) > \omega$, and choose a set $\mathcal{C} = \{C_\alpha : \alpha < \lambda\}$ satisfying the following requirements:

(1) $C_\alpha \subset \alpha$, C_α is cofinal and closed in α, and $0 \in C_\alpha$.

Later further requirements will be imposed upon $\mathcal{C}$, but, for now, the definitions below depend only on Condition (1).

Definition 20.3. 1. *Let $\alpha < \beta < \lambda$. Define an ordinal $n(\alpha, \beta) < \omega$ and a decreasing sequence of ordinals $\rho_n(\alpha, \beta)$ for $n \leq n(\alpha, \beta)$ as follows: $\rho_0(\alpha, \beta) = \beta$. If $\rho_n(\alpha, \beta) = \alpha$, then let $n = n(\alpha, \beta)$. If $\rho_n(\alpha, \beta) > \alpha$, then let*

$$\rho_{n+1}(\alpha, \beta) = \min(C_{\rho_n(\alpha, \beta)} \setminus \alpha).$$

Then we have $\alpha \leq \rho_{n+1}(\alpha, \beta) < \rho_n(\alpha, \beta)$ in view of (1), and so our definition is sound.

2. *For $n \leq n(\alpha, \beta)$, define $\rho_n^-(\alpha, \beta)$ as follows: $\rho_0^-(\alpha, \beta) = 0$, and if $n + 1 < n(\alpha, \beta)$, then put*

$$\rho_{n+1}^-(\alpha, \beta) = \sup(C_{\rho_n(\alpha, \beta)} \cap \alpha).$$

Observe that we have $\rho_n^-(\alpha, \beta) < \alpha$ for $n < n(\alpha, \beta)$.

Following Todorčević, the sequence

$$\rho(\alpha, \beta) = \langle \rho_n(\alpha, \beta) : n \leq n(\alpha, \beta) \rangle$$

is called the C-walk from β to α. The sequence of the ordinals $\rho_n^-(\alpha, \beta)$ plays an important role in the proof. For this reason, we will introduce a notation for this sequence.

Definition 20.4. *For $\alpha < \beta < \lambda$ and $n \leq n(\alpha, \beta)$, put*

$$\underline{\epsilon}_n(\alpha, \beta) = \langle \rho_m^-(\alpha, \beta) : m \leq n \rangle$$

and

$$\epsilon_n(\alpha, \beta) = \max\{\rho_m^-(\alpha, \beta) : m \leq n\} \dotplus 1$$

Lemma 20.2. *If $\alpha < \beta < \lambda$, $n \leq n(\alpha, \beta)$, and $\epsilon_n(\alpha, \beta) \leq \alpha' \leq \alpha$ then $\rho_m(\alpha', \beta) = \rho_m(\alpha, \beta)$ and $\rho_m^-(\alpha', \beta) = \rho_m^-(\alpha, \beta)$ for $m \leq n$, and so $\underline{\epsilon}_n(\alpha', \beta) = \underline{\epsilon}_n(\alpha, \beta)$.*

Proof. The first claim follows from Definition 20.3 by induction on m, and the second claim is a straightforward consequence of the first one.

Definition 20.5. *We call the set $S \subset \lambda$ a nonreflecting subset of λ if $S \subset \mathrm{Lim}(\lambda)$ and for every $\alpha \in \mathrm{Lim}(\lambda)$, we have $S \cap \alpha \notin \mathrm{Stat}(\alpha)$. If S is also stationary in λ, then we call it a nonreflecting stationary subset of λ.*

This is the one place in the book where the concept of α-stationary set is useful even if $\mathrm{cf}(\alpha) = \omega$. As we remarked after Definition 12.9 above, a subset of α is α-stationary in this case if and only if its complement is not cofinal in α.

Lemma 20.3. *If $\kappa = \mathrm{cf}(\kappa) \geq \omega$, then*

$$S = \{\alpha < \kappa^+ : \mathrm{cf}(\alpha) = \kappa\}$$

is a nonreflecting stationary subset of $\lambda = \kappa^+$.

Proof. According to Problem 1 of Section 12, S is λ-stationary. For each $\alpha \in \mathrm{Lim}(\lambda)$, let $B_\alpha \subset \alpha$ be a set cofinal in α consisting of successor ordinals such that $\mathrm{type}\, B_\alpha(<) \leq \kappa$. Let

$$C_\alpha = (B_\alpha \cup B'_\alpha) \cap \alpha.$$

Then we have $C_\alpha \cap S = \emptyset$, since each limit element of C_α has cofinality less than κ.

*   *
*

Theorem 20.4 *(S. Todorčević; S. Shelah).* Assume $\lambda = \mathrm{cf}(\lambda) > \omega$. *If λ has a nonreflecting stationary subset, then*

$$\lambda \nrightarrow [\lambda]^2_\lambda.$$

We would like to point out that Theorem 20.3 follows from Theorem 20.4 in view of Lemma 20.3.

Proof of Theorem 20.4. Let S be a nonreflecting stationary subset of λ. It is sufficient to establish the following assertion:

(∗) There is a function $g : [\lambda]^2 \to \lambda$ such that for each $X \subset \lambda$ with $|X| = \lambda$ we have

$$S \setminus g\text{“}[X]^2 \notin \mathrm{Stat}(\lambda).$$

The assertion of the theorem can be derived from (∗) as follows. According to Solovay's Theorem 12.5, there is a decomposition $S = \bigcup_{\xi < \lambda} S_\xi$ of S into pairwise disjoint λ-stationary sets S_ξ. Define the partition $f : [\lambda]^2 \to \lambda$ by the stipulation

$$f(\{\alpha, \beta\}) = \xi \iff g(\{\alpha, \beta\}) \in S_\xi.$$

Then f verifies the assertion of the theorem, since for every $X \in [\lambda]^\lambda$ and every $\xi < \lambda$, we have $g\text{“}[X]^2 \cap S_\xi \neq \emptyset$.

In what follows we will be concerned with establishing Assertion (∗). According to our assumptions on S, the set C described in (1) can be chosen such that the following condition is also satisfied:

(2) $C_\alpha \cap S = \emptyset$ if $\alpha \in \mathrm{Lim}(\lambda)$, and $C_\alpha = \{0, \alpha\}$ if $\alpha = \beta \dotplus 1$ is a successor ordinal.

Next we will give the definition of the function g that will witness Assertion (∗).

(3) For every α and β with $\alpha < \beta < \lambda$ put $g(\{\alpha, \beta\}) = \rho_n(\{\alpha, \beta\})$ if n
is the unique integer satisfying the followign conditions:

$$n < n(\alpha, \beta), \quad \varepsilon_n(\alpha, \beta) = \epsilon\,(= \max R(\underline{\epsilon}) \dotplus 1), \quad \underline{\varepsilon}_n(\alpha, \beta) = \underline{\varepsilon}_n(\epsilon, \beta),$$
$$\text{but} \quad \rho^-_{n+1}(\alpha, \beta) \neq \rho^-_{n+1}(\epsilon, \beta).$$

Let $g(\{\alpha, \beta\}) = 0$ in every other case.

Now let $X \subset \lambda$ with $|X| = \lambda$ be given. For every $\underline{\epsilon} \in {}^{<\omega}\lambda$, we define a
relation $R_{\underline{\epsilon}}$ with the following stipulation:

(4) For every ξ and η with $\xi < \eta < \lambda$ and for each $\underline{\epsilon} \in {}^{<\omega}\lambda$, put $R_{\underline{\epsilon}}(\xi, \eta)$
if and only if $\xi \in S$, $\eta \in X$, $\underline{\epsilon} = \underline{\varepsilon}_n(\xi, \eta)$ for $n = n(\xi, \eta)$, and
$\epsilon = \max R(\underline{\epsilon}) \dotplus 1 < \xi$.

Let $D = D(X)$ be a set consisting of limit ordinals and satisfying the
requirements of Lemma 20.1 for the relations $R_{\underline{\epsilon}}$ so defined. For the proof of
Assertion $(*)$ it is sufficient to show the following.

(5) If $\delta \in S \cap D$, then $\delta = g(\{\alpha, \beta\})$ for some α and β with $\alpha < \beta$.

Thus let $\delta \in S \cap D$. Choose ordinals $\delta_1 > \delta$ with $\delta_1 \in D$ and $\beta \in X$ with
$\beta \geq \delta_1$, and put $n = n(\delta, \beta)$ and $\underline{\epsilon} = \underline{\varepsilon}_n(\delta, \beta)$. We are going to show that

(6) $$R_{\underline{\epsilon}}(\delta, \beta)$$

holds.

To this end, we need to show only that $\epsilon < \delta$. As $n(\delta, \beta) = n$, the relation
$\rho^-_i(\delta, \beta) < \delta$ holds for $i < n$. Furthermore, since $\delta \in S$ and

$$\delta = \rho_n(\delta, \beta) \in C_{\rho_{n-1}(\delta, \beta)},$$

the equation $\rho_{n-1}(\delta, \beta) = \delta \dotplus 1$ follows in view of (2), and so $\rho^-_n(\delta, \beta) = 0$,
again in view of (2). Thus

$$\epsilon = \max R(\underline{\epsilon}) \dotplus 1 < \delta,$$

as δ is a limit ordinal.

By Lemma 20.1, we can pick an ordinal δ' with $\epsilon < \delta' < \delta$ such that

$$\{\gamma < \delta : R_{\underline{\epsilon}}(\delta', \gamma)\}$$

is cofinal in δ. Next, we can choose an ordinal γ with $\epsilon < \gamma < \delta$ and $\gamma \in C_\delta$,
and, finally, we can choose an α with $\gamma < \alpha < \delta$ for which $R_{\underline{\epsilon}}(\delta', \alpha)$ holds.
We claim that

(7) $$g(\{\alpha, \beta\}) = \delta \quad \text{and} \quad \alpha, \beta \in X;$$

this is sufficient to establish Assertion (5).

To establish this claim, note that $R_{\underline{\epsilon}}(\delta', \alpha)$ implies $\alpha \in X$ and

$$\underline{\epsilon} = \underline{\epsilon}_n(\delta', \alpha) = \underline{\epsilon}_n(\delta, \beta).$$

According to Lemma 20.2 we have $\underline{\epsilon}_n(\delta', \alpha) = \underline{\epsilon}_n(\epsilon, \alpha)$ and $\underline{\epsilon}_n(\delta, \beta) = \underline{\epsilon}_n(\alpha, \beta)$, and so $\underline{\epsilon}_n(\epsilon, \alpha) = \underline{\epsilon}_n(\alpha, \beta)$. We have $n < n(\alpha, \beta)$, as $\alpha < \delta$. On the other hand, $\rho_n(\epsilon, \alpha) = \rho_n(\delta', \alpha) = \delta'$ (the latter equality holds in view of $R_{\underline{\epsilon}}(\delta', \alpha)$), and so $\rho_{n+1}^-(\epsilon, \alpha) \leq \delta'$, while we have

$$\rho_{n+1}^-(\alpha, \beta) = \sup(C_\delta \cap \alpha) \geq \gamma > \delta'$$

in view of the choice of γ; note that to justify the equality here we used the fact that $\rho_n(\alpha, \beta) = \rho_n(\delta, \beta) = \delta$.

Thus $n = n(\delta, \beta)$ satisfies the requirements stipulated in (3) above; therefore we have $g(\{\alpha, \beta\}) = \rho_n(\alpha, \beta) = \delta$, and so (7) holds. This completes the proof of Theorem 20.4

*   *
*

The smallest cardinal for which Theorem 20.4 leaves the problem open whether the assumption $2^\kappa = \kappa^+$ is needed in Theorem 20.1 is $\aleph_{\omega+1}$. It was first proved by S. Shelah that $\aleph_{\omega+1} \not\to [\aleph_{\omega+1}]^2_{\aleph_{\omega+1}}$ also holds in ZFC, and he generalized this result to a wide class of cardinals λ. It is worth studying the paper [B, M] in connection with these results.

In 1998, the status of Erdős's original question is as follows. Elementary examples involving forcing show that it is possible that $2^{\aleph_0} =$ anything reasonable, but $2^{\aleph_0} \not\to [\aleph_1]^2_{\aleph_1}$ holds. It was an open question whether it is consistent with ZFC that there is a cardinal κ that is not excessively large such that $\kappa = 2^{\aleph_0}$ and $2^{\aleph_0} \to [\aleph_1]^2_3$ holds. In [Sh; 3] Shelah showed that the answer is affirmative. In his model, however, $2^{\aleph_0}$ is greater than $\aleph_2$; thus, it is not known whether

$$2^{\aleph_0} = \aleph_2 \quad \text{and} \quad 2^{\aleph_0} \to [\aleph_1]^2_3$$

is consistent with ZFC.

Problems

1. Prove that if $\lambda \subset M_{\alpha+1} \setminus M_{\alpha+2}$, then λ has a nonreflecting stationary subset. For the definition of Mahlo cardinals see Definitions 15.3–15.5.

2.[*] Prove that for $\kappa \geq \omega$, we have

$$\kappa \not\to [\aleph_0]^\omega_{2^{\aleph_0}}.$$

3. Prove that for $\kappa \geq \omega$, there is a mapping $\mathcal{G} : [\kappa]^\omega \to [[\kappa]^\omega]^{2^{\aleph_0}}$ such that we have $\mathcal{G}(A) \subset [A]^\omega$ for each $A \in [\kappa]^\omega$ and $\mathcal{G}(A) \cap \mathcal{G}(B) = \emptyset$ for any two distinct $A, B \in [\kappa]^\omega$.

4.* Prove that for $\kappa \geq \omega$, we have

$$\kappa \not\to [\kappa]^\omega_\kappa.$$

5.* Prove that if $2^{\aleph_0}$ is real-valued measurable, then

$$2^{\aleph_0} \to [2^{\aleph_0}]^2_{\omega_1}$$

holds.

6.* Prove that $2^{\aleph_0} \not\to [2^{\aleph_0}]^2_{\aleph_0}$.

21. PROPERTIES OF THE POWER OPERATION.
RESULTS ON THE SINGULAR CARDINAL PROBLEM

In Section 11, we touched upon some independence results related to the Generalized Continuum Hypothesis. Using the concepts described in the Appendix, we will now formulate a very general independence result concerning this hypothesis.

Easton's Theorem 21.1. *Let M be a countable transitive model of set theory (that is, ZFC) such that $M \models$ GCH. Let $f \in M$ be a function such that*

$$M \models \{f \text{ is a cardinal valued function defined}$$
$$\text{on cardinals} \wedge (1) \wedge (2)\},$$

where

   (1) $f(\sigma) \leq f(\tau)$ *whenever* $\omega \leq \sigma < \tau$;
   (2) $\mathrm{cf}\, f(\tau) > \tau$ *for each* $\tau \geq \omega$.

Then there is a transitive model $N \supset M$ such that $N \models$ ZFC, the cardinals and the cofinalities are the same in N as in M, and

$$N \models \forall \tau \geq \omega \; (\tau \text{ is regular} \implies 2^\tau = f(\tau)).$$

W. B. Easton proved his theorem not long after P. J. Cohen's discovery of the method of forcing. This theorem in effect says that it is not possible to prove a result stronger than Corollary 11.3 for 2^τ in the case when τ is a regular cardinal. As a particular case of the theorem, it is, for example, possible that $2^{\aleph_0} = \aleph_1$, $2^{\aleph_1} = \aleph_2, \ldots 2^{\aleph_4} = \aleph_5$, but $2^{\aleph_5}$ is "arbitrarily large."

After this theorem had been established, R. Solovay formulated the following problem.

The singular cardinal problem.
a) Is it possible that for some singular cardinal λ, we have

$$(\forall \tau : \omega \leq \tau < \lambda)(2^\tau = \tau^+), \text{ but } 2^\lambda > \lambda^+?$$

b) *If λ is a singular strong limit cardinal, is it then possible to give a bound in ZFC for 2^λ?*

Strong limit cardinals were defined after the proof of Corollary 10.5. Back in the early 1970s, M. Magidor proved that if we assume the consistency of the existence of a certain large cardinal that is "even larger than a measurable cardinal," then one can prove the assertion

$$\text{Con}(\text{ZFC} \cup \{\forall n < \omega \, 2^{\aleph_n} = \aleph_{n+1} \wedge 2^{\aleph_\omega} = \aleph_{\omega+2}\}).$$

It was, however, not possible to generalize his method to arbitrary singular cardinals. As a great surprise to many mathematicians pursuing generalizations of Magidor's Theorem, J. Silver showed in 1974 that such a generalization is not possible, by establishing the following result.

Silver's Theorem 21.2. *If*

$$\lambda > \text{cf}(\lambda) = \kappa > \omega$$

and

$$\{\tau < \lambda : 2^\tau = \tau^+\} \in \text{Stat}(\lambda)$$

then

$$2^\lambda = \lambda^+.$$

This shows that the answer to Solovay's Problem *a)* is negative if the cofinality of λ is greater than ω. Not much later, it turned out that the answer to *b)* is affirmative in certain circumstances.

Galvin–Hajnal Theorem 21.3. *If*

$$\aleph_\alpha = \lambda > \text{cf}(\lambda) = \kappa > \omega$$

is a strong limit cardinal then

$$2^{\aleph_\alpha} = 2^\lambda < \aleph_{(|\alpha|^\kappa)^+}.$$

Thus, for example, if $\aleph_{\omega_1}$ is a strong limit cardinal, then $2^{\aleph_{\omega_1}} < \aleph_{(2^{\aleph_1})^+}$.

In the original proof of Theorem 21.2, Silver used model-theoretic methods. Soon after the appearance of his proof, a number of people found a combinatorial proof of this theorem. In the remaining part of this section, we will give a proof of a common generalization of these two theorems given in Theorem 21.4. This proof uses a number of tools described in this book, and thus it gives a prominent illustration of the significance and strength of these tools.

After the appearance of Silver's result, R. B. Jensen proved that if $2^\lambda > \lambda^+$ for some singular strong limit cardinal, then there is a model in which λ is a measurable cardinal. The proof used methods from model theory, and so its presentation is beyond the scope of this book. This result had a basic influence on the further development of the theory of large cardinals. The reader interested in these developments may consult A. Kanamori's book [Ka]; we will not discuss these developments. We would, however, like to point out the significance of this result for the philosophy of mathematics. It shows that if one decides not to be concerned with measurable cardinals since "they are too large," then one is not even able to investigate the possible values of $2^{\aleph_\omega}$.

The result of Magidor mentioned above shows that Silver's Theorem does not remain valid for cardinals of cofinality ω. This made it even more surprising that in 1980 S. Shelah proved that the analogue of the Galvin–Hajnal Theorem remains valid for $\aleph_\omega$; namely, he showed that if $\aleph_\omega$ is a strong limit cardinal then

$$2^{\aleph_\omega} < \aleph_{(2^{\aleph_0})^+}.$$

This was the situation in 1983, when the first edition of this book was prepared. The result mentioned last above was soon improved by S. Shelah, who proved the following.

Theorem (Shelah). *For every limit ordinal α we have*

$$\aleph_\alpha^{\mathrm{cf}(\alpha)} < \aleph_{(|\alpha|^{\mathrm{cf}(\alpha)})^+}.$$

This theorem is a generalization of Theorem 21.3, but it does not imply Silver's Theorem or Theorem 21.4 of Galvin and Hajnal. We will present the proof of this latter result in its original form.

Shelah's approach may be sketched as follows. Let a be a set of regular cardinals such that $\min a > |a|$. Put

$$\mathrm{pcf}(a) = \big\{ \mathrm{cf}(\underset{\lambda \in a}{\times}\, \lambda / \mathcal{U}, <_{\mathcal{U}}) : \mathcal{U} \text{ is an ultrafilter on the set } a \big\};$$

here pcf stands for *possible cofinalities*. The cardinality of a cardinal power can be estimated by the cardinalities of certain sets of form $\mathrm{pcf}(a)$, and then the inequality $|\mathrm{pcf}(a)| \le 2^{|a|}$ is proved.

The following problem is unsolved at the time of this writing (1998).

Problem (Shelah). *Is it true that $|\mathrm{pcf}(a)| \le |a|$?*

If the answer to this question is affirmative then $\aleph_\omega^{\aleph_0} < (2^{\aleph_0})^+ \cdot \aleph_{\omega_1}$ holds. This appears to be the strongest possible result, since, using large cardinal assumptions, Magidor and Shelah showed that for each $\alpha < \omega_1$ it is consistent with ZFC that

$$2^{\aleph_0} < \aleph_\omega \wedge \aleph_\omega^{\aleph_0} > \aleph_\alpha$$

holds. Through a detailed study of the operation pcf, Shelah also proved the following:

Theorem *(Shelah).*

$$\aleph_\omega^{\aleph_0} < \aleph_{\omega_4} \cdot (2^{\aleph_0})^+,$$
$$\aleph_{\omega_1}^{\aleph_0} < \aleph_{\omega_5} \cdot (2^{\aleph_1})^+.$$

Here the subscripts 4 and 5 are not misprints; they describe the best results known to date. A description of this new theory is beyond the scope of the present book. Shelah collected his results in this area in the book [Sh; 2]. Before studying this book, the reader interested in the area may consult the survey article [B, M], and even before that, Section 22 of the present book. In that section, we describe the proof that was presented by Shelah at the 1980 ASL Summer Meeting in Patras, Greece. We expect that this proof can be followed with relative ease in the light of the material contained in this book, and the proof will be helpful for studying further results in this area.

In the remaining part of the present section, κ will denote a fixed regular cardinal greater than ω, and f, g, h, k, l will denote functions in the class $^\kappa \mathrm{On}$. We will use the symbols $\equiv$, $\prec$, $\preceq$, respectively, to abbreviate the symbols $\equiv_{\mathcal{F}}$, $\prec_{\mathcal{F}}$, $\preceq_{\mathcal{F}}$, introduced in Definition 16.3, whenever $\mathcal{F}$ stands for the filter $\mathcal{C}(\kappa)$, usually called the *club filter* on κ. If $A \in \mathrm{Stat}(\kappa)$ and $\mathcal{F}$ is the normal filter $\mathrm{co}\big(\mathrm{NS}(\kappa) + (\kappa \setminus A)\big)$, then the corresponding symbols will be abbreviated as $\equiv_A$, $\prec_A$, $\preceq_A$, respectively. We will use Assertions 1–5 of Lemma 16.2 without explicitly saying so.

Definition 21.1. 1. *Let $f \in {}^\kappa \mathrm{On}$, $\mathcal{F} \subset {}^\kappa \mathrm{On}$, $A \in \mathrm{Stat}(\kappa)$. We say that $\mathcal{F}$ is a system of almost disjoint transversals on f with respect to A, shortly $\mathcal{F}$ is an SADT for f on A if for each function $g \in \mathcal{F}$, we have*

$$g \prec_A f$$

and for every $g, h \in \mathcal{F}$, the set

$$A(g = h) \overset{def}{=} \{\xi \in A : g(\xi) = h(\xi)\}$$

has cardinality less than κ. If $A = \kappa$, then we will say SADT instead of SADT on κ.

2. *$T_A(f) = \sup\{|\mathcal{F}| : \mathcal{F}$ is an SADT for f on $A\}$. We will write $T(f)$ for $T_\kappa(f)$.*

3. *$\tilde{T}(f) = \sup\{T_A(f) : A \in \mathrm{Stat}(\kappa)\}$.*

4. *$\prod(f) = \prod_{\xi < \kappa} |f(\xi)|$.*

The reason for the introduction of systems of almost disjoint transversals is explained by the following fundamental observation.

Lemma 21.1. *Let $f \in {}^\kappa \mathrm{On}$. Define the function $\tilde{f} \in {}^\kappa \mathrm{On}$ as follows:*

$$\tilde{f}(\xi) = \Big| \underset{\eta < \xi}{\times} f(\eta) \Big| \qquad for \qquad \xi < \kappa.$$

Then
$$\prod(f) \le T(\tilde{f}).$$

Proof. Let
$$\{A_\nu^\xi : \nu < \tilde{f}(\xi)\}$$
be a wellordering of type $\tilde{f}(\xi)$ of $\bigtimes_{\eta<\xi} f(\eta)$. Write
$$P = \bigtimes_{\xi<\kappa} f(\xi).$$
For each $h \in P$, define the function g_h by stipulating
$$g_h(\xi) = \nu \iff h|\xi = A_\nu^\xi \qquad \text{for} \qquad \xi < \kappa.$$
Put
$$\mathcal{F} = \{g_h : h \in P\}.$$
It is clear that $g \prec \tilde{f}$ for $h \in P$. If $h_0, h_1 \in P$ with $h_0 \ne h_1$, then there is a $\xi < \kappa$ such that $h_0(\xi) \ne h_1(\xi)$, and so $g_{h_0}(\eta) \ne g_{h_1}(\eta)$ whenever $\xi < \eta < \kappa$. Hence $\mathcal{F}$ is an SADT for $\tilde{F}$. As the correspondence $h \mapsto g_h$ is clearly one-to-one, we have
$$\prod(f) = |P| = |\mathcal{F}| \le T(\tilde{f}).$$

* *
*

The lemma thus enables us to estimate the product of κ cardinals from above by some $T_A(f)$. While there seems to be no direct way to estimate products of cardinals in general, the following lemmas make it possible for us to estimate the cardinals $T_A(f)$.

We start our discussion with a simple lemma:

Lemma 21.2. *If for $f, g \in {}^\kappa On$, $A \in \mathrm{Stat}(\kappa)$, and $\xi \in A$, we have $|f(\xi)| \le |g(\xi)|$, then*
$$T_A(f) \le T_A(g).$$

Proof. Let $\mathcal{F}$ be an SADT for f on A. For each $\xi < \kappa$, choose a mapping ϕ_ξ such that
$$f(\xi) \sim_{\phi_\xi} A_\xi \subset g(\xi)$$
holds for each $\xi \in A$ with some A_ξ. It is clear that if $\phi(h)(\xi) = \phi_\xi(h(\xi))$ for each $\xi \in A$, then
$$\mathcal{G} = \{\phi(h) : h \in \mathcal{F}\}$$
is an SADT for g on A and $|\mathcal{F}| = |\mathcal{G}|$. Hence
$$T_A(f) \le T_A(g).$$

* *
*

The following definition of a technical nature allows us to obtain further estimates.

Definition 21.2. *For each $f \in {}^{\kappa}\mathrm{On}$, we define the rank $\|f\|$ of the function f by*

$$\|f\| = \sup\{\|g\| \dotplus 1 : g \prec f\}.$$

The definition is sound, since $\prec$ is well-founded on the class ${}^{\kappa}\mathrm{On}$; cf. Problem 2 of Section 16.

We point out that $\|f\| = 0$ is true only if

$$\{\xi < \kappa : f(\xi) = 0\} \in \mathrm{Stat}(\kappa).$$

It is also easy to show by transfinite induction that for each $\nu < \kappa$, we have

$$\|f\| \leq \nu \iff \{\xi < \kappa : f(\xi) \leq \nu\} \in \mathrm{Stat}(\kappa).$$

This observation will be generalized in Lemma 21.4. Before that, however, we summarize the basic facts about the rank operation.

Lemma 21.3.
1. $f \prec g \implies \|f\| < \|g\|$.
2. $f \preceq g \implies \|f\| \leq \|g\|$.
3. If $\|f\| = \nu$, then for each $\mu < \nu$ there is a $g \prec f$ such that $\|g\| = \mu$.
4. If $f \equiv g$, then $\|f\| = \|g\|$.
5. If $f(\xi) > 0$ for each $\xi < \kappa$, then

$$\|\|f\|\| \leq \prod(f).$$

Proof. Assertions 1 and 2 are clear from the definition. Note, however, that the converses of these assertions are not true.

3. We prove the assertion by transfinite induction on ν. Assume that the assertion is true for every $\mu < \nu$, and let

$$\nu = \|f\| = \sup\{\|g\| \dotplus 1 : g \prec f\}.$$

Let $\mu < \nu$. Then there is a $g \prec f$ such that

$$\mu \leq \|g\| = \mu' < \nu.$$

If $\mu = \mu'$, then there is nothing to prove. If $\mu < \mu'$, then, by the induction hypothesis, there is an $h \prec g$ such that

$$\|h\| = \mu;$$

as $h \prec g \prec f$, we have

$$h \prec f.$$

4. The assertion immediately follows from Assertion 2.

5. Assume $\|f\| = \nu$. In view of Assertion 3, for each $\mu < \nu$, we can choose an $f_\mu \prec f$ such that

$$\|f_\mu\| = \mu.$$

By possibly changing f_μ to zero on a nonstationary set, in view of our assumption and Assertion 4, we may suppose that

$$f_\mu \in \underset{\xi < \kappa}{\times} f(\xi).$$

Clearly, $f_\mu \neq f_{\mu'}$ whenever $\mu \neq \mu'$. Hence

$$|\nu| = |\{f_\mu : \mu < \nu\}| \leq \prod(f).$$

$$* \quad *$$
$$*$$

As a generalization of the assertion described before Lemma 21.3, for each $\nu < \kappa^+$ we define the functions $h_\nu \in {}^\kappa \mathrm{On}$.

Definition 21.3. 1. *Let*

$$\{g_\mu : \mu < \phi\} \subset {}^\kappa \mathrm{On}, \qquad \phi \leq \kappa.$$

The function g is said to be the diagonal supremum of the functions g_μ, $\mu < \phi$, if $g(\xi) = \sup\{g_\mu(\xi) : \mu < \xi\}$.

2. *For each ν with $0 < \nu < \kappa^+$, let*

$$\{\nu_\alpha : \alpha < \mathrm{cf}(\nu)\}$$

be a strictly increasing sequence of ordinals such that

$$\nu = \sup\{\nu_\alpha \dotplus 1 : \alpha < \mathrm{cf}(\nu)\}.$$

3. *Define the function h_ν for $\nu < \kappa^+$ by transfinite recursion on ν:*

$$\forall \xi < \kappa\big(h_0(\xi) = 0\big),$$

and for $\nu > 0$, let h_ν be the diagonal supremum of the system

$$\{h_{\nu_\alpha} \dotplus 1 : \alpha < \mathrm{cf}(\nu)\}$$

of functions.

In this definition, $g \dotplus 1$ is the function such that

$$(g \dotplus 1)(\xi) = g(\xi \dotplus 1) \qquad \text{for} \qquad \xi < \kappa.$$

If $\nu = \mu \dotplus 1$ is a successor ordinal, then

$$\mathrm{cf}(\nu) = 1, \qquad \nu_0 = \mu,$$

and so

$$h_\nu = h_\mu \dotplus 1.$$

Lemma 21.4.

1. $h_\nu(\xi) < \kappa$ for $\xi < \kappa$ and $\nu < \kappa^+$,

and

$$|\{\xi < \kappa : h_\mu(\xi) \geq h_\nu(\xi)\}| < \kappa \qquad \text{whenever} \qquad \mu < \nu < \kappa^+.$$

2. If $\nu < \kappa^+$, then $\|h_\nu\| = \nu$ and

$$\forall g \left(\|g\| \leq \nu \iff \exists A \in \text{Stat}(\kappa) \, (g \preceq_A h_\nu) \right).$$

Proof. 1. The first assertion immediately follows from the definition of the diagonal supremum, and the second one can be proved by transfinite induction on ν.

2. The assertion is proved by transfinite induction on ν. For $\nu = 0$, we have already verified the assertion. Assume that $\nu > 0$ and that the assertion holds for each $\mu < \nu$. Then, by Assertion 1 of Lemma 21.3 and the already proven Assertion 1 of the present lemma, we have

$$\|h_\nu\| \geq \nu.$$

Assume that we have $g \preceq_A h$ for some $A \in \text{Stat}(\kappa)$, and

$$\|g\| > \nu,$$

contrary to what is claimed. By Assertion 3 of Lemma 21.3, there is a $g' \prec g$ such that

$$\|g'\| = \nu.$$

Then there is a set
$$A' \subset A, \qquad A' \in \text{Stat}(\kappa),$$

such that
$$g'(\xi) < h_\nu(\xi) \qquad \text{for all} \qquad \xi \in A'.$$

By the definition of diagonal supremum, for each $\xi \in A'$ there is an

$$\alpha(\xi) < \xi \cap \text{cf}(\nu)$$

such that
$$g'(\xi) < h_{\alpha(\xi)}(\xi).$$

By Fodor's Theorem, there is a set

$$B \subset A', \qquad B \in \text{Stat}(\kappa),$$

and an ordinal
$$\alpha < \mathrm{cf}(\nu)$$
such that
$$\alpha(\xi) = \alpha \qquad \text{for} \qquad \xi \in B.$$
Hence
$$g' \preceq_B h_{\nu_\alpha},$$
and so, by the induction hypothesis we have
$$\|g'\| \le \nu_\alpha < \nu.$$

This is a contradiction. Thus, if $g \preceq_A h_\nu$ holds for some set $A \in \mathrm{Stat}(\kappa)$ then
$$\|g\| \le \nu.$$

This also proves the assertion $\|h_\nu\| = \nu$.

We have yet to show that if
$$\{\xi \in \kappa : g(\xi) \le h_\nu(\xi)\} \notin \mathrm{Stat}(\kappa)$$
then
$$\|g\| > \nu.$$

Indeed, if the assumption of this assertion holds, then $h_\nu \prec g$, and so, by Assertion 1 of Lemma 21.3 we have
$$\|h_\nu\| < \|g\|, \qquad \text{that is,} \qquad \nu < \|g\|.$$

$$* \qquad *$$
$$*$$

Lemma 21.5. *Assume that the function $f \in {}^\kappa \mathrm{On}$ is nondecreasing, continuous (i.e., $f(\xi) = \sup\{f(\eta) : \eta < \xi\}$ holds for each limit ordinal), the values of f are cardinals, and $f(0) \ge \omega$. If*
$$\rho = \sum_{\eta < \kappa} f(\eta)^\kappa$$
then
$$\tilde{T}(\kappa) \le \rho.$$

Proof. It is clearly sufficient to prove that for each set $A \in \mathrm{Stat}(\kappa)$, we have
$$T_A(f) \le \rho.$$

Let $\mathcal{F}$ be an SADT for f on A. Given $B \subset A$ with $B \in \text{Stat}(\kappa)$ and $\eta < \kappa$, write

$$\mathcal{F}_{B,\eta} = \{g \in \mathcal{F} : \forall \xi \in B \left(g(\xi) < f(\eta)\right)\}.$$

As $\mathcal{F}$ is an SADT for f, we have

$$g_0 | B \neq g_1 | B \qquad \text{whenever} \qquad g_0, g_1 \in \mathcal{F} \quad \text{and} \quad g_0 \neq g_1.$$

Hence

$$|\mathcal{F}_{B,\eta}| \leq |\{g | B : g \in \mathcal{F}_{B,\eta}\}| \leq f(\eta)^{\kappa} \leq \rho.$$

We claim that

$$\mathcal{F} = \bigcup\{\mathcal{F}_{B,\eta} : B \in \text{Stat}(\kappa) \wedge \eta < \kappa\}.$$

Let $g \in \mathcal{F}$, and put

$$A' = \{\xi \in A : \xi \text{ is a limit ordinal} \wedge g(\xi) < f(\xi)\}.$$

As $g \prec_A f$, we have $A' \in \text{Stat}(\kappa)$. We define the mapping ϕ_g on the set A' by stipulating

$$\phi_g(\xi) = \min\{\eta : g(\xi) < f(\eta)\}.$$

In view of the continuity of f, the function ϕ_g is regressive on A'. Hence there is a set $B \subset A'$ with $B \in \text{Stat}(\kappa)$ and an $\eta < \kappa$ such that

$$\phi_g(\xi) = \eta \qquad \text{for} \qquad \xi \in B.$$

Then

$$g(\xi) < f(\eta) \qquad \text{for} \qquad \xi \in B,$$

and so

$$g \in \mathcal{F}_{B,\eta}.$$

This proves the claimed equality. Hence

$$|\mathcal{F}| = |\bigcup\{\mathcal{F}_{B,\eta} : B \in \text{Stat}(\kappa) \wedge \eta < \kappa\}| \leq 2^{\kappa} \cdot \rho = \rho.$$

$$* \qquad *$$
$$*$$

Definition 21.4. For each cardinal $\kappa \geq \omega$, let $\kappa^{+\alpha}$ be the αth successor of κ, that is, if $\kappa = \aleph_\xi$, then

$$\kappa^{+\alpha} = \aleph_{\xi+\alpha}.$$

In particular,

$$\kappa^{+0} = \kappa \qquad \text{and} \qquad \kappa^{+1} = \kappa^{+}.$$

Theorem 21.4 *(F. Galvin–A. Hajnal). Let $\kappa > \omega$ be a regular cardinal. Assume that the values of g are infinite cardinals, and $f, g \in {}^{\kappa}\mathrm{On}$. Let*

$$\lambda = \max(\tilde{T}(g), 2^{\kappa}).$$

Then

$$T(g^{+f}) \leq \lambda^{+\|f\|}.$$

Proof. Let $h = g^{+f}$, that is,

$$h(\xi) = g(\xi)^{+f(\xi)}$$

for each $\xi < \kappa$, and put

$$\nu = \|f\|.$$

We prove the assertion by transfinite induction on ν. Let $\mathcal{F}$ be an SADT for h.

(I) $\nu = 0$. Put

$$A = \{\xi < \kappa : f(\xi) = 0\}.$$

According to Lemma 21.4, we have

$$A = \mathrm{Stat}(\kappa).$$

Then $\mathcal{F}$ is an SADT for g on A, and so

$$|\mathcal{F}| \leq T_A(g) \leq \tilde{T}(g) = \tilde{T}(g)^{+0}.$$

(II) Assume $\nu > 0$ and assume that the assertion is true for each ordinal $\mu < \nu$. Then for each $k \in \mathcal{F}$, we have

$$\{\xi : k(\xi) < g(\xi)^{+f(\xi)}\} \in \mathcal{C}(\kappa).$$

Let

$$\phi_k(\xi) = \min\{\eta : |k(\xi)| \leq g^{+\eta}\} \quad \text{and} \quad \xi < \kappa,$$

provided the minimum is meaningful and $\phi_k(\xi) = 0$ otherwise (i.e., when the set after the min sign is empty). We then have $\phi_k \prec f$ for each $k \in \mathcal{F}$, since

$$\{\xi < \kappa : \phi_k(\xi) \geq f(\xi)\} \subset$$
$$\{\xi < \kappa : k(\xi) \geq h(\xi)\} \cup \{\xi < \kappa : f(\xi) = 0\} \in \mathrm{NS}(\kappa),$$

as $\|f\| = \nu > 0$. Thus $\|\phi_k\| < \nu$ for every $k \in \mathcal{F}$.

Let

$$\mathcal{F}_\mu = \{k \in \mathcal{F} : \|\phi_k\| = \mu\}.$$

Then, by what we just said, we have

$$\mathcal{F} = \bigcup \{\mathcal{F}_\mu : \mu < \nu\}.$$

Taking into account that $\lambda^{+\nu} \geq |\nu|$, as $\lambda \geq \omega$, it is sufficient to prove that

$$|\mathcal{F}_\mu| \leq \lambda^{+\mu+1} \leq \lambda^{+\nu}$$

for each $\mu < \nu$. Assume now, on the contrary, that

$$|\mathcal{F}_\mu| \geq \lambda^{+\mu+2}$$

for some $\mu < \nu$.

For an arbitrary $k \in \mathcal{F}_\mu$, consider the set

$$\mathcal{H}(k) = \{l \in \mathcal{F}_\mu : l \prec k\}.$$

We claim that

$$|\mathcal{H}(k)| \leq \lambda^{+\mu}.$$

Indeed, $\mathcal{H}(k)$ is an SADT for k, and so

$$|\mathcal{H}(k)| \leq T(k).$$

By Lemma 21.2

$$T(k) \leq T(g^{+\phi_k})$$

holds. By the induction hypothesis, we however have

$$T(g^{+\phi_k}) \leq \lambda^{+\mu}.$$

This proves our claim.

Thus $\mathcal{H}$ is a set mapping of order $\lambda^{+\mu+1}$ on the set $\mathcal{F}_\mu$. In view of our assumption on the cardinality of $\mathcal{F}_\mu$, by Theorem 19.2 there exists a set $S \subset \mathcal{F}_\mu$ of cardinality $\lambda^{+\mu+2}$ that is free with respect to $\mathcal{H}$. As

$$\lambda^{+\mu+2} \geq (2^\kappa)^+,$$

we can pick a subset

$$S' \subset S, \qquad |S'| = (2^\kappa)^+,$$

and a wellordering

$$S' = \{k_\alpha : \alpha < (2^\kappa)^+\}.$$

Define a 2-partition ϕ of $(2^\kappa)^+$ in κ colors as follows:

Given $\alpha < \beta < (2^\kappa)^+$, put

$$\phi(\{\alpha, \beta\}) = \min_{<}\{\xi : k_\alpha(\xi) > k_\beta(\xi)\}.$$

This definition is sound; indeed, S is free with respect to $\mathcal{H}$, and so $k_\alpha \notin \mathcal{H}(k_\beta)$. Thus

$$\{\xi < \kappa : k_\alpha(\xi) \geq k_\beta(\xi)\} \in \mathrm{Stat}(\kappa);$$

on the other hand, the set

$$\{\xi < \kappa : k_\alpha(\xi) = k_\beta(\xi)\}$$

has cardinality less than κ, and so

$$\{\xi < \kappa : k_\alpha(\xi) > k_\beta(\xi)\} \neq \emptyset.$$

By Corollary 14.1 to the Erdős–Rado Theorem, $(2^\kappa)^+ \to (\omega)^2_\kappa$ holds, and so there is a sequence

$$\{\alpha_n : n < \omega\}, \qquad \alpha_0 < \cdots < \alpha_n < \cdots < (2^\kappa)^+,$$

and an ordinal $\xi < \kappa$ such that

$$\{\alpha_n : n < \omega\}$$

is homogeneous in color ξ with respect to ϕ. We then have

$$k_{\alpha_n}(\xi) > k_{\alpha_{n+1}}(\xi)$$

for each $n < \omega$; this is a contradiction.

$$* \quad *$$
$$*$$

Corollary 21.1. *Let*

$$\lambda > \mathrm{cf}(\lambda) = \kappa > \omega, \qquad \nu < \kappa.$$

Assume that

$$\{\tau < \lambda : 2^\tau \leq \tau^{+\nu}\} \in \mathrm{Stat}(\lambda),$$

where τ runs over cardinals. Then

$$2^\lambda \leq \lambda^{+\nu}.$$

Silver's Theorem 21.2 given above is the special case of this for $\nu = 1$.

Proof. Let

$$\{g(\xi) : \xi < \kappa\}$$

be a strictly increasing continuous sequence of cardinals such that

$$g(0) \geq \kappa \qquad \text{and} \qquad \lambda = \sup\{g(\xi) : \xi < \kappa\}.$$

Then the set $\{g(\xi) : \xi < \kappa\}$ is a λ-club, and so

$$\{\xi < \kappa : 2^{g(\xi)} \leq g(\xi)^{+\nu}\} \in \text{Stat}(\kappa).$$

We will prove the stronger assertion that if

$$\{\xi < \kappa : 2^{g(\xi)} \leq g(\xi)^{+h_\nu(\xi)}\} \in \text{Stat}(\kappa)$$

for some $\nu < \kappa^+$, then $2^\lambda < \lambda^{+\nu}$. We have

$$2^\lambda = 2^{\sum_{\xi < \kappa} g(\xi)} = \prod_{\xi < \kappa} 2^{g(\xi)},$$

$$\prod_{\eta < \xi} 2^{g(\eta)} \leq 2^{g(\xi) \cdot \kappa} = 2^{g(\xi)} \qquad \text{for} \qquad \xi < \kappa.$$

Thus, by Lemmas 21.1 and 21.2, we have

$$2^\lambda = \prod (2^g) \leq T(2^g) \leq T(g^{+h_\nu}).$$

By Theorem 21.4, we have

$$T(g^{+h_\nu}) \leq \max(\tilde{T}(g), 2^\kappa)^{+\|h_\nu\|}.$$

Furthermore, by Lemma 21.5, we have

$$\tilde{T}(g) \leq \sum_{\tau < \lambda} \tau^\kappa \leq \sum_{\tau < \lambda} 2^{\tau\kappa} \leq \lambda,$$

since λ is a strong limit cardinal in view of Assertion 1 of Lemma 21.4, according to the assumption

$$\{\xi < \kappa : 2^{g(\xi)} \leq g(\xi)^{+h_\nu(\xi)}\} \in \text{Stat}(\kappa)$$

made above. Finally, we have $\|h_\nu\| = \nu$ by Assertion 2 of Lemma 21.4. Thus

$$2^\lambda \leq T(g^{+h_\nu}) \leq \lambda^{+\nu}.$$

$$* \qquad *$$
$$*$$

Corollary 21.2. *If*

$$\lambda = \aleph_\alpha > \mathrm{cf}(\lambda) = \kappa > \omega$$

and λ is a strong limit cardinal then

$$2^\lambda < \aleph_{(|\alpha|^\kappa)^+}.$$

Proof. Let

$$\{\lambda_\xi : \xi < \kappa\}$$

be a nondecreasing continuous sequence of cardinals such that

$$\lambda_0 \geq \kappa \qquad \text{and} \qquad \lambda = \sum_{\xi<\kappa} \lambda_\xi.$$

Then

$$2^\lambda = 2^{\sum_{\xi<\kappa} \lambda_\xi} = \prod_{\xi<\kappa} 2^{\lambda_\xi}.$$

Noting that λ was assumed to be strong limit, we have

$$\prod_{\eta<\xi} 2^{\lambda_\eta} \leq 2^{\lambda_\xi} < \lambda.$$

Defining f by the equation

$$2^{\lambda_\xi} = \omega^{+f(\xi)} \qquad (\xi < \kappa),$$

we have

$$0 < f(\xi) < \alpha,$$

for each $\xi < \kappa$, since

$$\omega < 2^{\lambda_\xi} < \aleph_\alpha = \lambda.$$

By Lemma 21.1, we have

$$2^\lambda = \prod(\langle 2^{\lambda_\xi} : \xi < \kappa \rangle) \leq T(\omega^{+f}),$$

where ω denotes the constant function with value ω on κ. By Theorem 21.4, we have

$$T(\omega^{+f}) \leq \max(2^\kappa, \tilde{T}(\omega))^{+\|f\|}.$$

As $\lambda = \aleph_\alpha$ is strong limit, we have

$$2^\kappa = \aleph_\beta$$

for some $\beta < \alpha$; furthermore,

$$\tilde{T}(\omega) \leq \aleph_0^\kappa = 2^\kappa.$$

Hence

$$T(\omega^{+f}) \leq \aleph_{\beta \dotplus \|f\|}.$$

By Assertion 5 of Lemma 21.3, we have

$$\|\|f\|\| \leq \prod(f) \leq |\alpha|^\kappa,$$

and so

$$\beta \dotplus \|f\| < (|\alpha|^\kappa)^+.$$

$$* \qquad *$$
$$*$$

Corollary 21.2 is equivalent to Theorem 21.3.

Problems

1. Prove that if

$$\{\nu < \omega_1 : \aleph_\nu^{\aleph_0} \leq \aleph_{\nu \dotplus 1} \wedge 2^\aleph < \aleph_{\omega_1}\} \in \mathrm{Stat}(\omega_1),$$

then

$$\aleph_{\omega_1}^{\aleph_1} = \aleph_{\omega_1 \dotplus 1}.$$

2.[*] Prove that if $2^{\aleph_0} = \aleph_1$ and $2^{\aleph_1} > \aleph_2$, then there is no ω_1-complete ideal in $\mathrm{P}(\omega_1)$ that is ω_2-saturated in the sense of Definition 12.16 and that includes $[\omega_1]^{<\omega_1}$.

3.[*] Prove that if $2^{\aleph_0}$ is a real-valued measurable cardinal then

$$2^{\aleph_0} = 2^{\aleph_1}.$$

22. POWERS OF SINGULAR
CARDINALS. SHELAH'S THEOREM

In this section, $\mathcal{F}$ will always denote an ultrafilter consisting of subsets of ω, and f, g, h, k, l, will denote functions in the class ${}^\omega\mathrm{On}$.

Without further reference, we will use the fact that, for the properties

$$\prec_{\mathcal{F}}, \quad \preceq_{\mathcal{F}}, \quad \equiv_{\mathcal{F}},$$

introduced in Definition 16.3, we have

$$\preceq_{\mathcal{F}} \iff \prec_{\mathcal{F}} \vee \equiv_{\mathcal{F}},$$

and $\prec_{\mathcal{F}}$ is an ordering on the class of equivalence classes $[f]_{\mathcal{F}}$.

Theorem 22.1 *(S. Shelah). If $\aleph_\omega$ is a strong limit cardinal, then*

$$2^{\aleph_\omega} < \aleph_{(2^{\aleph_0})^+}.$$

As we already mentioned in the preceding section, our goal here is to establish S. Shelah's Theorem. For the proof, we will need a number of lemmas.

Definition 22.1. 1. $[\![f]\!]_{\mathcal{F}} \overset{def}{=} \{[g]_{\mathcal{F}} : g \prec_{\mathcal{F}} f\}$.

2. $\mathrm{cf}_{\mathcal{F}}(f) \overset{def}{=} \mathrm{cf}(\langle [\![f]\!]_{\mathcal{F}}, \prec_{\mathcal{F}} \rangle)$.

3. $f \prec g \overset{def}{\iff} \forall n \in \omega \big(f(n) < g(n)\big)$.

4. $f \preceq g \overset{def}{\iff} \forall n \in \omega \big(f(n) \leq g(n)\big)$.

5. $\mathcal{P}(f) \overset{def}{=} \mathsf{X}_{n<\omega} f(n) = \{g \in {}^\omega\mathrm{On} : g \prec f\}$.

6. The system $\mathcal{H} \subset \mathcal{P}(f)$ is called a cover of f if it is cofinal in $\mathcal{P}(f)$ with respect to the partial ordering, that is, if

$$\forall g \in \mathcal{P}(f) \exists h \in \mathcal{H} \, (g \preceq h).$$

Lemma 22.1. *If $\lambda > 2^{\aleph_0}$ is a regular cardinal, and*

$$\{f_\alpha : \alpha < \lambda\} \in {}^\omega \mathrm{On}$$

is strictly increasing in the ordering $\prec_{\mathcal{F}}$, then there is a function

$$f \in {}^\omega \mathrm{On}$$

such that $[f]_{\mathcal{F}}$ is the least upper bound of the set $\{[f_\alpha]_{\mathcal{F}} : \alpha < \lambda\}$.

We will use the notation

$$\sup_{\mathcal{F}}\{f_\alpha : \alpha < \lambda\}$$

for this least upper bound.

Proof. Let

$$\mathcal{K} = \{g : \forall \alpha < \lambda \, (f_\alpha \prec_{\mathcal{F}} g)\}.$$

We claim that there is a set $\mathcal{L} \subset \mathcal{K}$ such that

$$|\mathcal{L}| \leq 2^{\aleph_0} \qquad \text{and} \qquad \forall g \in \mathcal{K} \, \exists h \in \mathcal{L} \, (h \preceq_{\mathcal{F}} g);$$

we may express the latter relation here by saying that $\mathcal{L}$ is *downward cofinal* in the set of upper bounds $\mathcal{K}$. Assume, on the contrary, that there is no such a set $\mathcal{L}$. Define the sequence

$$\{g_\nu : \nu < (2^{\aleph_0})^+\} \subset \mathcal{K}$$

by transfinite recursion as follows: Let $g_0 \in \mathcal{K}$ be arbitrary, and if the sequence $\{g_\mu : \mu < \nu\}$ has already been defined for some $\nu < (2^{\aleph_0})^+$, then let $g_\nu \in \mathcal{K}$ be a function such that

$$g_\nu \prec_{\mathcal{F}} g_\mu$$

for each $\mu < \nu$; such a g_ν exists by the assumption made above. Then, in exactly the same way as in the proof of Theorem 21.4, we arrive at a contradiction by using the particular case $(2^{\aleph_0})^+ \to (\omega)^2_{\aleph_0}$ of the Erdős–Rado Theorem.

In fact, we want to show that the set $\mathcal{L}$ can even be chosen as a one-element set. To this end, we may assume that there is an element g in the set $\mathcal{L}$ just described such that

$$\forall \alpha < \lambda \, (f_\alpha \preceq g).$$

Let

$$C = \{g(n) : n < \omega \wedge g \in \mathcal{L}\}.$$

Then
$$|C| \le |^\omega C| \le 2^{\aleph_0},$$

as $|C| \le \mathcal{L}$. For each $\alpha < \lambda$, define the function $f_\alpha^* \in {}^\omega C$ by putting

$$f_\alpha^*(n) = \min(C \setminus f_\alpha(n)) \qquad \text{for each} \qquad n < \omega.$$

Then
$$f_\alpha \preceq f_\alpha^*$$

for each $\alpha < \lambda$. By our assumption that $\lambda > 2^{\aleph_0}$ is regular, there is an $f \in {}^\omega C$ and an $L \subset \lambda$ such that

$$|L| = \lambda$$

and
$$f_\alpha^* = f \qquad \text{for each} \qquad \alpha \in L.$$

We claim that this f satisfies the requirements of the theorem. Indeed, for each $\beta < \lambda$, there is an $\alpha \in L$ with $\beta < \alpha$, and so

$$f_\beta \prec_{\mathcal{F}} f_\alpha \preceq f_\alpha^* = f.$$

Thus $[f]_{\mathcal{F}}$ is an upper bound, i.e., $f \in \mathcal{K}$. Assuming that $[f]_{\mathcal{F}}$ is not the least upper bound, there is an $h \in \mathcal{K}$ with $h \prec_{\mathcal{F}} f$ and a $g \in \mathcal{L}$ with $g \preceq_{\mathcal{F}} h$, i.e., there is a $g \in \mathcal{L}$ with

$$g \prec_{\mathcal{F}} f.$$

As $g \in {}^\omega C$ and g is an upper bound,

$$f_\alpha \preceq f_\alpha^* \prec_{\mathcal{F}} g$$

holds for each $\alpha < \lambda$, since by the definition of f_α^*, we have

$$f_\alpha(n) \le g(n) \implies f_\alpha^*(n) \le g(n).$$

This is impossible for any $\alpha \in L$, and so we indeed have

$$f = \sup_{\mathcal{F}} \{ f_\alpha : \alpha < \lambda \}.$$

$$* \qquad *$$
$$*$$

Lemma 22.2. *For each f, we have*

$$\mathrm{cf}_{\mathcal{F}}(f) = \mathrm{cf}_{\mathcal{F}}(\mathrm{cf} \circ f).$$

Proof. For each $n < \omega$, let $A_n \subset f(n)$ be a set that is cofinal in $f(n)$ and for which

$$\text{type } A_n(<) = \text{cf}(f(n)).$$

If $g \prec_{\mathcal{F}} f$, then

$$X = \{n : g(n) < f(n)\} \in \mathcal{F}.$$

For each $n \in X$, there is an $h(n) \in A_n$ such that

$$g(n) \in h(n).$$

For this h, we have

$$g \prec_{\mathcal{F}} h,$$

and

$$[h]_{\mathcal{F}} \in \underset{n \in \omega}{\times} A_n / \mathcal{F} \overset{def}{=} \{[h]_{\mathcal{F}} : \{n < \omega : h(n) \in A_n\} \in \mathcal{F}\}.$$

Hence the set $\times_{n \in \omega} A_n / \mathcal{F}$ is cofinal in the ordered set $\langle [\![f]\!]_{\mathcal{F}}, \prec_{\mathcal{F}} \rangle$, and so their cofinalities are equal according to Theorem 10.10. On the other hand, the ordered set

$$\underset{n \in \omega}{\times} A_n / \mathcal{F}$$

is clearly similar to the set

$$\langle [\![\text{cf} \circ f]\!]_{\mathcal{F}}, \prec_{\mathcal{F}} \rangle$$

$$* \qquad *$$
$$*$$

Lemma 22.3. If $f = \aleph \circ g$ (i.e., $f(\alpha) = \aleph_{g(\alpha)}$ for any $\alpha \in D(g)$) and $g \in {}^{\omega}(2^{\aleph_0})^+$, then

$$\text{cf}_{\mathcal{F}}(f) < \aleph_{(2^{\aleph_0})^+}.$$

Proof. Let

$$\rho = \aleph_{(2^{\aleph_0})^+}.$$

Assuming $\text{cf}_{\mathcal{F}}(f) \geq \rho$, there is sequence

$$\{f_{\xi} : \xi < \rho\}$$

strictly increasing in $\prec_{\mathcal{F}}$ such that

$$f_{\xi} \prec_{\mathcal{F}} f.$$

Clearly, we have

$$(2^{\aleph_0})^+ \leq \aleph_{(2^{\aleph_0})^+}.$$

Even

$$(2^{\aleph_0})^+ < \aleph_{(2^{\aleph_0})^+}$$

holds, since the subscript $(2^{\aleph_0})^+$ on the right-hand side is a limit ordinal.

Write

$$\Gamma = \{\gamma < (2^{\aleph_0})^+ : (2^{\aleph_0})^+ \le \aleph_{\gamma+1}\}.$$

Then, by the inequality just stated, we have

$$|\Gamma| = (2^{\aleph_0})^+.$$

For each $\gamma \in \Gamma$, let

$$f^\gamma = \sup_{\mathcal{F}}\{f_\xi : \xi < \aleph_{\gamma+1}\}.$$

This definition is sound in view of Lemma 22.1, as

$$\mathrm{cf}(\aleph_{\gamma+1}) = \aleph_{\gamma+1} \ge (2^{\aleph_0})^+ \qquad \text{for} \qquad \gamma \in \Gamma.$$

We have $f^\gamma \preceq_{\mathcal{F}} f_{\aleph_{\gamma+1}} \prec_{\mathcal{F}} f$, and so $f^\gamma \prec_{\mathcal{F}} f$, for each $\gamma \in \Gamma$. It is clear that

$$\mathrm{cf}_{\mathcal{F}}(f^\gamma) = \aleph_{\gamma+1} \qquad \text{for} \qquad \gamma \in \Gamma,$$

since

$$\{[f_\xi]_{\mathcal{F}} : \xi < \aleph_{\gamma+1}\}$$

is cofinal in $\langle [\![f^\gamma]\!]_{\mathcal{F}}, \prec_{\mathcal{F}} \rangle$. Thus we certainly have

$$\mathrm{cf}_{\mathcal{F}}(f^\gamma) \ne \mathrm{cf}_{\mathcal{F}}(f^\delta) \qquad \text{whenever} \qquad \gamma, \delta \in \Gamma \quad \text{and} \quad \gamma \ne \delta.$$

By Lemma 22.2, we have

$$\mathrm{cf}_{\mathcal{F}}(f^\gamma) = \mathrm{cf}_{\mathcal{F}}(\mathrm{cf} \circ f^\gamma);$$

here we may assume that

$$\mathrm{cf} \circ f^\gamma \subset \bigtimes_{n<\omega} (\{0,1\} \cup \{\aleph_\nu : \nu < g(n)\}).$$

The cardinality of the set on the right-hand side is at most

$$\prod_{n<\omega} |g(n) \dotplus 2| \le 2^{\aleph_0}.$$

This is a contradiction, since we have

$$|\Gamma| = (2^{\aleph_0})^+,$$

as we saw above.

$$* \qquad *$$
$$*$$

The next lemma is the first major step leading to the proof of the main result.

Lemma 22.4. *If*

$$f = \aleph \circ g \quad \text{and} \quad g \in {}^{\omega}(2^{\aleph_0})^+,$$

then there is a cover $\mathcal{H} \subset \mathcal{P}(f)$ *of* f *such that*

$$|\mathcal{H}| \leq \aleph_{(2^{\aleph_0})^+} \cdot 2^{2^{\aleph_0}}.$$

Proof. For an arbitrary ultrafilter $\mathcal{F} \in \mathrm{P}(\omega)$, let $\mathcal{H}_{\mathcal{F}}$ be a set such that the set

$$\{[f]_{\mathcal{F}} : f \in \mathcal{H}_{\mathcal{F}}\},$$

is cofinal in

$$\langle [\![f]\!]_{\mathcal{F}}, \prec_{\mathcal{F}} \rangle,$$

$$|\mathcal{H}_{\mathcal{F}}| < \aleph_{(2^{\aleph_0})^+}$$

holds, and we have

$$g \prec f$$

for every $g \in \mathcal{H}_{\mathcal{F}}$. Such a set $\mathcal{H}_{\mathcal{F}}$ exists according to Lemma 22.3. Let

$$\mathcal{H}' = \bigcup \{\mathcal{H}_{\mathcal{F}} : \mathcal{F} \subset \mathrm{P}(\omega) \wedge \mathcal{F} \text{ is an ultrafilter}\},$$

and put

$$\mathcal{H} = \{\max \mathcal{H}'' : \mathcal{H}'' \in [\mathcal{H}']^{<\omega}, \mathcal{H}'' \neq \emptyset\}.$$

Here $\max \mathcal{H}''$ is the function k such that

$$k(n) = \max\{g(n) : g \in \mathcal{H}''\}.$$

Then

$$|\mathcal{H}| = \max(|\mathcal{H}'|, \omega) \leq \aleph_{(2^{\aleph_0})^+} \cdot 2^{2^{\aleph_0}}.$$

We claim that $\mathcal{H}$ is a cover of f. Assume, on the contrary, that, for some $h \prec f$, none of the elements of the set

$$D = \{[g < h] : g \in \mathcal{H}\}$$

are empty, where

$$[g < h] = \{n < \omega : g(n) < h(n)\};$$

the set $[h \leq g]$, to be used below, can be defined analogously. As, for each $\mathcal{H}'' \in [\mathcal{H}]^{<\omega}$, we have $\max \mathcal{H}'' \in \mathcal{H}$, and, further,

$$\bigcap \{[g < h] : g \in \mathcal{H}''\} = [\max \mathcal{H}'' < h],$$

the set system D has the Finite Intersection Property. Let $\mathcal{F} \subset P(\omega)$ be an ultrafilter that includes D. Then

$$\mathcal{H}_{\mathcal{F}} \subset \mathcal{H}' \subset \mathcal{H},$$

and by the choice of $\mathcal{H}_{\mathcal{F}}$ there is a $g \in \mathcal{H}$ such that

$$[h \le g] \in \mathcal{F}.$$

This is a contradiction, since $[g < h] \in D \subset \mathcal{F}$ holds.

$$* \qquad *$$
$$*$$

Lemma 22.5. *Let*

$$\sigma = (2^{\aleph_0})^+.$$

For each $\gamma < \sigma^+$, there is a set system

$$\mathcal{A}_\gamma \subset [\aleph_\gamma]^{\sigma^+}$$

such that

$$|\mathcal{A}_\gamma| \le \aleph_\gamma \cdot 2^{\sigma^+}$$

and

$$\forall X \in [\aleph_\gamma]^{\sigma^+} \exists A \in \mathcal{A}_\gamma \, (A \subset X).$$

Proof. We proceed by transfinite induction on γ. The assertion is obvious if $\aleph_\gamma \le \sigma^+$, since then

$$[\aleph_\gamma]^{\sigma^+} = \emptyset \qquad \text{for} \qquad \aleph_\gamma < \sigma^+$$

and

$$|[\aleph_\gamma]^{\sigma^+}| = (\sigma^+)^{\sigma^+} = 2^{\sigma^+} \qquad \text{for} \qquad \aleph_\gamma = \sigma^+.$$

Assume therefore that

$$\sigma^+ < \aleph_\gamma < \aleph_{\sigma^+}$$

and, further, that the assertion is true for every $\delta < \gamma$ replacing γ. We distinguish two cases.

   (I) $\gamma = \delta + 1$.
   (II) γ is a limit ordinal.

(I) For each $\xi < \aleph_{\delta+1}$ we have

$$|\xi| \leq \aleph_\delta,$$

and so there is an

$$\mathcal{A}'_\xi \subset [\xi]^{\sigma^+} \qquad \text{with} \qquad |\mathcal{A}'_\xi| \leq \aleph_\delta \cdot 2^{\sigma^+}$$

such that

$$\forall X \in [\xi]^{\sigma^+} \exists A \in \mathcal{A}'_\xi \, (A \subset X).$$

Put

$$\mathcal{A}_\gamma = \bigcup \{\mathcal{A}'_\xi : \xi < \aleph_{\delta+1}\}.$$

Then

$$|\mathcal{A}_\gamma| \leq \aleph_\delta \cdot 2^{\sigma^+} \cdot \aleph_{\delta+1} = 2^{\sigma^+} \cdot \aleph_\gamma,$$

and for each

$$X \in [\aleph_{\delta+1}]^{\sigma^+},$$

there is a $\xi < \aleph_{\delta+1}$ such that

$$X \in [\xi]^{\sigma^+},$$

since

$$\mathrm{cf}(\aleph_{\delta+1}) = \aleph_{\delta+1} > \sigma^+.$$

Hence $\mathcal{A}_\gamma$ satisfies the requirements.

(II) Let

$$\mathcal{A}_\delta \subset [\aleph_\delta]^{\sigma^+}$$

be a set system satisfying the requirements for each $\delta < \gamma$, and put

$$\mathcal{A}_\gamma = \bigcup \{\mathcal{A}_\delta : \delta < \gamma\}.$$

Then

$$|\mathcal{A}_\gamma| \leq \left(\sum_{\delta<\gamma} \aleph_\delta\right) \cdot 2^{\sigma^+} \leq \aleph_\gamma \cdot 2^{\sigma^+}.$$

Let

$$X \in [\aleph_\gamma]^{\sigma^+}.$$

As γ is a limit ordinal, we have

$$\aleph_\gamma = \bigcup \{\aleph_\delta : \delta < \gamma\}$$

and

$$X = \bigcup \{X \cap \aleph_\delta : \delta < \gamma\}.$$

As σ^+ is a regular cardinal and $\gamma < \sigma^+$, we have

$$|X \cap \aleph_\delta| = \sigma^+$$

for some $\delta < \gamma$. In this case, however, there is an $A \subset \mathcal{A}_\delta$ such that

$$A \subset X \cap \aleph_\delta \subset X$$

holds, and so $\mathcal{A}_\gamma$ satisfies the requirements of the theorem.

$$* \qquad *$$
$$*$$

We now turn to a detailed study of the set $\aleph_\omega$. In what follows, we assume that $2^{\aleph_0} < \aleph_\omega$, and so

$$\sigma^+ = (2^{\aleph_0})^{++} = \aleph_{n_0}$$

for some $n_0 < \omega$ (here κ^{++} abbreviates $(\kappa^+)^+$). Put

$$S = \sigma^+, \qquad S_n = \aleph_{n_0+n+1} \setminus \aleph_{n_0+n} \quad \text{for} \quad n < \omega.$$

Then $\aleph_\omega = S \cup \bigcup_{n<\omega} S_n$ is a partition of $\aleph_\omega$ into pairwise disjoint sets. Put

$$f = \aleph \circ g,$$

where $g(n) = n_0 + n + 1$. Then, by Lemma 22.4, there is a cover $\mathcal{H}$ of f such that

$$|\mathcal{H}| \overset{\text{def}}{=} \tau \leq \aleph_\sigma \cdot 2^{2^{\aleph_0}}.$$

Let

$$\mathcal{H} = \{f_\alpha : \alpha < \tau\}$$

be a wellordering of order type τ of $\mathcal{H}$. By Lemma 22.5, we may pick a set system

$$\mathcal{A} \subset [\tau]^{\sigma^+}, \qquad |\mathcal{A}| \leq \tau \cdot 2^{\sigma^+},$$

such that for each $X \in [\tau]^{\sigma^+}$, there is a set $A \in \mathcal{A}$ with $A \subset X$.

For each $Y \subset \aleph_\omega$, define the function $g_Y \in {}^\omega \mathrm{On}$ by stipulating

$$g_Y(n) = \sup S_n \cap Y \quad \text{for} \quad n < \omega.$$

Then we have $g_Y \dot{+} 1 \prec f$ for $Y \in [\aleph_\omega]^{\sigma^+}$, since

$$f(n) = \aleph_{n_0+n+1} > \sigma^+.$$

For each $\xi < \aleph_\omega$, let ϕ_ξ be a function such that

$$\xi \sim_{\phi_\xi} |\xi|.$$

The set $Y \subset \aleph_\omega$ is said to be ϕ-*closed* if $S \subset Y$ and, for each $\xi, \eta \in Y$, we have $\phi_\xi(\eta) \in Y$ whenever $\eta < \xi$, and, further, we have

$$\phi_\xi^{-1}(\{\eta\}) \subset Y$$

whenever $\eta < |\xi|$.

As we saw earlier, several times, in similar situations, for each $Y \subset \aleph_\omega$ there is a smallest set $\tilde{Y}$ that is ϕ-closed; furthermore, we have $|Y| = |\tilde{Y}|$ if $|Y| \geq \sigma^+$ (cf., e.g., the proof of the existence of transitive closure in the proof of Theorem A7.2). $\tilde{Y}$ will be called the ϕ-*closure* of Y.

The key to the rest of the proof is the following definition.

Definition 22.2. *We call the set $B \in [\aleph_\omega]^{\sigma^+}$ neat if it satisfies the following three conditions:*

1. *B is ϕ-closed.*
2. *There is a set $X \in [\tau]^{\sigma^+}$ with type $X(<) = \sigma^+$ such that*

$$\{f_\alpha : \alpha \in X\}$$

is a sequence that is increasing in the partial ordering $\prec$ and, for each $n < \omega$, the set

$$\{f_\alpha(n) : \alpha \in X\} \subset S_n \cap B$$

is cofinal in $g_B(n)$.

3. *If $n < \omega$, then $S_n \cap B$ includes a subset that is club in $g_B(n)$.*

The set of neat sets will be denoted by $\mathcal{B}$.

Cofinality Lemma 22.6. *For each $Y \in [\aleph_\omega]^{\sigma^+}$, there is a neat set $B \in \mathcal{B}$ such that $Y \subset B$.*

Consequently,

$$\aleph_\omega^{\sigma^+} = \left| [\aleph_\omega]^{\sigma^+} \right| \leq 2^{\sigma^+} \cdot |\mathcal{B}|.$$

Proof. Define the sequence

$$\{B_\nu : \nu < \sigma^+\} \subset \mathrm{P}(\aleph_\omega)$$

of sets and the sequence

$$\{\alpha_\nu : \nu < \sigma^+\} \subset \tau$$

of ordinals by transfinite recursion as follows. Let $\nu < \sigma^+$ and assume that the sets B_μ have been defined for $\mu < \nu$ such that

$$|B_\mu| = \sigma^+.$$

Let

$$B'_\nu = \bigcup_{\mu < \nu} B_\mu \cup S \cup Y \cup \{\omega_n : n < \omega\}.$$

Then

$$|B'_\nu| = \sigma^+.$$

Put

$$\alpha_\nu = \min\{\alpha < \tau : g_{B'_\nu} \dotplus 1 \preceq f_\alpha\}.$$

The definition of α_ν is sound, as $\mathcal{H}$ is a cover of f and we have $g_{B'_\nu} \dotplus 1 \prec f$ in view of $|B'_\nu| = \sigma^+$ (cf. the inequality $g_Y \dotplus 1 \prec f$ mentioned at the point the function g_Y was defined above). Let

$$B''_\nu = B'_\nu \cup \mathrm{R}(g_{B'_\nu}) \cup \mathrm{R}(f_{\alpha_\nu})$$

and

$$B_\nu = \widetilde{B''_\nu},$$

where the tilde on the right-hand side indicates the ϕ-closure of B''_ν described above. Then

$$|B_\nu| = |B''_\nu| = |B'_\nu| + \omega = \sigma^+.$$

This completes the definition of the sequence $\{B_\nu : \nu < \sigma^+\}$. Write

$$B = \bigcup_{\nu < \sigma^+} B_\nu.$$

It is clear that $Y \subset B$ and $|B| = \sigma^+$. We are going to show that B is neat. Indeed, B is ϕ-closed, since it is a nondecreasing union of ϕ-closed sets. Put

$$X' = \{\alpha_\nu : \nu < \sigma^+\}.$$

By the Erdős–Dushnik–Miller Theorem 14.6, there is a set $X \subset X'$ with $|X| = \sigma^+$ such that

$$\alpha_\nu < \alpha_{\nu'} \quad \text{whenever} \quad \nu < \nu' \quad \text{and} \quad \alpha_\nu, \alpha'_\nu \in X.$$

Thus $\{f_\alpha : \alpha \in X\}$ is increasing in the ordering $\prec$, as

$$f_{\alpha_\mu} \prec f_{\alpha_\nu} \quad \text{whenever} \quad \mu < \nu.$$

Given $n < \omega$ and $x \in B \cap S_n$, we have $x \in B'_\nu$ for some $\nu < \sigma^+$; thus

$$x < f_{\alpha_\nu}(n) < \aleph_{n_0 + n + 1},$$

and so $\{f_{\alpha_\nu}(n) : \alpha_\nu \in X\}$ is a cofinal subset of $B \cap S_n$.

Finally, the set $\{g_{B'_\nu} : \nu < \sigma^+\}$ is a subset of $B \cap S_n$ that is a club in $g_B(n)$.

$$* \qquad *$$
$$*$$

First Counting Lemma 22.7. *If B_0 and B_1 are neat sets and $g_{B_0} = g_{B_1}$, then $B_0 = B_1$. Thus, writing $C = \{g_B : B \in \mathcal{B}\}$, we have $|\mathcal{B}| = |\mathcal{C}|$.*

Proof. It is sufficient to prove that

$$B_0 \cap \omega_m = B_1 \cap \omega_m$$

holds for each $m < \omega$. We prove this by induction. If $m = n_0$, then

$$B_0 \cap \omega_{n_0} = B_1 \cap \omega_{n_0} = S.$$

Assume that $m = n_0 + n + 1$ and the assertion is true for ω_{n_0+n}. Let $\eta \in B_0 \cap \omega_m$. Then

$$\eta < g_{B_0}(m) = g_{B_1}(m);$$

to see this, observe that Condition 2 in Definition 22.2 implicitly says that $g_{B_0}(m) \neq 0$. By Condition 3 of neat sets, there is an $\eta < \xi < g_{B_0}(m)$ such that $\xi \in B_0 \cap B_1$, since, in view of $\mathrm{cf}(g_{B_0}(m)) = \sigma^+ > \omega$, the intersection of the two closed unbounded sets whose existence is asserted by Condition 3 in Definition 22.2 of the neat sets B_0 and B_1 with m replacing n is a closed unbounded set. We have $\phi_\xi(\eta) \in B_0$ in view of the properties of neat sets. By the choice of ϕ_ξ

$$\phi_\xi(\eta) < |\xi| \leq \omega_{n_0+n}$$

holds, and so

$$\phi_\xi(\eta) \in B_1$$

by the induction hypothesis. Then

$$\{\eta\} = \phi^{-1}(\{\phi_\xi(\eta)\}) \subset B_1,$$

as B_1 is also ϕ-closed. Thus $B_0 \cap \omega_m \subset B_1 \cap \omega_m$. The reverse inclusion follows the same way, completing the proof.

$$* \qquad *$$
$$*$$

Second Counting Lemma 22.8.

$$|\mathcal{C}| \leq |\mathcal{A}|.$$

Proof. Let B be a neat set, let $X \in [\tau]^{\sigma^+}$ be a set satisfying Condition 2 of being a neat set, and let $\Phi(B) = A$ for some set $A \in \mathcal{A}$ with $A \subset X$.

Then, for each $n < \omega$,

$$g_B(n) = \sup\{f_\alpha(n) : \alpha \in A\}$$

holds, since A is a subset of cardinality σ^+ of the set X of order type σ^+. Thus, if B_0 and B_1 are neat sets with $\Phi(B_0) = \Phi(B_1)$, then

$$g_{B_0} = g_{B_1}.$$

$$* \quad *$$
$$*$$

Putting Lemmas 22.6, 22.7, and 22.8 together, we obtain that

$$\aleph_\omega^{\sigma^+} \le 2^{\sigma^+} \cdot |\mathcal{B}| = 2^{\sigma^+} \cdot |\mathcal{C}| \le 2^{\sigma^+} \cdot |\mathcal{A}| \le 2^{\sigma^+} \cdot \tau \le 2^{\sigma^+} \cdot \aleph_\sigma.$$

Thus $\aleph_\omega^{\sigma^+} \le 2^{\sigma^+} \cdot \aleph_\sigma$. In the course of the proof we used only the assumption

$$2^{\aleph_0} < \aleph_\omega.$$

The conclusion is, however, true even without this assumption, since if $\aleph_\omega < 2^{\aleph_0}$, then

$$\aleph_\omega^{\sigma^+} = 2^{\aleph_0 \cdot \sigma^+} = 2^{\sigma^+}.$$

We have thus proved the following.

Theorem 22.2 *(S. Shelah).*

$$\aleph_\omega^{\sigma^+} \le 2^{\sigma^+} \cdot \aleph_\sigma,$$

where $\sigma = (2^{\aleph_0})^+$.

If $\aleph_\omega$ is a strong limit cardinal, then

$$2^{\aleph_\omega} = 2^{\sum_{n<\omega} \aleph_n} = \prod_{n<\omega} 2^{\aleph_n} \le \aleph_\omega^{\aleph_0} \le \aleph_\omega^{(2^{\aleph_0})^+} \le \aleph_{(2^{\aleph_0})^+}.$$

Equality here cannot hold, since

$$\mathrm{cf}(2^{\aleph_\omega}) > \aleph_\omega$$

by Corollary 11.3 to König's Theorem, but

$$\mathrm{cf}(\aleph_{(2^{\aleph_0})^+}) = (2^{\aleph_0})^+ < \aleph_\omega.$$

Hence, in this case we have

$$2^{\aleph_\omega} < \aleph_{(2^{\aleph_0})^+}.$$

HINTS FOR SOLVING PROBLEMS OF PART II

Section 12

1. Let $B \in \mathcal{C}(\kappa)$. By transfinite recursion, one can pick a strictly increasing sequence $\{\alpha_\nu : \nu < \lambda\} \subset B$ with $\sup\{\alpha_\nu : \nu < \lambda\} \in S_{\lambda,\kappa}$.

2. Use the idea of Problem 1.

3. Assertions $a)$ and $c)$ can be obtained by refining the proofs in the text.

$b)$ Let $\kappa > \mathrm{cf}(\kappa)$ and let $\{\alpha_\nu : \nu < \mathrm{cf}(\kappa)\}$ be the strictly increasing continuous sequence of ordinals obtained in Problem 2. We may assume that $\alpha_0 = 0$ and $\alpha_1 = \mathrm{cf}(\kappa)$. Define f such that $f(\alpha) = \alpha_\nu$ if $\alpha_\nu < \alpha < \alpha_{\nu+1}$, $f(\alpha_\nu) = \nu$ if $\nu < \mathrm{cf}(\kappa)$. [F].

4. Otherwise, by Fodor's Theorem, we would have $B = \{\alpha < \kappa : \mathrm{type}(A \cap \alpha) = \beta\}$ for some $B \in \mathrm{Stat}(\kappa)$ and $\beta < \kappa$. Then, however, $A \subset \beta_0 = \min B$ would hold.

5. Otherwise, by Fodor's Theorem, there is a $\beta < \omega_1$ and a stationary set $B \subset \omega_1$ such that, for every γ with $\beta < \gamma < \omega_1$ and for every $\alpha \in B$, the set $A \cap \{\langle \gamma, \beta \rangle : \beta < \alpha\}$ is not cofinal in α. According to Problem 4, we then have $|A \cap \{\langle \gamma, \beta \rangle : \beta < \omega_1\}| < \omega_1$ for every γ with $\beta < \gamma < \omega_1$, and so the order type of A is not ω_1^2. [H; 2].

6. If $A_\alpha \in \mathrm{NS}(\kappa)$ for each $\alpha < \kappa$ and $B \in \mathrm{NS}(\kappa)$, then $A \setminus B \in \mathrm{NS}(\kappa)$, as confirmed by the regressive function f defined by $f(\xi) = \min_< A_\alpha$ for $\xi \in A_\alpha \setminus B$.

7. Assume that $\{C_\alpha : \alpha < \kappa^+\} \subset \mathrm{Stat}(\kappa)$ is a sequence of order type κ^+ with $C_\alpha \cap C_\beta \in \mathrm{NS}(\kappa)$ whenever $\alpha \neq \beta$ and $\alpha, \beta < \kappa^+$. For each $\alpha < \kappa^+$, let $\{g(\alpha, \nu) : \nu < \kappa\}$ be an enumeration of order type κ of α. Let $B_\alpha = \Delta_{\nu<\kappa}(C_\alpha \setminus C_{g(\alpha,\nu)})$. Then $B_\alpha \in \mathrm{Stat}(\kappa)$ and $|B_\alpha \cap B_\beta| \leq |B_\alpha \cap C_\beta| < \kappa$ whenever $\beta < \alpha < \kappa^+$.

8. It is enough to find a sequence $A_\alpha'' \subset A_\alpha$ with $A_\alpha'' \in \mathrm{Stat}(\kappa)$ for every $\alpha < \kappa$ that consists of pairwise disjoint sets. In view of Problem 7 and of the assumptions, for each $\alpha < \kappa$ we can find a system consisting of pairwise almost disjoint κ-stationary sets $B_{\alpha,\xi} \subset A_\alpha$, $\xi < \kappa^+$. Let

$$X_0 = \left\{ \alpha < \kappa : \left| \{\xi < \kappa^+ : B_{0,\xi} \cap A_\alpha \in \mathrm{Stat}(\kappa)\} \right| = \kappa^+ \right\}.$$

The system $\{A_\alpha : \alpha \in X_0\}$ may be made almost disjoint by picking sets $A_\alpha'' \subset A_\alpha$ with $A_\alpha'' \in \mathrm{Stat}(\kappa)$ for every $\alpha \in X_0$ such that $A_\beta \cap A_\alpha'' \in \mathrm{NS}(\kappa)$ holds for

$\beta \in \kappa \setminus X_0$. Indeed, we may first pick $A''_\alpha = B_{0,\xi} \cap A_\alpha$ for an appropriately chosen $\xi < \kappa^+$ such that $B_{0,\xi} \cap A_\alpha \in \mathrm{NS}(\kappa)$ whenever $\alpha \in \kappa \setminus X_0$. Using the result of Problem 6, we may shrink A''_α so that $\bigcup \{A''_\alpha : \alpha \in X_0\} \cap B_\beta \in \mathrm{NS}(\kappa)$ holds as well whenever $\beta \in \kappa \setminus X_0$. Repeat this procedure for the remaining set system $\{A_\alpha : \alpha \in \kappa \setminus X_0\}$ (transfinitely). [B, H, M].

We remark that it is relatively consistent with ZFC that the condition of the present problem is satisfied for every regular $\kappa > \omega$; for example, the condition holds on the model on constructible sets, briefly mentioned in Section A9. Shelah proved in ZFC that $\mathrm{NS}(\kappa)$ cannot be κ^+ saturated on $S_{\lambda,\kappa}$ if $\omega \le \lambda = \mathrm{cf}(\lambda) < \kappa = \mathrm{cf}(\kappa)$.

9. It is obvious that if $\mathrm{NS}(\omega_1)$ is ω_1-dense on A, then the ω_1-stationary sets A_α, $\alpha < \omega_1$, showing this cannot be made disjoint.

Assume now that for every stationary $A \subset \omega_1$, the nonstationary ideal on A is not ω_1-dense. This means that

(*) for every sequence $\{B_\alpha : \alpha < \omega_1\} \subset \mathrm{Stat}(\omega_1)$, there is a set $B \subset B_0$ with $B \in \mathrm{Stat}(\omega_1)$ and $B_\alpha \setminus B \in \mathrm{Stat}(\omega_1)$ for each $\alpha < \omega_1$.

According to the proof of Theorem 12.5.A there are functions f_n, $n < \omega_1$, regressive on ω_1 such that for each $A \in \mathrm{Stat}(\omega_1)$, there is an $n < \omega$ for which f_n is not essentially bounded on A. Assume we are given the ω_1-stationary sets $\{A_\alpha : \alpha < \omega_1\}$. Let $n(\alpha)$ denote the smallest n for which f_n is not essentially bounded on A_α.

Let $X_0 = \{\alpha : n(\alpha) = 0\}$. The set system $\{A_\alpha : \alpha \in X_0\}$ can be made disjoint by choosing sets $A'_\alpha \subset A_\alpha$ with $A'_\alpha \in \mathrm{Stat}(\omega_1)$ such that f_0 is constant on each A'_α and such that $\{\min A'_\alpha : \alpha \in X_0\} \in \mathrm{NS}(\omega_1)$. Using (*), we can then choose sets $A^0_\alpha \subset A'_\alpha$ with $A^0_\alpha \in \mathrm{Stat}(\omega_1)$ for each $\alpha \in X_0$ such that for every n with $1 \le n \le \omega$, for every $\beta < \omega_1$, and for every $\gamma \in \omega_1 \setminus X_0$ we have

$$f_n^{-1}(\{\beta\}) \cap A_\gamma \cap A'_\alpha \setminus A^0_\alpha \in \mathrm{Stat}(\omega_1)$$
$$\text{whenever} \quad f_n^{-1}(\{\beta\}) \cap A_\gamma \cap A'_\alpha \in \mathrm{Stat}(\omega_1).$$

Let $A^0 = \bigcup \{A^0_\alpha : \alpha \in X_0\}$ and $A^1_\alpha = A_\alpha \setminus A^0$ for $\alpha \in \omega_1 \setminus X_0$. For $\alpha \in \omega_1 \setminus X_0$ we have $n(\alpha) \ge 1$, $f_{n(\alpha)}$ is not essentially bounded on A^1_α, and, now continuing with $\{A^1_\alpha : \alpha \in \omega_1 \setminus X_0\}$ in place of $\{A_\alpha : \alpha < \omega_1\}$, the procedure can be repeated countably many times. [B, H, M].

10. From the assumptions on the cardinals, it can be easily seen that there is an $\alpha < \kappa^+$ such that $|\{\beta < \kappa^+ : A_\beta \cap (\xi \dotplus 1) = A_\alpha \cap (\xi \dotplus 1)\}| = \kappa^+$ for every $\xi < \kappa$. Hence there is a sequence $\{\alpha_\xi : \xi < \kappa\} \subset \kappa^+$ of pairwise distinct ordinals such that $A_{\alpha_\xi} \cap (\xi \dotplus 1) = A_\alpha \cap (\xi \dotplus 1)$ for each $\xi < \kappa$. Let $A = \bigcup \{A_{\alpha_\xi} : \xi < \kappa\}$. Then $A \setminus A_\alpha \subset \bigcup \{A_{\alpha_\xi} \setminus (\xi \dotplus 1) : \xi < \kappa\} \in \mathrm{NS}(\kappa)$.

The proof can of course be carried out for an arbitrary normal ideal so as to obtain the corresponding assertion. The result is due to F. Galvin, and the proof can be found in [B, H, M].

Section 13

1. *a)* If $\lambda = \kappa$, then the sets $A_\alpha = \alpha < \kappa$ form a set system of cardinality κ that does not include a Δ-system of three elements.

b) Let $\lambda = \min_< \{\lambda' : \exists \tau < \kappa\, (\tau^{\lambda'} \geq \kappa)\}$ and $\mathcal{F} = \{\{f|\xi : \xi < \lambda\} : f \in {}^\tau\lambda\}$. Then $|\mathcal{F}| \geq \kappa$, the elements of $\mathcal{F}$ have cardinality λ, and $\mathcal{F}$ does not include a Δ-system of cardinality τ^+.

c) If κ is singular, $\mathrm{cf}(\kappa) < \kappa_0 < \cdots < \kappa_\xi < \ldots$ is an increasing sequence of cardinals with $\sup\{\kappa_\xi : \xi < \mathrm{cf}(\kappa)\} = \kappa$, then

$$\left\{\{\xi,\alpha\} : \xi < \mathrm{cf}(\kappa) \wedge \kappa_\xi \leq \alpha < \kappa_{\xi+1}\right\}$$

is a sequence of two-element sets that does not include any Δ-system of cardinality κ. [E, R; 2].

2. See the hints for Problem 3.

3. By induction. If $\mathcal{F}$ consists of n-element sets and it does not include a Δ-system having three elements, then there is a set of at most $2n$ elements that intersects every element of $\mathcal{F}$. Hence $d(n,3) - 1 \leq 2n\,(d(n-1,3) - 1)$.

4. Let $\mathbb{Q} = \{r_n : n < \omega\}$ be a one-to-one enumeration of the rationals. Let $F_x = \{n_k(x) : k < \omega\}$ be such that $r_{n_k(x)} \to x$ as $k \to \infty$. Then $\mathcal{F} = \{F_x : x \in \mathbb{R}\}$, where $\mathbb{R}$ is the set of real numbers, satisfies the requirements.

5. See the solution of Problem 1*b)*.

6. Proof in the case when $\lambda > \mathrm{cf}(\kappa)$:

Let λ_ξ form an increasing sequence of regular cardinals with

$$\lambda = \sup\{\lambda_\xi : \xi < \mathrm{cf}(\kappa),\ \lambda_0 > cf(\kappa)\}.$$

We construct a set system $\mathcal{F} = \{F_\alpha : \alpha < \lambda^+\}$ by transfinite recursion on α. Assuming that $F_\beta \subset [\lambda]^{\mathrm{cf}(\kappa)}$ has been given for $\beta < \alpha$, let $h_\alpha : \alpha \to \lambda$ be one-to-one. Put

$$x_\xi^\alpha = \min\left(\lambda_\xi \setminus \left(\bigcup\{F_\beta : h_\alpha(\beta) < \lambda_\xi\} \cup \{x_\eta^\alpha : \eta < \xi\}\right)\right)$$

and

$$F_\alpha = \{x_\xi^\alpha : \xi < \mathrm{cf}(\kappa)\}.$$

[T].

7. Let $A \subset \lambda$ be cofinal in λ with type $A = \mathrm{cf}(\lambda)$ and put $X_\xi = X \cap \xi$ for $\xi \in A$. Assume, by reductio ad absurdum, that $|X_\xi| < \kappa$ for $\xi \in A$. We have $X = \bigcup\{X_\xi : \xi \in A\}$; thus $\mathrm{cf}(\kappa) \leq \mathrm{cf}(\lambda)$, and so $\mathrm{cf}(\kappa) < \mathrm{cf}(\lambda)$. Therefore, the sequence $|X_\xi|$ must be eventually constant; hence $|X_\xi| \leq \tau$ for some $\tau < \kappa$, i.e., we must have $\kappa = \tau^+$. The sequence of the X_ξ's cannot be eventually constant, as this would imply $X = X_\xi$ for some ξ; thus $\kappa \geq \mathrm{cf}(\lambda)$. Consequently, $\mathrm{cf}(\kappa) = \kappa \geq \mathrm{cf}(\lambda)$, which is a contradiction. [T].

8. For an arbitrary $F \in \mathcal{F}$, let F' be a bounded subset of F having cardinality κ; the existence of such an F' is guaranteed by Problem 7. As $F_0' \neq F_1'$ whenever $F_0 \neq F_1$ and $F_0, F_1 \in \mathcal{F}$, putting $\mathcal{F}' = \{F' : F \in \mathcal{F}\}$, we have $|\mathcal{F}'| = |\mathcal{F}|$. Now $\mathcal{F}' \subset \bigcup\{[\xi]^\kappa : \xi < \lambda\}$, and the latter set has cardinality κ according to the assumptions. [T].

9. Let $\mathcal{F} = \{F_\alpha : \alpha < \kappa\}$. As $\kappa = \mathrm{cf}(\kappa)$, we have $|F_\alpha \setminus \bigcup\{\beta < \alpha : F_\beta\}| = \kappa$ for an arbitrary $\alpha < \kappa$. Let $x_\alpha = \min\left(F_\alpha \setminus \left(\bigcup\{F_\beta : \beta < \alpha\} \cup \{x_\beta : \beta < \alpha\}\right)\right)$ and $F_\alpha = \{x_\alpha : \alpha < \kappa\}$.

10. If $\kappa = \mathrm{cf}(\kappa)$, then the assertion is identical to that of Problem 8. If $\mathrm{cf}(\kappa) < \kappa$, let κ_ξ form an increasing sequence of cardinals such that $\sup\{\kappa_\xi : \xi < \mathrm{cf}(\kappa)\} = \kappa$ and write $\mathcal{F} = \{F_\alpha : \alpha < \kappa\}$. Define the sequence x_ξ, $\xi < \mathrm{cf}(\kappa)$, by transfinite recursion by putting $x_\xi = \min\left(\kappa \setminus \left(\bigcup\{F_\alpha : \alpha < \kappa_\xi\} \cup \{x_\eta : \eta < \xi\}\right)\right)$. The set $X = \{x_\xi : \xi < \mathrm{cf}(\kappa)\}$ is almost disjoint to each element of $\mathcal{F}$; to show this, we do not need to use the assumption that the elements of $\mathcal{F}$ are pairwise almost disjoint. [T].

11. Let $\omega = A_0 \cup A_1$, $A_0 \cap A_1 = \emptyset$, and $A_0 \sim_f A_1$. Put

$$\mathcal{F} = \{A \cup f``(A_0 \setminus A) : A \subset A_0\}.$$

12. Let $\mathcal{F} = \{F_n : n < \omega\}$. By induction, define pairwise distinct integers $b_n \in F_n$, $a_n \notin F_n$. Put $B = \{b_n : n < \omega\}$. [Be].

13. By Problem 4, there is a system $\mathcal{F}_0 \subset [\omega]^\omega$ with $|\mathcal{F}_0| = 2^{\aleph_0}$ of almost disjoint sets. Let $\bigcup_{n \in \omega} A_n$ be a disjoint partition with $|A_n| = \aleph_0$ for $n \in \omega$. Let $\mathcal{F}_1$ be the set of all transversals of the system of the sets A_n, i.e.,

$$\mathcal{F}_1 = \{\mathrm{R}(f) : f \in \underset{n \in \omega}{\times} A_n\}.$$

Let Φ be such that $\mathcal{F}_1 \sim_\Phi \mathcal{F}_0$. Finally, put

$$\mathcal{F} = \{A_n : n \in \omega\} \cup \left\{F \cap \bigcup\{A_n : n \in \Phi(F)\} : F \in \mathcal{F}_1\right\}.$$

The elements of $\mathcal{F}$ are pairwise disjoint, since the same is true for the elements of $\mathcal{F}_0$.

Assume B meets each A_n. Then $\mathrm{R}(f) \subset B$ for some $f \in \times_{n \in \omega} A_n$. We have $\mathrm{R}(f) = F \in \mathcal{F}_1$ and $F \cap \bigcup\{A_n : n \in \Phi(F)\} \in \mathcal{F}$. [Mi].

Section 14

1. If $f, g, h \in P$, then the set $\{\delta(f,g), \delta(f,h), \delta(g,h)\}$ has at most two elements, and if $f \prec g \prec h$ or $f \succ g \succ h$, then $\delta(f,h) = \min\{\delta(f,g), \delta(g,h)\}$. The assertion easily follows from this relation; this relation will also be frequently used in the solutions of some problems below.

2. Let

$$\delta_0 = \min\{\delta(f_\alpha, f_\beta) : \alpha, \beta < \lambda \wedge \alpha \neq \beta\},$$

$$\alpha_0 = \min\{\alpha < \lambda : \exists \beta (\alpha < \beta < \lambda \land \delta(\alpha, \beta) = \delta_0)\},$$
$$\bar{\alpha}_0 = \min\{\beta < \lambda : \alpha_0 < \beta \land \delta(f_{\alpha_0}, f_\beta) = \delta_0\}.$$

If $\bar{\alpha}_0 \leq \beta < \lambda$, then $\delta(f_{\alpha_0}, f_\beta) = \delta_0$, and if $\bar{\alpha}_0 \leq \beta, \gamma < \lambda$, then $\delta(f_\beta, f_\gamma) \geq \delta_0$. Choose α_0 as the first element of L, and pick the other elements of L from the set $L_0 = \{f_\alpha : \bar{\alpha}_0 \leq \alpha < \lambda\}$. As $|L \setminus L_0| < \lambda$ and λ is a regular cardinal, the process of picking the elements of L appropriately can be continued up to λ steps by transfinite recursion.

3. Let $P = {}^\kappa 2$, let $\prec$ be the ordering of P described in Problem 1, and let $\prec_0$ be an arbitrary wellordering of P. Define the disjoint partition $[P]^2 = K_0 \cup K_1$ with the stipulation $\{f, g\} \in K_0 \iff f \prec_0 g \land f \prec g$. If there is a set of cardinality κ homogeneous in, say, color 0, then there is a set $\{f_\alpha : \alpha < \kappa^+\} \subset P$ such that $f_\alpha \prec f_\beta$ whenever $\alpha < \beta < \kappa^+$.

According to Problem 2, then there is a set $L \subset \kappa^+$ and a sequence $\{\delta_\alpha : \alpha < \kappa^+\} \subset \kappa$ of ordinals such that $\delta(f_\alpha, f_\beta) = \delta_\alpha$ and $\delta_\alpha \leq \delta_\beta$ whenever $\alpha < \beta$ and $\alpha, \beta \in L$. In view of the choice of P, $\delta_\alpha < \delta_\beta$ would hold in this case; this is, however, impossible.

The assertion for color 1 can be established similarly.

4. Let $P = {}^\kappa 2$ and define the 2-partition $\Phi : [P]^2 \to \kappa$ by stipulating that $\Phi(\{f, g\}) = \delta(f, g)$ for $\{f, g\} \in [P]^2$.

5. Let $P = {}^\omega \kappa$ and let $\prec$ be the lexicographic ordering defined in Problem 1. Let $\prec_0$ be an arbitrary wellordering of P and define the 2-partition $\Psi : [P]^2 \to 2$ by stipulating that $\Phi(\{f, g\}) = 0$ if and only if $f \prec g$ and $f \prec_0 g$. If there is a set of cardinality κ^+ that is homogeneous in color 0, then according to Problem 2, we obtain that there is a set $\{f_\alpha : \alpha < \kappa^+\} \subset P$ and a sequence $\{\delta_\alpha : \alpha < \kappa^+\}$ such that $\delta_\alpha \leq \delta_\beta < \omega$, $f_\alpha \prec f_\beta$, and $\delta(f_\alpha, f_\beta) = \delta_\alpha$ whenever $\alpha < \beta < \kappa^+$.

Then, for some δ and some $\alpha_0 < \kappa^+$, we have $\delta = \delta_{\alpha_0} = \delta_\alpha$ whenever $\alpha_0 \leq \alpha < \kappa^+$, and so $f_\alpha(\delta) < f_\beta(\delta)$ whenever $\alpha_0 \leq \alpha < \beta < \kappa^+$. This is impossible; a contradiction.

If there is a homogeneous set of cardinality ω_1 in color 1, then there are sequences $\{f_\alpha : \alpha < \omega_1\} \subset P$ and $\{\delta_\alpha : \alpha < \omega_1\} \subset \omega$ such that

$$f_\alpha \succ f_\beta, \quad \delta_\alpha \leq \delta_\beta \qquad \text{whenever} \quad \alpha < \beta < \omega_1.$$

Then we again obtain a δ and an $\alpha_0 < \omega_1$ such that $\delta_\alpha = \delta$ whenever $\alpha_0 \leq \alpha < \omega_1$. In this case, however, $\{f_\alpha(\delta) : \alpha_0 \leq \alpha < \omega_1\}$ would be a decreasing sequence of ordinals.

6. The assertion is a special case of Theorem 14.5.

7. Let $P = {}^{\mathrm{cf}(\kappa)}\kappa$, let $\prec$ be the lexicographic ordering defined in Problem 1, and let $\prec_0$ be an arbitrary wellordering of P. Define a 2-partition Φ of P with $\mathrm{cf}(\kappa)$ colors by stipulating that for each $\{f, g\} \in [P]^2$ with $f \prec g$ we have

$$\Phi(\{f, g\}) = \begin{cases} 0 & \text{if } f \prec_0 g, \\ 1 + \delta & \text{if } g \prec_0 f \text{ and } \delta(f, g) = \delta \quad (\delta < \mathrm{cf}(\kappa)). \end{cases}$$

We can see that there is no homogeneous set of cardinality κ^+ in color 0 as we saw this in Problem 3. Assume that $\delta < \mathrm{cf}(\kappa)$ and there is an infinite homogeneous set in color $1 \dotplus \delta$. Then we obtain a sequence $\{f_n : n < \omega\}$ such that

$$f_m \succ f_n, \quad 1 \dotplus \delta = \delta(f_m, f_n) \qquad \text{whenever} \quad m \le n \le \omega.$$

Thus we have $f_m(1 \dotplus \delta) > f_n(1 \dotplus \delta)$ whenever $m < n < \omega$, which is impossible.

8. According to the assumptions, $(\kappa^+)^\kappa = \kappa^{\mathrm{cf}(\kappa)} = \kappa^+$. In view of what we proved in Problem 4, we may assume that $\kappa > \mathrm{cf}(\kappa)$. Let $\{\kappa_\xi : \xi < \mathrm{cf}(\kappa)\} \subset \kappa$ be a sequence of cardinals such that $\omega \le \kappa_\xi < \kappa$ and $\kappa = \bigcup_{\xi < \mathrm{cf}(\kappa)} \kappa_\xi$. Let $[\kappa^+]^\kappa = \{A_\alpha : \alpha < \kappa^+\}$ be a wellordering of $[\kappa^+]^\kappa$ of type κ^+, $S_\alpha = \{A_\beta : A_\beta \subset \alpha \wedge \beta < \alpha\}$, and put

$$S_{\alpha,\xi} = \{A_\beta \in S_\alpha : \beta < \kappa_\xi\} \qquad \text{whenever} \quad \alpha < \kappa^+, \ \xi < \mathrm{cf}(\kappa).$$

We will define a 2-partition $\Phi : [\kappa^+]^2 \to \mathrm{cf}(\kappa)$ of κ^+ as follows: For each $\alpha < \kappa^+$ and each $\xi < \mathrm{cf}(\kappa)$ we are going to define a set $\Phi_\xi(\alpha) \subset \alpha$ with $|\Phi_\xi(\alpha)| \le \kappa_\xi$, and then we will specify Φ with the stipulation that, for $\{\alpha, \beta\} \in [\kappa^+]^2$ with $\beta < \alpha$ and for $\xi < \mathrm{cf}(\kappa)$,

$$\Phi(\{\alpha, \beta\}) = 1 \dotplus \xi \qquad \text{holds if and only if} \qquad \beta \in \Phi_\xi(\alpha).$$

(Then, clearly, $\Phi(\{\alpha, \beta\}) = 0$ holds if and only if $\beta \notin \bigcup_{\xi < \mathrm{cf}(\kappa)} \Phi_\xi(\alpha)$.)

If the following requirements are satisfied for $\alpha < \kappa^+$ and $\xi < \mathrm{cf}(\kappa)$, then Φ will be a witness to the assertion to be proved.

(1) If $A \in S_{\alpha,\xi}$, then $\Phi_\xi(\alpha) \cap A \ne \emptyset$;

(2) If $\beta, \gamma \in \Phi_\xi(\alpha)$, then $\beta \notin \Phi_\xi(\gamma)$ and $\gamma \notin \Phi_\xi(\beta)$.

We are going to carry out transfinite recursion on α. Assume that $\beta < \alpha$ and (1) and (2) are satisfied and $|\Phi_\xi(\beta)| < \kappa_\xi$ holds for each $\xi < \mathrm{cf}(\kappa)$; then it is sufficient to show that there is a set $B = \Phi_\xi(\alpha) \subset \alpha$ that satisfies Conditions (1) and (2). We will omit the proof of this; however, there will be hints for the proof among the hints for the problems of Section 19.

9. The assertion is a special case of Theorem 14.7.

10. We may assume that $\kappa > \mathrm{cf}(\kappa)$. Let $\kappa = \bigcup_{\xi < \mathrm{cf}(\kappa)} A_\xi$ be a disjoint partition of κ such that $|A_\xi| < \kappa$ for each ξ. Let $[\mathrm{cf}(\kappa)]^2 = I_0 \cup I_1$ be a 2-partition verifying the assumption. Define a 2-partition $[\kappa]^2 = J_0 \cup J_1$ such that

$$\{\alpha, \beta\} \in J_0 \text{ if and only if there are } \xi, \eta < \mathrm{cf}(\kappa) \text{ for which}$$
$$\alpha \in A_\xi \wedge \beta \in A_\eta \wedge \{\xi, \eta\} \in I_0 \text{ holds.}$$

The partition $\{J_0, J_1\}$ is called a *canonical expansion* of the partition $\{I_0, I_1\}$.

11. Assume $f : [\kappa]^2 \to 2$ is given. From the assumptions, it follows that $\lambda \le \mathrm{cf}(\kappa) < \kappa$, and so, according to Theorem 14.7, for each $\mu < \kappa$ there is a $\nu < \kappa$ (in fact, we can take $\nu = (2^{\max(\mu,\lambda)})^+$) for which $\nu \to (\mu,\lambda)^2$. Hence we may choose a sequence of pairwise disjoint sets $A_\xi \subset \kappa$, $\xi < \mathrm{cf}(\kappa)$, such that, for each ξ with appropriate κ_ξ and B_ξ, we have $|A_\xi| = \kappa_\xi^+ < \kappa$, $\bigcup_{\eta<\xi} A_\eta = B_\xi$, $2^{|B_\xi|} \leq \kappa_\xi < \kappa_{\xi+1}$, $\kappa_0 \geq 2^{\mathrm{cf}(\kappa)}$, $\kappa = \sup\{\kappa_\xi : \xi < \mathrm{cf}(\kappa)\}$, and A_ξ is homogeneous in color 0 with respect to f.

In view of the assumption $2^{|B_\xi|} \leq \kappa_\xi$ for each $\xi < \mathrm{cf}(\kappa)$ we can pick a set $A_\xi^1 \subset A_\xi$ such that $|A_\xi^1| = |A_\xi| = \kappa_\xi^+$ and

(1)   if $x,y \in A_\xi^1$, $z \in B_\xi$, then $f(\{x,z\}) = f(\{y,z\})$.

Using (1) and the inequality $\kappa_0 > 2^{\mathrm{cf}(\kappa)}$, we may choose sets $A_\xi^2 \subset A_\xi^1$, $|A_\xi^2| = \kappa_\xi^+$ such that

(2)   for each $x,y \in A_\eta^1$ and $z \in A_\xi^2$ we have $f(\{x,z\}) = f(\{y,z\})$.

whenever $\eta < \xi < \mathrm{cf}(\kappa)$.

The partition f restricted to the set $A \overset{def}{=} \bigcup_{\xi<\mathrm{cf}(\kappa)} A_\xi^2$ of cardinality κ is *canonical* in the same sense as the partition J defined in the solution of Problem 10 was canonical. Thus we can define the 2-partition $[\mathrm{cf}(\kappa)]^2 = I_0 \cup I_1$ of the set $\mathrm{cf}(\kappa)$ by stipulating that for each $\{\xi,\eta\} \in [\mathrm{cf}(\kappa)]^2$ we have
$\{\xi,\eta\} \in I_0$ if and only if there are $x \in A_\xi^2$, $y \in A_\eta^2$ such that $f(\{\xi,\eta\}) = 0$, that is, if and only if each $x \in A_\xi^2$ and $y \in A_\eta^2$ satisfy $f(\{\xi,\eta\}) = 0$.

The assumption $\mathrm{cf}(\kappa) \to (\mathrm{cf}(\kappa),\lambda)^2$ applied with this partition gives the required conclusion. The assertion presented here is a special case of the general canonization lemmas, about which more information can be found in [E, H, M, R].

12. Let $X = \{f_\alpha : \alpha < \kappa\}$ be such that $f_\alpha \prec_0 f_\beta$ whenever $\alpha < \beta < \kappa$. If we have $f_\beta \prec f_\alpha$ whenever $\alpha < \beta$, then the second assertion of the problem holds with $Y = X$. Assume therefore that α_0 and β_0 are such that $\alpha_0 < \beta_0 < \kappa$ and $f_{\alpha_0} \prec f_{\beta_0}$. As $[X]^3 \cap K = \emptyset$ by our assumptions, we must have $f_{\beta_0} \prec f_\beta$ whenever $\beta_0 < \beta < \kappa$. For the same reason, we must have $f_\beta \prec f_\gamma$ whenever $\beta_0 \leq \beta < \gamma < \kappa$, and so $Y = \{f_\alpha : \beta_0 \leq \alpha < \kappa\}$ satisfies the requirements.

13. Let $P = {}^\kappa 2$, where $\kappa = 2^{\aleph_0}$, let $\prec$ be the lexicographic ordering of P defined in Problem 1, and let $\prec_0$ be a wellordering of P. Let $[\kappa]^2 = J_0 \cup J_1$ be a 2-partition verifying the relation $2^{\aleph_0} \nrightarrow (\aleph_1)_2^2$ (see Problem 3).

Let $\{f,g,h\} \in [P]^3$ be such that $f_0 \prec_0 g \prec_0 h$. Put

$$\{f,g,h\} \in K_0 \overset{def}{\Longleftrightarrow} f \prec g \succ h,$$

$$\{f,g,h\} \in K_1 \overset{def}{\Longleftrightarrow} f \succ g \prec h,$$

$$\{f,g,h\} \in S_0 \overset{def}{\Longleftrightarrow} f \prec g \prec h,$$

$$\{f, g, h\} \in S_1 \overset{def}{\Longleftrightarrow} f \succ g \succ h.$$

Put $S = S_0 \cup S_1$. Finally, define $[P]^3 = I_0 \cup I_1$ as follows. If $\{f, g, h\} \in K_j$ for some $j < 2$, then put $\{f, g, h\} \in I_j$. If $\{f, g, h\} \in S$, then let $\Delta(f, g, h) = \{\delta(f, g), \delta(g, h)\}$. Clearly, we have $\Delta(f, g, h) \in [\kappa]^2$; hence the following definition is sound:

$$\{f, g, h\} \in I_j \qquad \text{if and only if} \qquad \Delta(f, g, h) \in J_j \quad (j < 2).$$

For reasons of symmetry, it is sufficient to show that there is no homogeneous set of cardinality $\aleph_1$ in color 0. Proceeding via reductio ad absurdum, assume there is $X \subset P$ with $|X| = \aleph_1$ and $[X]^3 \subset I_0$. As we have $[X]^3 \cap K_1 = \emptyset$ in this case, according to Problem 12 we have a $Y \subset X$ with $|Y| = \aleph_1$ for which we may assume that it has the form $Y = \{f_\alpha : \alpha < \omega_1\}$ such that $f_\alpha \prec_0 f_\beta \wedge f_\alpha \prec f_\beta$ whenever $\alpha < \beta < \omega_1$ or $f_\alpha \prec_0 f_\beta \wedge f_\alpha \succ f_\beta$ whenever $\alpha < \beta < \omega_1$.

We may assume that the first of these alternatives holds. According to the assertion of Problem 2, there is an $L \subset \omega_1$, $|L| = \aleph_1$, and $\{\delta_\alpha : \alpha < \omega_1\} \subset \kappa = 2^{\aleph_0}$ such that $\delta(f_\alpha, f_\beta) = \delta_\alpha$ and $\delta_\alpha \leq \delta_\beta$ whenever $\alpha, \beta \in L$ and $\alpha < \beta$. In view of the choice of P, if we have $\alpha, \beta \in L$ and $\alpha < \beta$, then even $\delta_\alpha < \delta_\beta$ holds, and

$$\{\delta_\alpha, \delta_\beta\} = \Delta(f_\alpha, f_\beta, f_\gamma),$$

where $\gamma = L \setminus (\beta + 1)$. Hence, writing $D = \{\delta_\alpha : \alpha \in L\}$, we have $|D| = \aleph_1$ and $[D]^2 \subset J_0$, which is a contradiction.

The assertion just proven is a special case of the so-called Negative Stepping-up Lemma.

14. We give two solutions of the problem. The first one is an ad hoc approach. The second one can be used to establish a number of generalizations of the assertion.

a) Let $\{A_\alpha : \alpha < \kappa^{\aleph_0}\}$ be a wellordering of the set $[\kappa]^{\aleph_0}$. Define a 2-partition $I_0 \cup I_1$ of $[\kappa]^{\aleph_0}$ by stipulating that $A_\alpha \in I_0$ if and only if $A_\beta \not\subset A_\alpha$ for every $\beta < \alpha$.

If $X \subset [\kappa]^{\aleph_0}$ and $\alpha = \min\{\beta : A_\beta \subset X\}$, then $A_\alpha \in [X]^{\aleph_0}$ and $A_\alpha \in I_0$. If $\langle X_n : n < \omega \rangle$ is an arbitrary infinite sequence of infinite subsets of X increasing with respect to inclusion, then we have $X_n \in I_1$ for some $n < \omega$.

b) First, it is easy to prove the assertion for $\kappa = \aleph_0$ by transfinite induction. Now, assuming $\kappa > \aleph_0$, consider a maximal system $\mathcal{F} \subset [\kappa]^{\aleph_0}$ of almost disjoint sets. For each $F \in \mathcal{F}$, we choose a partition $[F]^{\aleph_0} = I_0^F \cup I_1^F$ that verifies the relation $\aleph_0 \not\to (\aleph_0)_2^{\aleph_0}$ on $[F]^{\aleph_0}$. Let $\{F_\alpha : \alpha < |\mathcal{F}|\}$ be a wellordering of $\mathcal{F}$. For $A \in [\kappa]^{\aleph_0}$, write $\alpha(A) = \min\{\beta : |A \cap F_\beta| = \aleph_0\}$. Put $A \in I_0$ if and only if $\alpha = \alpha(A)$ and $A \cap F_\alpha \in I_0^{F_\alpha}$.

For all the problems of this section, see [E, H, M, R].

Section 16

1. The proof is similar to the the proof of Theorem A7.2. For each element u of X, we define a set $T^R(u) \subset X$ such that $u \in T^R(u)$ and for $v \in T^R(u)$, we have $[v]_R \subset T^R(u)$. (If $T_0^R = \{u\}$ and $T_{n+1}^R = \bigcup\{[v]_R : vRu\}$, then $T^R = \bigcup_{n<\omega} T_n^R$ satisfies these requirements – this argument is closely related to the proof of the existence of transitive closure given in the proof of Theorem A7.2.) If there is a u such that $\neg\Phi(u)$ holds, then let $T = \{v \in T^R(u) : \neg\Phi(v)\}$. T is not empty, and in view of the well-foundedness of R, there is a $v \in T$ such that $[v]_R \cap T = \emptyset$. As $[v]_R \subset T^R(u)$, for every w with wRv, we have $\Phi(w)$; hence $\Phi(v)$ holds; this contradicts the relation $v \in T$.

2. The proof is similar to the proof of the Transfinite Recursion Theorem 9.2 (cf. also Theorem A5.1). Using the statement proved in Problem 1, we first show that for every $u \in X$, there is at most one function f that satisfies the functional equation restricted to $T^R(u)$, where $T^R(u)$ is the set defined in the solution of the preceding problem. Then, again using the statement of Problem 1, we conclude that for each $u \in X$, there is exactly one such function f_u. The operation $\mathcal{F}$ can then be defined as $\mathcal{F} = \bigcup\{f_u : u \in X\}$.

3. Using the Theorem of Well-Founded Recursion of Problem 2, we define the operation $\mathcal{F}(u) = \{\mathcal{F}(v) : vRu\}$. All the assertions are direct consequences of the Theorem of Well-Founded Induction given in Problem 1.

4. Let $\mathcal{F}$ be a nonprincipal normal ultrafilter on κ. For each α with $\omega \leq \alpha < \kappa$, let ϕ_α be such that $P(\alpha) \sim_{\phi_\alpha} |\alpha|^+$ (except that $\phi_\alpha = \mathrm{Id}_\alpha$, the identity function on α, for $\alpha < \omega$). For each $A \subset \kappa$, let $\Phi_A \in \bigtimes_{\alpha<\kappa} \alpha$ be a function such that

$$\Phi_A(\alpha) = \phi_\alpha(A \cap \alpha) \qquad \text{whenever} \qquad \omega \leq \alpha < \kappa.$$

We claim that if $A \neq B$ and $A, B \subset \kappa$, then $[\Phi_A]_{\mathcal{F}} \neq [\Phi_A]_{\mathcal{F}}$. Indeed, for some $\alpha_0 < \kappa$, we have $A \cap \alpha_0 \neq B \cap \alpha_0$, and so $A \cap \alpha \neq B \cap \alpha$ whenever $\alpha_0 \leq \alpha < \kappa$; thus $\Phi_A(\alpha) \neq \Phi_B(\alpha)$ whenever $\alpha_0 \leq \alpha < \kappa$. It is therefore sufficient to prove that the set $P = \{[\Phi]_{\mathcal{F}} : \Phi \in \bigtimes_{\alpha<\kappa} \alpha^+\}$ has cardinality at most κ^+. As the property $\prec_{\mathcal{F}}$ given in Definition 16.3 is a wellordering of P, it is sufficient to establish that every proper initial segment of $\langle P, \prec_{\mathcal{F}} \rangle$ has cardinality at most κ. For an arbitrary element Φ of P, we have $|\Phi(\alpha)| \leq |\alpha| \leq \alpha$ whenever $\omega \leq \alpha < \kappa$. Let $\Psi_\alpha : \Phi(\alpha) \to |\alpha|$ be one-to-one for each α with $\omega \leq \alpha < \kappa$. The assertion now follows, since, given an arbitrary h with $[h]_{\mathcal{F}} < [\Phi]_{\mathcal{F}}$, the function g defined by $g(\alpha) = \Psi(h(\alpha))$ is regressive, and so it is constant almost everywhere. [Sc].

5. First assume that κ is smaller than the first measurable cardinal greater than ω. As a discrete space of cardinality $\aleph_0$ can be embedded in $\mathbb{R}$ as a closed subspace, it is sufficient to prove that κ can be embedded into a power of C, where C is a discrete space of cardinality $\aleph_0$. Let P be the set of 1-partitions of κ with $\aleph_0$ (pairwise disjoint, possibly empty) classes (i.e., with $\aleph_0$ colors), and for $p \in P$, denote by p_c, $c \in C$, the classes of p ($\kappa = \bigcup\{p_c : c \in C\}$).

For each $\alpha < \kappa$ define an element f_α of $^P C$ by stipulating that $f_\alpha(p) = c$ if and only if $\alpha \in p_c$. We claim that $\{f_\alpha : \alpha < \kappa\}$ is a discrete closed subspace of $^P C$. Assume, on the contrary, that $f \in {}^P C$ is a limit point of the set $\{f_\alpha : \alpha < \kappa\}$. For each $p \in P$, put $F_p = p_{f(p)}$.

It is easy to see that $\mathcal{F} = \{F_p : p \in P\}$ is a filter on κ. An additional important property of the filter $\mathcal{F}$ can be seen immediately: For any partition p of κ into $\aleph_0$ classes, there is at least one class p_c of p that belongs to $\mathcal{F}$. This property implies that $\mathcal{F}$ is in fact an ω_1-complete ultrafilter. Finally, for $\alpha < \kappa$, we have $\{\alpha\} \notin \mathcal{F}$, since otherwise f would have a neigborhood not containing any f_β with $\alpha \neq \beta < \kappa$; thus $\mathcal{F}$ is nonprincipal. Then, according to Theorem 16.1, there is a measurable cardinal that is greater than ω but is not greater than κ.

Conversely, assume that there is a measurable cardinal λ with $\omega < \lambda < \kappa$. It is sufficient to show that a discrete space of cardinality λ cannot be embedded as a closed subspace into a power of $\mathbb{R}$. Assume that for some cardinal σ, the set $\{f_\alpha : \alpha < \lambda\}$ is a subspace of $^\sigma\mathbb{R}$, and let $\mathcal{I}$ be a nonprincipal λ-complete prime ideal on λ. For each $\xi < \sigma$ there is an $f(\xi)$ such that $\{\alpha < \lambda : f_\alpha(\xi) = f(\xi)\} \notin \mathcal{I}$. Then f is a limit point of the set $\{f_\alpha : \alpha < \lambda\}$. [K, T].

Section 17

1. Assume $\{f_\alpha : \alpha < \kappa\}$ is a κ-scale and μ is a nontrivial real-valued measure on κ. Write $A_{n,m} = \{\alpha < \kappa : f_\alpha(n) = m\}$. Then $\bigcup_{m=0}^\infty A_{n,m} = \kappa$ for each $n \in \omega$. For every $n \in \omega$, there is a $g(n) < \omega$ such that $\mu\left(\bigcup_{m=g(n)}^\infty A_{n,m}\right) < 1/2^{n+1}$. Thus

$$\mu\left(\{\alpha < \kappa : \forall n \in \omega \, f_\alpha(n) < g(n)\}\right) > 0,$$

and yet $|\{\alpha < \kappa : \forall n \in \omega \, f_\alpha(n) < g(n)\}| < \kappa$; this is a contradiction.

2. It is sufficient to show that if $f_n \in {}^\omega\omega$, then there is an $f \in {}^\omega\omega$ such that $f_n \prec f$ for each $n < \omega$. In fact, $f(n) = \max\{f_i(n) : i \leq n\} + 1$ is such a function.

3. We have to show that, given $F_\alpha \in \mathrm{P}(\kappa)\backslash\mathcal{I}$ for $\alpha < \kappa$ such that $F_\alpha \cap F_\beta \in \mathcal{I}$ whenever $\alpha < \beta < \kappa$, then there are sets $F_\alpha' \subset F_\alpha$ with $F_\alpha' \notin \mathcal{I}$ such that

$$F_\alpha' \cap F_\beta' = \emptyset \qquad \text{whenever} \qquad \alpha < \beta < \kappa.$$

The sets $F_\beta' = F_\beta \setminus \bigcup\{F_\alpha : \alpha < \beta\}$ satisfy these requirements.

Section 18

1. If the partitions $\bigcup_{\xi<\gamma} I_\xi^n$ for $n < \omega$ witness $\lambda \nrightarrow (\kappa)^{<\omega}_\gamma$, then define J_ξ^n for $\xi < \gamma$ and n with $1 \leq n < \omega$ as follows. If $X = \{x_0, \ldots, x_{n-1}\} \in [\lambda]^n$ with $x_0 < \cdots < x_{n-1}$, then put

$$X \in J_\xi^n \iff n = 2^k(2l+1) \wedge \{x_0, \ldots, x_{k-1}\} \in I_\xi^n.$$

Then, for each $A \subset \lambda$ with $|A| = \kappa$, there are infinitely many n's such that A is not homogeneous with respect to $\bigcup_{\xi < \gamma} J_\xi^n$.

2. *a)* For $X = \{x_0, \ldots, x_{n-1}\} \in [\omega]^n$, $x_0 < \cdots < x_{n-1}$, $n > 0$, put $X \in I_0^n \iff x_0 \geq n$.

b) Let $\kappa_0 = \min\{\lambda > \omega : \lambda \text{ is strongly inaccessible}\}$. Given λ with $\omega < \lambda < \kappa_0$, it is sufficient to show that $\lambda \nrightarrow (\omega)_2^{<\omega}$ holds if $\lambda' \nrightarrow (\omega)_2^{<\omega}$ holds for each λ' with $\omega \leq \lambda' < \lambda$. Assuming the latter, according to the assertion of Problem 1, we can choose partitions $[\lambda']^n = I_0^n(\lambda') \cup I_1^n(\lambda')$ for each $n < \omega$ such that there is no infinite set that is homogeneous for all but finitely many n's.

We distinguish two cases: (i) λ is singular, or (ii) $2^{\lambda'} \geq \lambda$ for some $\lambda' < \lambda$.

Ad (i). Let $\lambda = \bigcup_{\xi < \mathrm{cf}(\lambda)} A_\xi$ be a disjoint partition such that $|A_\xi| = \lambda_\xi < \lambda$. We are going to define the partitions $[\lambda]^n = I_0^n(\lambda) \cup I_1^n(\lambda)$. For $X = \{x_0, \ldots, x_{n-1}\} \in [\lambda]^n$, put $\bar{X} = \{\xi_0, \ldots, \xi_{n-1}\}$, where ξ_i is determined such that $x_i \in A_{\xi_i}$. If $X \subset A_\xi$ for some ξ, then put $X \in I_0^n(\lambda) \iff X \in I_0^n(A_\xi)$, where the partition $[A_\xi]^n = I_0^n(A_\xi) \cup I_1^n(A_\xi)$ is obtained from the partition $[\lambda_\xi]^n = I_0^n(\lambda_\xi) \cup I_1^n(\lambda_\xi)$ by mapping λ_ξ onto A_ξ in a one-to-one way. If $|\bar{X}| = |X| = n > 1$ then $X \in I_0^n(\lambda) \iff \bar{X} \in I^n(\mathrm{cf}(\lambda))$. We put $X \in I_0^n(\lambda)$ in all other cases. As for each $A \subset \lambda$ with $|A| = \omega$, we either have $|A \cap A_\xi| = \omega$ for some ξ or $|\{\xi : A \cap A_\xi \neq 0\}| \geq \omega$, this partition verifies the assertion.

Ad (ii). Write $P = {}^{\lambda'}2$, let $\prec$ be the lexicographic ordering defined in Problem 1 of Section 14, and let $\prec_0$ be a wellordering of P. Define the partitions $[P]^n = I_0 \cup I_1$ by the following stipulations. Given $\{f_0, \ldots, f_{n-1}\} \in [P]^n$ with $f_0 \prec_0 \cdots \prec_0 f_{n-1}$:

If $n \geq 3$ and $f_0 \prec f_1 \succ f_2$, then $\{f_0, \ldots, f_{n-1}\} \in I_0^n$.

If $n \geq 3$ and $f_0 \succ f_1 \prec f_2$, then $\{f_0, \ldots, f_{n-1}\} \in I_1^n$.

If $n \geq 3$, $f_0 \prec \cdots \prec f_{n-1}$ or $f_0 \succ \cdots \succ f_{n-1}$, and the sequence $\Delta(f_0, \ldots, f_{n-1}) = \{\delta(f_0, f_1), \ldots, \delta(f_{n-2}, f_{n-1})\} \in [\lambda']^{n-1}$ is strictly increasing, then we put

$$\{f_0, \ldots, f_{n-1}\} \iff \Delta(f_0, \ldots, f_{n-1}) \in I_0^n(\lambda').$$

We put $\{f_0, \ldots, f_{n-1}\} \in I_0^n$ in all other cases.

The lemmas described in the hints for the solutions of the problems of Section 14 show that the partitions I_0^n, I_1^n satisfy the requirements.

c) It follows from the assumption that, for each $\alpha < \lambda$ and $n \geq 1$, there is a partition $[\alpha]^2 = I_0^n(\alpha) \cup I_1^n(\alpha)$ such that each set $A \subset \alpha$ with $|A| = \omega$ is not homogeneous with respect to $I_0^n(\alpha) \cup I_1^n(\alpha)$ for infinitely many n. For each $n \geq 2$ and each $\{x_0, \ldots, x_{n-1}\} \in [\lambda]^n$ with $x_0 < \cdots < x_{n-1}$, put $\{x_0, \ldots, x_{n-1}\} \in I_0^n \iff \{x_0, \ldots, x_{n-2}\} \in I_0^{n-1}(x_{n-1})$. It is clear that this partition verifies the conclusion. [Si; 1].

3. Assume $f : [\kappa]^i \to {}^\omega 2$ is given for each i with $1 \leq i < \omega$. Define $g : [\kappa]^n \to 2$ for n with $1 \leq n < \omega$ by stipulating that if $n = 2^i(2j+1)$, $X \subset [\kappa]^n$, $X = \{x_0, \ldots, x_{n-1}\}$, and $x_0 < \cdots < x_{n-1} < \kappa$, then $g(x) =$

$f(\{x_0, \ldots, x_{j-1}\})(j)$; the latter denotes the value of the function $f(\{x_0, \ldots, x_{j-1}\}) \in {}^\omega 2$ at j. According to the assumptions, there is an $A \subset \kappa$ with type $A(<) = \alpha$ that is homogeneous with respect to g for each n. We claim that A is also homogeneous with respect to f. To see this, let $X = \{x_0, \ldots, x_{i-1}\}$, $Y = \{y_0, \ldots, y_{i-1}\}$, $X, Y \in [A]^i$, $x_0 < \cdots < x_{i-1}$, $y_0 < \cdots < y_{i-1}$, and $j < \omega$ be arbitrary. Choose a $Z = \{z_0, \ldots, z_{n-1}\} \subset A$ such that $x_{i-1}, y_{i-1} < z_0 < \cdots < z_{n-1}$ and $n = 2^i(2j+1)$. Then $g(X \cup Z) = g(Y \cup Z)$, and so $f(X)(j) = f(Y)(j)$ for each j. [Si; 1].

4. If κ is weakly compact, then $\kappa \to (\kappa, 4)^3$ holds according to Theorem 18.2c). Now assume that κ is not weakly compact, and let the 2-partition $[\kappa]^2 = J_0 \cup J_1$ witness this. Define the 3-partition $[\kappa]^3 = I_0 \cup I_1$ by stipulating that if $\{\alpha, \beta, \gamma\} \in [\kappa]^3$ with $\alpha < \beta < \kappa$, then

$$\{\alpha, \beta, \gamma\} \in I_1 \iff \{\alpha, \beta\} \in J_0 \wedge \{\beta, \gamma\} \in J_1.$$

Let $A \subset \kappa$ with $|A| = \kappa$. There are $\alpha, \beta \in A$ with $\alpha < \beta$ such that $\{\alpha, \beta\} \in J_0$; otherwise A would be homogeneous of color 1 with respect to the partition J. If there is a $\gamma \in A$ with $\beta < \gamma$ such that $\{\beta, \gamma\} \in J_1$, then A includes a triple of color 1; we may assume therefore that $\{\beta, \gamma\} \in J_0$ for each $\gamma \in A$ with $\beta < \gamma$. In this case, however, there are $\gamma, \delta \in A$ with $\gamma < \delta$ such that $\{\gamma, \delta\} \in J_1$; otherwise the set $A \setminus \beta$ would be homogeneous of color 0 with respect to the partition J. Now let $\alpha < \beta < \gamma < \delta < \kappa$ be arbitrary. If $\{\alpha, \beta, \gamma\} \in I_1$, then $\{\beta, \gamma, \delta\} \notin I_1$. The 3-partition I witnesses $\kappa \nrightarrow (\kappa, 4)^3$. [H; 1].

5. First assume that κ is weakly compact, $\mathcal{S}$ is a κ-complete field of sets generated by at most κ elements, $|\mathcal{S}| \geq \kappa$, and $\mathcal{F} \subset \mathcal{S}$ is a κ-complete filter in $\mathcal{S}$. According to the assumptions we have $\kappa^{<\kappa} = \kappa$, and so $\mathcal{S}$ has cardinality κ. Let $\{A_\alpha : \alpha < \kappa\} = \mathcal{S}$ be a wellordering of $\mathcal{S}$. For each $\alpha < \kappa$, we define a function $f_\alpha : \alpha \to 2$ as follows. Choose an $x_\alpha \in \bigcap\{A_\beta : \beta < \alpha \text{ and } A_\beta \in \mathcal{F}\}$, and put $f_\alpha(\beta) = 1$ if and only if $x_\alpha \in A_\beta$.

Let $T = \{f_\alpha | \beta : \beta \leq \alpha < \kappa\}$. T ordered by inclusion is clearly a tree. Let b be a branch of cardinality κ; such a branch exists according to Theorem 18.2. Put $f = \bigcup b$; then f is a function on κ. It is easy to see that $\mathcal{U} = \{A_\alpha : f(\alpha) = 1\}$ is an extension of $\mathcal{F}$ that is a κ-complete ultrafilter.

Assume now that the assertion on the right-hand side of the biconditional formulated in the problem is true. Following the proof of Theorem 16.1, it is easy to show that κ is strongly inaccessible.

Let T be a κ-tree. Let $\mathcal{S}$ be the κ-complete field of subsets of T generated by $[T]^{<\kappa} \cup \{T| \succeq x : x \in T\}$, where $T| \succeq x = \{y \in T : x \preceq y\}$. Let $\mathcal{I}$ be a prime ideal in $\mathcal{S}$ that is an extension of $[T]^{<\kappa}$. For each $\alpha < \kappa$, there is exactly one $x_\alpha \in T_\alpha$ such that $T| \succeq x_\alpha \notin \mathcal{I}$; hence $\{x_\alpha : \alpha < \kappa\}$ is a branch of T. Thus κ has the tree property. [K, T].

6. Assume first that κ is weakly compact. Let $\{X_\nu : \nu < \kappa\}$ be a sequence with the property described in the problem, and put $T = \times_{\nu < \kappa} X_\nu$. It is easy to see that the weight of T is also at most κ. Let $\mathcal{S}$ be the field of subsets of T generated by a base of cardinality $\leq \kappa$ of the space T. Let $\mathcal{O}$ be an open cover of T; without loss of generality, we may assume that $\mathcal{O} \subset \mathcal{S}$. Assume, contrary to the assertion to be established, that $\mathrm{co}(\mathcal{O})$ generates a κ-complete filter. According to Problem 5, this filter can be extended to a κ-complete ultrafilter $\mathcal{U}$ in $\mathcal{S}$. Let $\nu < \kappa$ be arbitrary. The set

$$\mathcal{U}_\nu = \big\{ Y_\nu \subset X_\nu : Y_\nu \times \underset{\substack{\mu<\kappa \\ \mu\neq\nu}}{\bigtimes} X_\mu \in \mathcal{U} \big\}$$

is a κ-complete ultrafilter in the κ-complete field of sets

$$\mathcal{S}_\nu = \big\{ Y_\nu \subset X_\nu : Y_\nu \times \underset{\substack{\mu<\kappa \\ \mu\neq\nu}}{\bigtimes} X_\mu \in \mathcal{S} \big\}.$$

In view of the κ-compactness of X_ν, $\mathcal{F}_\nu$ converges to a (single) point $x_\nu \in X_\nu$; that is, $\mathcal{F}_\nu$ contains all neighborhoods of x_ν that belong to $\mathcal{S}_\nu$. Indeed, assuming on the contrary that each $x_\nu \in X_\nu$ has a neighborhood $G_{x_\nu} \in \mathcal{S}_\nu \setminus \mathcal{F}_\nu$, fewer than κ of these would cover X_ν, contradicting the κ-completeness of $\mathcal{F}_\nu$. (As the space is Hausdorff, an ultrafilter cannot converge to more than one point.) Then, simply by the definition of topological product, and without the use of κ-completeness, we can conclude that $\mathcal{F}$ converges to $\langle x_\nu : \nu < \kappa \rangle$; i.e., that $\mathcal{F}$ contains all neighborhoods belonging to $\mathcal{S}$ of this point. On the other hand, $\mathcal{O} \cap \mathcal{F} = \emptyset$; hence $\mathcal{O}$ cannot cover this point, a contradiction.

Now assume, conversely, that the assertion in the problem about κ-compactness of products holds, and let $T = \langle \kappa, \prec \rangle$ be a κ-tree. Denote by T_α the α's level of this tree. Let $t \notin \kappa$, and put $D_\alpha = T_\alpha \cup \{t\}$. D_α is κ-compact in the discrete topology.

Let $X = \bigtimes_{\alpha<\kappa} D_\alpha$. According to the assumption, X is κ-compact. For each $\alpha < \kappa$, define $f_\alpha \in X$ as follows. Given $\alpha < \kappa$, fix $x_\alpha \in T_\alpha$; writing $b_\alpha = T| \preceq x_\alpha$, put

$$f_\alpha(\beta) = \begin{cases} x_\alpha|_T\beta & \text{if } \beta < \alpha, \\ t & \text{otherwise.} \end{cases}$$

The set $\{f_\alpha : \alpha < \kappa\}$, being of cardinality κ, has a κ-accumulation point f (i.e., a point every neighborhood of which contains at least κ elements of the above set) in X; to see this, one needs to use the regularity of κ, which follows from the assumptions. The set $\{f(\alpha) : \alpha < \kappa\}$ is a branch of the κ-tree T. Thus κ has the tree property. [K, T]

7. If κ is singular and $\kappa = \sum_{\xi<\mathrm{cf}(\kappa)} \kappa_\xi$ with $\kappa_\xi < \kappa$ for each $\xi < \mathrm{cf}(\kappa)$ then the sum of the κ_ξ's with respect to the ordering $>$ has the required property. If $\kappa \leq 2^\lambda$ for some $\lambda < \kappa$, then an arbitrary subset of cardinality κ of $\langle {}^\lambda 2, \prec \rangle$ has the required property according to Problem 3 in Section 14, with $\prec$ being the lexicographic ordering defined in Problem 1 of Section 14.

If κ is strongly inaccessible and $\langle T, \prec \rangle$ is a κ-tree that has no branch of length κ, then the squashing $\langle T, \prec^* \rangle$ of T satisfies the requirements, as was shown in the proof of Theorem 18.2.

8. In the preceding problem, we saw that there is an ordered set $\langle R, \prec \rangle$ of cardinality κ that has no increasing or decreasing wellordered subset of order type κ. We may assume that the ordering is dense, that is, for all $x, y \in R$ with $x \prec y$, there is a $z \in R$ such that $x \prec z \prec y$. All such orderings can be made complete: If $R^* = \{S : S$ is an initial segment of $R\}$ and $S \prec S' \iff S \subsetneq S'$ for any $S, S' \in R^*$, then $\langle R, \prec \rangle$ can be embedded into $\langle R^*, \prec \rangle$ in a natural way. As every initial segment includes a wellordered cofinal subset of order type less than κ, we have $|R^*| \leq \kappa^{<\kappa} = \kappa$. It is immediate that R^* does not include an increasing or decreasing wellordered set of type κ, and that R^* is κ-compact.

In order to verify the assertion of the problem, it is sufficient to show that there are two κ-compact spaces X_1 and X_2 of weight κ and cardinality κ such that $X_1 \times X_2$ is not κ-compact.

We claim that $\langle R^*, \prec \rangle$ is also κ-compact in the *Sorgenfrey topology*, defined as follows: The basic open sets in this topology are the intervals $[a, b)$ closed on the left and open on the right. Omitting the easy proof of the κ-compactness of the space so defined, let X_1 be the space with the Sorgenfrey topology generated by intervals closed on the left and open on the right, and let X_2 be the space with the similarly defined *reverse-Sorgenfrey topology* generated by intervals open on the left and closed on the right. In the product space $X_1 \times X_2$, the diagonal is a closed discrete subspace of cardinality κ. [H, J].

9. Assume we are given $f : [^2\kappa]^2 \to 2$. Let $a = \{\langle \alpha_0, \alpha_1 \rangle, \langle \alpha_2, \alpha_3 \rangle\}$ and $b = \{\langle \beta_0, \beta_1 \rangle, \langle \beta_2, \beta_3 \rangle\}$ be two elements of $[^2\kappa]^2$. Define the relation $a \leftrightsquigarrow b$ by stipulating that $a \leftrightsquigarrow b$ if and only if $\alpha_i < \alpha_j \iff \beta_i < \beta_j$ and $\alpha_i = \alpha_j \iff \beta_i = \beta_j$ whenever $i, j < 4$ and $i \neq j$. This relation splits $[^2\kappa]^2$ into finitely many equivalence classes. By repeated applications on Theorem 18.2, we obtain a set $A \subset \kappa$ with $|A| = \kappa$ such that for all $a, b \in [^2A]^2$ with $a \leftrightsquigarrow b$ we have $f(a) = f(b)$. In $^2\kappa$ we can choose a subset B of order type κ^2 in the lexicographic ordering of $^2\kappa$ such that only such pairs $a = \{\langle \alpha, \beta \rangle, \langle \gamma, \delta \rangle\}$ occur in B for which $\alpha > \beta$, $\gamma > \delta$, and which are one of the following six types:

(i)	$\beta < \alpha < \delta < \gamma,$		(i*)	$\delta < \gamma < \beta < \alpha,$
(ii)	$\beta < \delta < \alpha < \gamma,$		(ii*)	$\delta < \beta < \gamma < \alpha,$
(iii)	$\delta < \beta < \alpha < \gamma,$		(iii*)	$\beta < \delta < \gamma < \alpha.$

If we have $f(a) = 1$ for pairs a of one of these types, then for each $n < \omega$ there is a homogeneous set of n elements in color 1. If $f(a) = 0$ for pairs a of each of these types, then B is a homogeneous set of order type κ^2 in color 0. [Sp].

Section 19

1. Define the $k+1$-partition $[\omega]^{k+1} = I_0 \cup I_1$ with two colors by stipulating that $X \in I_0$ for $X \in [\omega]^{k+1}$ if and only if $x \in F(X \setminus \{x\})$ for some $x \in X$. If $Y \subset \omega$ with $|Y| = n > (l-1)k + 1$, then Y is not homogeneous in color 0 with respect to I, since a set of n elements has at most $l\binom{n}{k} < \binom{n}{k+1}$ subsets of $k+1$ elements in I_0. By Ramsey's Theorem, there is an infinite set $A \subset \omega$ such that $[A]^{k+1} \subset I_1$. A is free with respect to f. [E, H; 1].

2. We will apply Rado's Selection Lemma, described in Problem 8 of Section 9. For each $\alpha \in X$, let $A_\alpha = k$, and for each $V \in [X]^{<\omega}$, let $f_V \in {}^Vk$ be such that for every $i < k$ the set $f^{-1}(\{i\})$ is free with respect to F; this is possible, since V can be represented as the union of k free sets, according to the assumptions. Let $f \in {}^Xk$ be a function satisfying the conclusion of Rado's Selection Lemma. For each $i < k$, the set $X_i = f^{-1}(\{i\})$ is free with respect to F and $X = \bigcup_{i<k} X_i$. [B, E].

3. We will show that each finite subset V of X is the union of $2k$ free sets; we are going to use induction of $|V| = n$. For this, it is enough to establish that in each nonempty subset V there is an $x \in V$ for which

$$(*) \qquad |\{y \in V \setminus \{x\} : x \in F(y) \vee y \in F(x)\}| < 2k.$$

Writing $E = \big\{\{x, y\} \in [V]^2 : x \in F(y) \vee y \in F(x)\big\}$, we have $|E| \le n(k-1)$. If $(*)$ were not true, we would have $|E| \ge 2k$. [B, E].

4. We prove that there is even a free set of order type η_0. We choose pairwise disjoint intervals $\{I_n : n < \omega\}$ such that they form a set of order type η_0 in the ordering $<$ extended to intervals ($I < J$ if for each $x \in I$ and for each $y \in J$ we have $x < y$). Denote by $\mathcal{J}$ the σ-ideal of sets of first category, and by $\mathcal{S}$ the set of nowhere dense sets. Pick $A_n \subset I_n$ with $A_n \notin \mathcal{J}$. We are going to prove that there is a sequence $x_n \in A_n$ such that the set $\{x_n : n \in \omega\}$ is free with respect to F. Deviating from our standard notation, write $F^{-1}(y) = \{x : y \in F(x)\}$, and put $A_{0,m} = \{y \in A_0 : A_m \setminus F^{-1}(y) \in \mathcal{J}\}$ for each m with $0 < m < \omega$. We claim that $A_{0.m} \in \mathcal{S}$ for each m. Otherwise there would be a $B \subset A_{0,m}$ with $|B| = \omega$ that is dense in some interval. But then $\bigcap\{F^{-1}(y) : y \in B\} = \emptyset$, i.e., $\bigcup\{A_m \setminus F^{-1}(y) : y \in B\} = A_m \notin \mathcal{J}$, which is a contradiction. Choose $x_0 \in A_0 \setminus \bigcup_{m=1}^{\infty} A_{0,m}$ and put $A_m^1 = A_m \setminus F^{-1}(x_0) \setminus F(x_0)$. As $A_m^1 \notin \mathcal{J}$ whenever $0 < m < \omega$, we can continue this procedure. [M].

5. Let $\mathbb{R} = \{x_\alpha : \alpha < \omega_1\}$ and $[\mathbb{R}]^\omega = \{A_\alpha : \alpha < \omega_1\}$ be wellorderings of the set of the reals and of the set of all countably infinite subsets of the reals, respectively. Write $\mathcal{S}_\alpha = \{A_\beta : \beta < \alpha \wedge x_\alpha$ is a limit point of $A_\beta \cap (-\infty, x_\alpha)\}$. We choose $F(x_\alpha)$ to be an increasing sequence tending to x_α and satisfying $F(x_\alpha) \cap A_\beta \neq \emptyset$ whenever $A_\beta \in \mathcal{S}_\alpha$.

Assume $A \subset \mathbb{R}$ with $|A| = \omega_1$. Let $B \subset A$ be a countable set and $\alpha_0 < \omega_1$ be an ordinal such that x_α is a limit point of $B \cap (-\infty, x_\alpha)$ whenever $x_\alpha \in A$ and $\alpha_0 < \alpha < \omega_1$. We have $B = A_{\alpha_1}$ for some $\alpha_1 < \omega_1$. If $x_\alpha \in A$ and $\alpha > \max(\alpha_0, \alpha_1)$, then $F(x_\alpha) \cap A \neq \emptyset$. [He].

6. We may assume that the sets X_ξ are pairwise disjoint. We follow the proof of Theorem 19.2. The proof proceeds differently depending on whether

κ is regular or singular. We confine our comments to the more difficult case of κ being singular. Instead of the sets S_α and $S_{\alpha,\nu}$ there, we define the sets $S_\alpha^\xi, S_{\alpha,\nu}^\xi \subset X_\xi$ for $\xi < \mu$.

The key point is that for $y \in Y_\alpha$ we need to define $\nu(y)$ as

$$\nu(y) = \sup\{\nu < \tau : \exists \beta < \alpha \, \exists \xi < \mu \, (S_{\beta,\nu}^\xi \cap F(y) \neq \emptyset)\}.$$

7. We prove both assertions by induction on κ.

Ad 1. Let $F_0(\alpha) = \alpha$ for $\alpha < \omega_1$. Assume that F_k satisfies the assumption for some $k \geq 0$. Write $X \in [\omega_{k+2}]^{k+2}$ in the form $X = Y \cup \{\alpha\}$ where $Y < \alpha$, and let $\phi_\alpha : \alpha \to \omega_{k+1}$ be one-to-one. $F_{k+1}(X) = \phi_\alpha^{-1}(F_k(\phi_\alpha``Y))$ satisfies the requirements of the problem.

Ad 2. For $k = -1$, the assertion is straightforward, since every one-element set is free. Assume that the assertion is true for some $k \geq -1$, and let F be a set mapping of type $k+2$ and order ω_1 on ω_{k+3}. Let $\alpha \in \omega_{k+3} \setminus (\bigcup\{F(X) : X \in [\omega_{k+2}]^{k+2}\} \cup \omega_{k+2})$. Put $\bar{F}(Y) = F(Y \cup \{\alpha\}) \cap \omega_{k+2}$ for $Y \in [\omega_{k+2}]^{k+1}$. According to the induction hypothesis, there is a $Z \subset \omega_{k+2}$ with $|Z| = k+2$ that is free with respect to $\bar{F}$. Then $Z \cup \{\alpha\}$ is a set of $k+3$ elements that is free with respect to F. [Ku].

8. First we prove the assertion for $\kappa = \omega$. For each $X \in [\omega]^\omega$, the set $X \in [\omega]^\omega \setminus \{X\}$ has cardinality $2^{\aleph_0}$. Hence, according to Problem 1 of Section 10, for each $X \in [\omega]^\omega$ we can choose a set $Y_X \subsetneq X$ such that $Y_X \neq Y_{X'}$ whenever $X, X' \in [\omega]^\omega$ with $X \neq X'$. Pick $f(Y_X) \in X \setminus Y_X$ for each $X \in [\omega]^\omega$, and choose $f(Z)$ arbitrarily if $Z \notin \{Y_X : X \in [\omega]^\omega\}$. Clearly, there is no infinite free set with respect to f.

Now let $\kappa \geq \omega$ and let $\mathcal{F} \in [\kappa]^\omega$ be a system consisting of pairwise almost disjoint sets that is maximal with respect to inclusion. For each $F \in \mathcal{F}$ let $f_F : [F]^\omega \to F$ be a set mapping such that there is no infinite subset of F that is free with respect to f_F. Let $\mathcal{F} = \{F_\alpha : \alpha < \kappa^{\aleph_0}\}$ be a wellordering of $\mathcal{F}$. For each $X \in [\kappa]^\omega$, put $\alpha(X) = \min\{\alpha < \kappa^{\aleph_0} : |F_\alpha \cap X| = \omega\}$, and put $f(X) = f_F(X \cap F)$ for $F = F_{\alpha(X)}$. If $Y \in [\kappa]^\omega$ and $\alpha = \alpha(Y)$, then $\alpha(Z) = \alpha$ for each $Z \in [Y \cap F_\alpha]^\omega$, and so $f|[Y \cap F_\alpha]^\omega = f_{F_\alpha}|[Y \cap F_\alpha]^\omega$; hence Y cannot be free with respect to F.

With some care, the reader can check that these arguments work even if $[A]^\omega$ (for any $A \subset \kappa$) is taken to mean the set of all subsets of order type ω rather than of cardinality ω of the set A. [E, H; 4].

9. Write $\lambda = (2^\kappa)^+$. For each $\alpha < \lambda$ and for each $\beta \in \lambda \setminus \{\alpha\}$, put $F_\alpha(\beta) = F(\{\alpha, \beta\})$. F_α is a set mapping type 1 and order κ. According to Fodor's Theorem 19.1, there are pairwise disjoint sets $A_{\alpha,\xi}$ for $\xi < \kappa$ that are free with respect to F_α and $\lambda \setminus \{\alpha\} = \bigcup\{A_{\alpha,\xi} : \xi < \kappa\}$.

Define $f : [\lambda]^2 \to \kappa \times \kappa$ as follows. For each α and β with $\alpha < \beta < \lambda$ put $f(\{\alpha, \beta\}) = \langle \xi, \eta \rangle$ if and only if $\beta \in A_{\alpha,\xi}$ and $\alpha \in A_{\beta,\eta}$. According to Corollary 14.1, there is a set $X \subset \lambda$ with $|X| = \kappa^+$ that is homogeneous with respect to f in, say, color $\langle \xi, \eta \rangle$. Then X is also a free set with respect to F. Indeed, let $\alpha, \beta, \gamma \in X$ be such that $\alpha < \beta < \gamma$. Then we have $\beta, \gamma \in A_{\alpha,\xi}$; hence $\gamma \notin F(\{\alpha, \beta\})$ and $\beta \notin F(\{\alpha, \gamma\})$. Similarly, $\alpha, \beta \in A_{\gamma,\eta}$; hence $\alpha \notin F(\{\beta, \gamma\})$. [E, H; 1].

10. Let $\{A_\alpha : \alpha < \omega_1\} = [\omega_1]^\omega$ be a wellordering and put $S_\alpha = \{\beta < \alpha : A_\beta \subset \alpha\}$ for $\alpha < \omega_1$. Let $\alpha < \omega_1$ be arbitrary. According to Problem 1 of Section 10, there is an $x_\beta \in A_\beta$ for each $\beta \in S_\alpha$ such that $x_\beta \neq x_\gamma$ whenever $\beta \neq \gamma$ and $\beta, \gamma \in S_\alpha$. Pick $f(\{x_\beta, \alpha\}) \in A_\beta$ for each $\beta \in S_\alpha$, and choose $f(\{y, \alpha\})$ arbitrarily if $y \notin \{x_\beta : \beta \in S_\alpha\}$ and $y < \alpha$. [E, H; 1].

11. Write $\mathcal{F} = \{F_\alpha : \alpha < \lambda\}$. We may assume that, for some real number $\delta > 0$, we have $\mu(F_\alpha) > \delta$ for each $\alpha < \lambda$. Put $S_\alpha = \bigcup\{F_\beta : \alpha \leq \beta < \lambda\}$. and let $S = \bigcap_{\alpha<\lambda} S_\alpha$. Then $\mu(S) \geq \delta$; thus $|S| \neq 0$. Furthermore, $S = \{\xi < \kappa : |\{\alpha < \lambda : \xi \in F_\alpha\}| = \lambda\}$.

12. We may assume that $\mathrm{cf}(\lambda) = \lambda > \omega$. Put $F_0 = F$, and define the set mapping F_n of type $< \omega$ by induction on n as follows: If F_n has already been defined and $X \in [\kappa]^k$ for some k with $1 \leq k < \omega$, then put

$$y \in F_{n+1}(X) \quad \text{if and only if}$$
$$\mu(\{\alpha < \kappa : X < \alpha \wedge y \in F_n(X \cup \{\alpha\})\}) > 0.$$

We claim that F_n is a set mapping of order λ for each $n < \omega$. For $n = 0$, this is true according to our assumptions. If we had $\{y_\xi : \xi < \lambda\} \subset F_{n+1}(X)$ with pairwise distinct y_ξ's for some $X \in [\kappa]^k$, then, according to Problem 11, there would be an α with $X < \alpha < \kappa$ and an $L \subset \lambda$ with $|L| = \lambda$ such that

$$\{y_\xi : \xi \in L\} \subset F_n(X \cup \{\alpha\}).$$

Put $\bar{F}(\alpha) = \bigcup_{n<\omega} F_n(\{\alpha\})$ for each $\alpha < \kappa$. Then $\bar{F}$ is a set mapping of type 1 and of order λ on κ. According to Theorem 19.1, there is a set $S \subset \kappa$ with $\mu(S) > 0$ that is free with respect to $\bar{F}$. Define the sequence $\{\alpha_\nu : \nu < \kappa\} \subset S$ by transfinite recursion on ν. Given $\nu < \kappa$, assume that α_μ has already been defined for $\mu < \kappa$. Write $A_\nu = \{\alpha_\mu : \mu < \nu\}$, and for $X \in [A_\nu]^k$ and for $y \in A_\nu$, put

$$F_n^{-1}(X,y) \stackrel{\text{def}}{=} \begin{cases} \{\alpha : A_\nu < \alpha < \kappa \wedge y \notin F_n(X \cup \{\alpha\})\} \\ \qquad \text{if this set has measure 1,} \\ \kappa \setminus A_\nu \qquad \text{otherwise.} \end{cases}$$

Pick an α_ν with

$$\alpha_\nu = \min\Big(S \cap \bigcap\{F_n^{-1}(X,y) : X \in [A_\nu]^k \wedge y \in A_\nu \wedge n < \omega$$
$$(*) \qquad \wedge\, 1 \leq k < \omega\} \setminus \bigcup\{F_n(X) : X \in [A_\nu]^k \wedge n < \omega \wedge 1 \leq k < \omega\}\Big).$$

Let $A = \{\alpha_\nu : \nu < \kappa\}$. We claim that A is a set free with respect to F (and with respect to all of the F_n's).

Proceeding via reductio ad absurdum, assume, on the contrary, that, for some $X \in [A]^k$ with $1 \leq k < \omega$ and some $y \in A$, there is an $n < \omega$ such that

$$y \in F_n(X).$$

Let k be the least possible such that this assertion holds. Then we have $k > 1$ in view of $A \subset S$. Writing $\alpha = \max X$, we have $y < \alpha$ in view of (*). If $X' = X \setminus \{\alpha\}$, then, again by (*), we have $y \in F_{n+1}(X)$. This is in contradiction with k being least possible. [P; 1].

13. Let $T = \langle \kappa, \prec \rangle$ be a κ-tree and let μ be a real-valued measure on κ. Denote by T_α the α's level of the tree for each $\alpha < \kappa$. For each $\alpha < \kappa$, there is an $x_\alpha \in T_\alpha$ such that $\mu(T| \succeq x_\alpha) > 0$. Put $A_\alpha = T| \succeq x_\alpha$. There is an $A \subset \kappa$ with $|A| = \kappa$ and a real number $\delta > 0$ such that $\mu(A_\alpha) \geq \delta$ holds for each $\alpha \in A$. Writing $\lceil x \rceil$ for the least integer not smaller than the real number x, put $n = \lceil 1/\delta \rceil + 1$. Define the 2-partition $[A]^2 = I_0 \cup I_1$ by stipulating that for $\{\alpha, \beta\} \in [A]^2$, we have $\{\alpha, \beta\} \in I_0$ if and only if $A_\alpha \cap A_\beta = \emptyset$. As there is no set of cardinality n that is homogeneous in color 0 with respect to the partition I, by the Erdős–Dushnik–Miller Theorem 14.6, there is a set $B \subset A$ with $|B| = \kappa$ that is homogeneous in color 1 with respect to I. Then we have $A_\alpha \cap A_\beta \neq \emptyset$ whenever $\alpha, \beta \in B$, and so $\{y : \exists \alpha \in B\, (y \prec x_\alpha)\}$ is a branch of the κ-tree T. [Si; 1].

Section 20

1. Let $S_0 = \lambda \cap M$. According to the assumptions, $S_0 \in \mathrm{Stat}(\lambda)$. Let D be a λ-large set. $D \subset \lambda \setminus M_{\alpha+1}$, $S = S_0 \cap D \in \mathrm{Stat}(\lambda)$. If $\xi \in D$, then there is a ξ-club $C_\xi \subset \xi$ such that $C_\xi \cap S \subset C_\xi \cap M_\alpha = \emptyset$. If $\xi \notin D$ and $\xi \in \mathrm{Lim}(\lambda)$, then $S \cap \xi$ is bounded in ξ and $C_\xi = \xi \setminus \sup(S \cap \xi)$ satisfies the requirement.

2. This is a generalization of Problem 14 in Section 14. Solution *b)* given for that problem can be used, with slight modifications, to give a solution of the present problem. [E, H, R].

3. Given an arbitrary $F \in [\kappa]^\omega$, we can easily define a function $\mathcal{G}_F$ on $[F]^\omega$ satisfying the requirements of the problem. Let $\mathcal{F} \subset [\kappa]^\omega$ be a maximal system of almost disjoint sets in $[\kappa]^\omega$, and let $\mathcal{F} = \{F_\alpha : \alpha < |\mathcal{F}|\}$ be a wellordering. For each $X \in [\kappa]^\omega$, let $\alpha(X) = \min\{\alpha : |X \cap F_\alpha| = \omega\}$, and, with $\alpha = \alpha(X)$ put

$$\mathcal{G}(X) = \{Y \cup (X \setminus F_\alpha) : Y \in \mathcal{G}_{F_\alpha}(X \cap F_\alpha)\}.$$

[E, H; 4].

4. First we present a proof due to R. Solovay for regular $\kappa > \omega$; the second proof works for arbitrary $\kappa \geq \omega$.

a) Assuming $\kappa = \mathrm{cf}(\kappa) > \omega$, let $S = \bigcup_{\xi < \kappa} S_\xi$ be a decomposition of the set S of ordinals $< \kappa$ of cofinality ω as a union of κ pairwise disjoint stationary sets; this is possible according to Solovay's Theorem 12.5 – a full proof of this theorem is given after Corollary 17.2. For $X \subset \kappa$ with type $X(<) = \omega$ put

$$f(X) = \xi \iff \sup X \in S_\xi.$$

Given an arbitrary $A \in [\kappa]^\kappa$, the set A' of its limit points is κ-large, and so $A' \cap S_\xi \neq \emptyset$ for each $\xi < \kappa$.

b) In view of Problem 3, it is sufficient to show that there is a mapping $\mathcal{H} : [\kappa]^\omega \to [\kappa]^{2^{\aleph_0}}$ such that for each $A \in [\kappa]^\kappa$, we have

$(*)$ $$\bigcup \{\mathcal{H}(X) : X \in [A]^\omega\} = \kappa.$$

Indeed, the result of Problem 3 ensures that for each $\mathcal{H} : [\kappa]^\omega \to [\kappa]^{2^{\aleph_0}}$, there is an $\mathcal{H}_0 : [\kappa]^\omega \to \kappa$ such that for every $X \in [\kappa]^\omega$,

$$\mathcal{H}(X) = \{\mathcal{H}_0(Y) : Y \in \mathcal{G}(X)\}$$

with the mapping $\mathcal{G}$ in Problem 3; then for every $A \subset \kappa$, we have

$$\bigcup \{\mathcal{H}(X) : X \in [A]^\omega\} \subset \mathcal{H}_0 \text{``} [A]^\omega.$$

Thus the $\mathcal{H}_0$ corresponding to the $\mathcal{H}$ satisfying $(*)$ verifies the partition relation in the problem.

To establish the existence of an $\mathcal{H}$ satisfying $(*)$, let $f : [\kappa]^\omega \to \kappa$ be a set mapping of type ω and order 2 with respect to which there is no infinite free set; the existence of such a set mapping is guaranteed by Problem 8 of Section 19. For each $V \in [\kappa]^{<\omega}$, define the function $\mathcal{H}_V : [\kappa]^\omega \to [\kappa]^{<\omega} (\subset [\kappa]^{\leq 2^{\aleph_0}})$ by stipulating the equality $\mathcal{H}_V(X) = \{f(X \cup V') : V' \subset V\}$ for every $X \in [\kappa]^\omega$. Assume, by reductio ad absurdum, that no $\mathcal{H}_V$ satisfies the statement corresponding to $(*)$ on any set $Y \in [\kappa]^\kappa$, i.e., for every $V \in [\kappa]^{<\omega}$ and for every $Y \in [\kappa]^\kappa$, there is an $A \in [Y]^\kappa$ such that

$$\bigcup \{\mathcal{H}_V(X) : X \in [A]^\omega\} \not\supset Y.$$

Then, by recursion, we can define a sequence $\alpha_0 < \alpha_1 < \cdots < \alpha_n < \cdots < \kappa$ such that the set $\{\alpha_n : n < \omega\}$ is free with respect to f.

With some care, the reader can check that the arguments used to solve Problems 3 and 4 work even if $[X]^\omega$ (for any $X \subset \kappa$) is taken to mean the set of all subsets of order type ω rather than of cardinality ω of the set X. [E, H; 4].

5. We are going to prove the stronger assertion that if $f : [2^{\aleph_0}]^2 \to \omega_1$, then there is a $\nu < \omega_1$ and an $A \subset 2^{\aleph_0}$ with $|A| = 2^{\aleph_0}$ such that $f \text{``} [A]^2 \subset \nu$. So, writing $\kappa = 2^{\aleph_0}$, assume we are given $f : [\kappa]^2 \to \omega_1$, and denote by $\mathcal{I}$ the κ-complete ideal of sets of measure 0. Write $\mathcal{F} = \text{co}(\mathcal{I})$. We will only use the fact that $\mathcal{I}$ is ω_1-saturated. For each $\alpha < \kappa$, there is an ordinal $\nu(\alpha) < \omega_1$ such that, writing

$$f_\nu(\alpha) = \{\beta < \kappa : \alpha < \beta \wedge f(\{\alpha, \beta\}) = \nu\},$$

we have

$$\bar{f}(\alpha) \stackrel{def}{=} \bigcup\{f_\nu(\alpha) : \nu < \nu(\alpha)\} \in \mathcal{F}.$$

Using transfinite recursion, we may choose a sequence of ordinals α_ξ, $\xi < \kappa$ such that

$$\alpha_\xi \in \bigcap\{\bar{f}(\alpha_\eta) : \eta < \xi\} \qquad \text{for} \qquad \xi < \kappa.$$

As we have $\nu(\alpha_\xi) = \nu$ on a set of ξ's of cardinality κ, the assertion follows. [P; 2].

6. Write $\kappa = 2^{\aleph_0}$. We are going to define a sequence of functions $f_n : \kappa \to \mathbb{R}$ for $n \in \omega$ by recursion on n. Let f_0 be an arbitrary one-to-one function. Assume that f_i has already been defined for $i \leq n$. Given $\mu, \nu < \kappa$, we will write $f(\mu) < f(\nu)$ if $f_i(\mu) < f_i(\nu)$ holds for every $i \leq n$. Assume that for every $X \subset \kappa$ with $|X| = \kappa$, there are $\mu, \nu \in X$ such that $f(\mu) < f(\nu)$. We want to define f_n in such a way that conditions (i) and (ii) below are satisfied:

(i)   For each $X \in [\kappa]^\kappa$, there are $\mu, \nu \in X$ such that

$$f(\mu) < f(\nu) \quad \text{and} \quad f_{n+1}(\mu) > f_{n+1}(\nu);$$

(ii) For each $X \in [\kappa]^\kappa$, there are $\mu, \nu \in X$ such that

$$f(\mu) < f(\nu) \quad \text{and} \quad f_{n+1}(\mu) < f_{n+1}(\nu).$$

Let $[\kappa]^\omega = \{A_\alpha : \alpha < \kappa\}$ be a wellordering. We define $f_{n+1}(\alpha)$ by transfinite recursion on α in such a way that

$$f_{n+1}(\alpha) \neq \beta \quad \text{if} \quad \beta < \alpha$$
$$f_{n+1}(\alpha) \neq \sup\{f_{n+1}(\beta) : \beta \in A_\gamma \wedge f(\beta) < f(\alpha)\},$$
$$f_{n+1}(\alpha) \neq \inf\{f_{n+1}(\beta) : \beta \in A_\gamma \wedge f(\beta) < f(\alpha)\}$$

whenever $\gamma < \alpha$ and $A_\gamma \subset \alpha$.

We claim that with this definition, (i) and (ii) above are satisfied. Assume, on the contrary, first that (i) is not satisfied for some $X \in [\kappa]^\kappa$. Define $h : \kappa \to \mathbb{R}^{n+2}$ by stipulating that $h(\nu) = \langle f_0(\nu), \ldots, f_{n+1}(\nu) \rangle$ for each $\nu < \kappa$. Then there is an $A \in [X]^\omega$ such that $h``A$ is dense in $h``X$. Let $A = A_\gamma$ for some $\gamma < \kappa$. Taking $\mathrm{cf}(\kappa) > \omega$ into account, we can see that the set $X_0 = \{\alpha \in X : \gamma < \alpha \wedge A_\gamma \subset \alpha\}$ also has cardinality κ. By the assumption that (i) is not satisfied, we have

$$f_{n+1}(\alpha) > \sup\{f_{n+1}(\beta) : \beta \in A_\gamma \wedge f(\beta) < f(\alpha)\} \quad \text{whenever} \quad \alpha \in X_0.$$

For some real $\epsilon > 0$ and some set $X_1 \in [X_0]^\kappa$, we have

$$f_{n+1}(\alpha) - \epsilon > \sup\{f_{n+1}(\beta) : \beta \in A_\gamma \wedge f(\beta) < f(\alpha)\} \quad \text{whenever} \quad \alpha \in X_1.$$

Now there is an $X_2 \in [X_1]^\kappa$ such that $|f_{n+1}(\alpha_0) - f_{n+1}(\alpha_1)| < \epsilon/2$ whenever $\alpha_0, \alpha_1 \in X_2$. Choose $\alpha_0, \alpha_1 \in X_2$ with $\alpha_0 \neq \alpha_1$ such that $f(\alpha_0) < f(\alpha_1)$.

As $h``A_\gamma$ is dense in $h``X$, there is a $\beta \in A_\gamma$ such that $|f_{n+1}(\beta) - f_{n+1}(\alpha_0)| < \epsilon/2$ and

$$|f_i(\beta) - f_i(\alpha_0)| < f_i(\alpha_1) - f_i(\alpha_0) \quad \text{for} \quad i \leq n.$$

Clearly, $f(\beta) < f(\alpha_0)$ and $f_{n+1}(\beta) < f_{n+1}(\alpha_0)$, in contradiction to the assumption that (i) fails. Thus (i) has been established; (ii) can be established in the same way. The definition of the functions f_n is now complete.
  Put

$$\{\alpha, \beta\} \in I_n \iff f(\alpha) < f(\beta) \wedge f_{n+1}(\alpha) > f_{n+1}(\beta).$$

Then the I_n's are pairwise disjoint and, for every $X \in [\kappa]^\kappa$, we have

$$[X]^2 \cap I_n \neq \emptyset \quad \text{whenever} \quad n < \omega.$$

[G, S].

Section 21

1. Let $g(\nu) = \aleph_\nu$ and $f(\nu) = 1$ for $\nu < \omega_1$. We have $\aleph_{\omega_1+1} \leq \aleph_{\omega_1}^{\aleph_1} \leq T(g^{+f}) \leq \max(\tilde{T}(g), 2^{\aleph_1})^+ \leq \aleph_{\omega_1+1}$ according to Theorem 21.4 and Lemmas 21.1 and 21.5.

2. According to Lemma 21.1 and the assumptions on the cardinalities, we have $T(\boldsymbol{\omega}_1) \geq \aleph_3$, where $\boldsymbol{\omega}_1$ is the constant function with value ω_1 on ω_1. Let $\mathcal{F}$ be an SADT for $\boldsymbol{\omega}_1$ with $|\mathcal{F}| = \omega_3$. Let $\mathcal{I} \subset \mathrm{P}(\omega_1)$ be an ω_1-complete ideal on ω_1 with $[\omega_1]^{<\omega_1} \subset \mathcal{I}$. Define the set mapping $\mathcal{H}(g) = \{h \in \mathcal{F} : \{\xi < \omega_1 : h(\xi) < g(\xi)\} \notin \mathcal{I}\}$ on $\mathcal{F}$. If $g \neq h$ for $g, h \in \mathcal{F}$, then either $g \in \mathcal{H}(h)$ or $h \in \mathcal{H}(g)$. According to Lemma 19.1, we must have $|\mathcal{H}(g)| \geq \omega_2$ for some $g \in \mathcal{F}$. In the same way as in Lemma 21.2, it follows that there is a SADT $\mathcal{H}'$ for the function $\boldsymbol{\omega}$ (the constant function with value ω) such that $|\mathcal{H}'| = \omega_2$ and

$$A_g \overset{def}{=} \{\xi < \omega_1 : g(\xi) < \omega\} \notin \mathcal{I}$$

holds for $g \in \mathcal{H}'$. Now there is an $n < \omega$ and a set $\mathcal{H}'' \in \mathcal{H}'$ with $|\mathcal{H}''| = \omega_2$ such that

$$A_g^n \overset{def}{=} \{\xi < \omega_1 : g(\xi) = n\} \notin \mathcal{I}$$

for $g \in \mathcal{H}''$. The set $\{A_g^n : g \in \mathcal{H}''\}$ consists of almost disjoint sets. [Ke].

3. Suppose that $2^{\aleph_0} < 2^{\aleph_1}$. Instead of the real-valued measurablity of $\kappa \overset{def}{=} 2^{\aleph_0}$, we will use only the weaker assumption that on κ there is a κ-complete ω_1-saturated ideal $\mathcal{I} \subset \mathrm{P}(\kappa)$ such that $[\kappa]^{<\kappa} \in \mathcal{I}$. According to

Lemma 21.1 and the assumption on cardinalities, there is an SADT $\mathcal{F}$ with $|F| = \kappa^+$ for the function assuming the constant value κ on ω_1. In the same way as in Problem 2, from this we can conclude that for some $g \in \mathcal{F}'$ and for some $\mathcal{F}' \subset \mathcal{F}$ with $|\mathcal{F}'| = \kappa$, we have

$$A_h \stackrel{def}{=} \{\nu < \omega_1 : h(\nu) < g(\nu)\} \notin \mathcal{I} \qquad \text{whenever} \qquad h \in \mathcal{F}'.$$

As $\mathcal{F}$ is an SADT, for each $h, k \in \mathcal{F}$ with $h \neq k$ there is an ordinal $\nu(h, k) < \omega_1$ such that $h(\nu) \neq k(\nu)$ whenever $\nu(h, k) < \nu < \omega_1$. According to the main assertion established in the solution of Problem 5 of Section 20, there is an ordinal $\nu_0 < \omega_1$ and there is a set $\mathcal{F}'' \subset \mathcal{F}'$ with $|\mathcal{F}''| = \kappa$ such that $\nu(h, k) < \nu_0$ whenever $h, k \in \mathcal{F}''$ with $h \neq k$. Now there are a set $\mathcal{F}''' \subset \mathcal{F}''$ with $|\mathcal{F}'''| = \kappa$ and an ordinal ν_1 with $\nu_0 < \nu_1 < \omega_1$ such that $\nu_1 \in A_h$ whenever $h \in \mathcal{F}'''$. It follows that the values $h(\nu_1)$ are pairwise distinct for $h \in \mathcal{F}'''$, which is a contradiction, since $h(\nu_1) < g(\nu_1)$ for all such h. [P; 1].

BIBLIOGRAPHY

[B, H, M]   J. Baumgartner, A. Hajnal, and A. Máté, *Weak saturation properties of ideals*, in: Infinite and Finite Sets, Coll. Math. Soc. J. Bolyai **10** (1973), 137–158.

[Be]   F. Bernstein, *Zur Theorie der Trigonometrischen Reichen*, Leipz. Ber. **60** (1908), 325–338.

[B, E]   N. G. de Bruijn and P. Erdős, *A color problem for infinite graphs and a problem in the theory of relations*, Indagationes Math. **13** (1951), 371–373.

[B, M]   M. R. Burke and M. Magidor, *Shelah's pcf theory and its applications*, Ann. Pure Appl. Logic **50** (1990), 207–254.

[C, K]   C. C. Chang and H. J. Keisler, *Model Theory*, second ed., North-Holland Publishing Co., Amsterdam–New York–Oxford, 1977.

[D, M]   B. Dushnik and E. W. Miller, *Partially ordered sets*, Amer. J. of Math. **63** (1941), 600–610.

[E, H; 1]   P. Erdős and A. Hajnal, *On the structure of set mappings*, Acta Acad. Sci. Math. Hungar. **9** (1958), 111–130.

[E, H; 2]   P. Erdős and A. Hajnal, *On a property of families of sets*, Acta Acad. Sci. Math. Hungar. **12** (1961), 277–298.

[E, H; 3]   P. Erdős and A. Hajnal, *Unsolved problems in set theory*, Proceedings of Symposia in Pure Mathematics **XIII** (1971), American Mathematical Society, Providence, R.I., pp. 17–48.

[E, H; 4]   P. Erdős and A. Hajnal, *On a problem of B. Jónsson*, Bulletin de l'Academie Polonaise de Sciences **14** (1966), 61–99.

[E, H, R]   P. Erdős, A. Hajnal, and R. Rado, *Partition relations for cardinal numbers*, Acta Acad. Sci. Math. Hungar. **16** (1965), 93–196.

[E, H, M, R]   P. Erdős, A. Hajnal, A. Máté, and R. Rado, *Combinatorial Set Theory: Partition Relations for Cardinals*, Disquisitiones Math. Hung. and Studies in Logic and Foundations of Mathematics, Vol. 106, Akadémiai Kiadó and North-Holland Publishing Co., Budapest, New York, 1984.

[E, R; 1]   P. Erdős and R. Rado, *Intersection theorems for systems of sets*, Journal of the London Math. Soc. **35** (1960), 85–90.

[E, R; 2]   P. Erdős and R. Rado, *Intersection theorems for systems of sets II*, Journal of the London Math. Soc. **44** (1969), 467–479.

[F]   G. Fodor, *Eine Bemerkung zur Theorie der regressiven Funktionen*, Acta Sci. Math. **17** (1956), 139–142.

[G, S]   F. Galvin and S. Shelah, *Some counterexamples in the partition calculus*, Journal of Combinatorial Theory, Ser. A **15** (1973), 167–174.

[H; 1]   A. Hajnal, *Remarks on a theorem of W. P. Hanf*, Fundamenta Math. **54** (1964), 109–113.

[H; 2]   A. Hajnal, *A negative partition relation*, Proceedings of the National Academy, USA **68** (1971), 142–144.

[H, J]   A. Hajnal and I. Juhász, *On square compact cardinals*, Periodica Math. Hungar. **3** (1972), 64–70.

[He]   S. H. Hechler, *Independence results concerning the number of nowhere dense sets necessary to cover the real line*, Acta Math. Acad. Sci. Hungar. **24** (1973), 27–32.

[J]   T. Jech, *Set Theory*, Academic Press, New York, 1978.

[Ka]   A. Kanamori, *The Higher Infinite. Large Cardinals in Set Theory from Their Beginnings*, Perspectives in Mathematical Logic, Springer-Verlag, Berlin, 1994.

[K, T] H. J. Keisler and A. Tarski, *From accessible to inaccessible cardinals*, Fundamenta Math. **53** (1964), 225–307.

[Ke]   J. Ketonen, *Some combinatorial principles*, Trans. Amer. Math. Soc. **188** (1974), 387–394.

[Ku]   K. Kuratowski, *Sur une caractérisation des alephs*, Fundamenta Math. **38** (1951), 14–17.

[M]   A. Máté, *On the theory of relations*, Magyar Tud. Akad. Mat. Fiz. Oszt. Közl. **9A** (1964), 331–333.

[Mi]   E. W. Miller, *On a property of families of sets*, Comptes Rendus Varsovie **30** (1937), 31–38.

[P; 1] K. Prikry, *Ideals and powers of cardinals*, Bull. Amer. Math. Soc. **81** (1975), 907–909.

[P; 2] K. Prikry, *Changing measurable into accessible cardinals*, Dissertationes Math. **68** (1970), 55 pp.

[Sc]   D. Scott, *Measurable cardinals and constructible sets*, Bulletin de l'Académie Polonaise des Sciences, Série des Sciences **9** (1961), 521–524.

[Sh; 1] S. Shelah, *Proper Forcing*, Lecture Notes in Mathematics, vol. 940, Springer-Verlag, Berlin, Heidelberg, New York, 1982.

[Sh; 2] S. Shelah, *Cardinal Arithmetic*, Clarendon Press, Oxford, 1994.

[Sh; 3] S. Shelah, *Was Sierpiński right?* Israel Journal of Mathematics **62** (1988), 355–380.

[Sh, S] S. Shelah and S. Stanley, *Ideals, Cohen sets, and consistent extensions of the Erdős–Dushnik–Miller theorem*, Journal of Combinatorial Theory (to appear).

[Si; 1] J. H. Silver, *Some applications of model theory in set theory*, Annals of Math. Logic **3** (1971), 45–110.

[Si; 2] J. H. Silver, *On the singular cardinal problem*, Proc. Internat. Congr. Mathematicians, vol. 1, Vancouver (1974), 1975, pp. 256–268.

[Sp]   E. Specker, *Teilmengen von Mengen mit Relationen*, Commentarii Math. Helvetici **31** (1957), 302–334.

[T]   A. Tarski, *Sur la décomposition des ensembles en sous ensembles presque disjoint*, Fundamenta Math. **14** (1929), 205–215.

[To]   S. Todorčević, *Partitioning pairs of countable ordinals*, Acta Math. **159** (1986), 357–372.

Page numbers usually refer to the place where the symbol in question was introduced. In certain cases, where important discussion concerning the symbol occurs at several places, the number of the page containing the main source of information is printed in italics.

M_α, 189
$\overline{M}_\alpha$, 189
M^∞, 189
$\overline{M}^\infty$, 189
$\min_<\{\alpha : \Phi(\alpha)\}$, 58
$\min_{\prec'} X'$, 50

$\underline{n}$, 12
$NL(x)$, 126
$\text{NS}(\xi)$, 148

ω, 127
ω, 5
ω_0, 44
ω_α, 84
ω^*, 53
On, 188
$\text{Ordinal}(x)$, 124

$P(A)$, 7, 117
$P^n(\omega)$, 26
$\Phi(x, y)$, 9

$\mathbb{Q}$, 5, 127

$\mathbb{R}$, 5, 127
$R(f)$, 8, 122
$\langle R_\kappa, \in \rangle \models \phi$, 138
$R \upharpoonright B$, 42
$R(x)$, 130
$\|f\|$, 248
$\text{Rel}(x)$, 122
$\text{rk}(x)$, 130

$\text{Singlevalued}(x)$, 122
$\text{Stat}(\xi)$, 148
$\sup A$, 65

$T(f)$, 246
T_α, 217
$T_A(f)$, 246
$\tilde{T}(f)$, 246
$\text{Term}(L)$, 109
$\Theta_0 \dotplus \Theta_1$, 48
$\Theta' \times \Theta''$, 50
Θ^*, 53
$\text{Trans}(x)$, 124
$\text{type}\langle A, \prec \rangle$, 44
$\text{type } A(\prec)$, 44, 125

$[u]_R$, 202

V, 188

$\text{Wff}(L)$, 109

$x(y)$, 123
$x \dotplus 1$, 125
^XOn, 193
$x \circ y$, 123

$\mathbb{Z}$, 127
ZF, 112
ZF′, 114
ZF*, 130

NAME INDEX

Names are indexed here if they are mentioned in the text or if they occur as authors of quoted works. The pages of the Bibliography are not covered by this index.

SUBJECT INDEX

Page numbers in italics refer to the main source of information about whatever is being indexed. Often, they refer to the statement of a theorem, while the other page numbers listed refer to various applications of the same.